Ai6X6 RESEARCH LAB

Transformation Blueprint Series

Published by EDGE Publish, www.EdgePublish.com

Generative AI tools were used during the research and creative process to support topic exploration, case study identification, content structuring, narrative refinement, and image generation. All AI-assisted outputs were reviewed, edited, validated, and finalized by the author in collaboration with the editorial, design, and operations teams of the Ai6x6 Research Lab across New York, London, and Kolkata. All interpretations, conclusions, structural decisions, and final narrative determinations represent the collective institutional judgment of the Ai6x6 Research Lab. No AI system holds rights to any portion of this work. All clipart and vector art in this book are AI-generated and included solely to enhance clarity and engagement. They are not intended to depict real individuals or groups.

Product Transformation for the AI Era, First Edition, 2026
ISBN-13: 978-1-971148-04-5, Printed in the United States of America.

AI6X6 RESEARCH LAB

Transformation Blueprint Series

PRODUCT TRANSFORMATION FOR THE AI ERA

Creating Value in a World of Intelligent Systems.
The 6x6 Blueprint for Adaptive Innovation and AI-First Design.

SUBRATA MUKHERJEE

EDGE PUBLISH
www.EdgePublish.com

"For the grace of God, who lights the path.
And, in loving memory of my mother Rupali and my father Dilip, who taught me the importance of kindness and perseverance. This work is born from the love you gave, the sacrifices you made, and the quiet wisdom that still echoes in my life. This one is for you."

Personal message from our Founder & Executive Director

SUBRATA MUKHERJEE

For over twenty-five years, I have sat at the heart of global digital transformations—from shifting a 180-year-old institution like *The Economist* into a digital powerhouse, to architecting hyper-automation for Fortune 500 leaders. I have seen the rise of the internet, the mobile revolution, and the cloud. But none of these shifts compare to the era we have just entered.

The Ai6x6 Research Lab was born out of a singular, urgent observation: Most organizations are trying to solve tomorrow's problems with yesterday's mental models. Traditional transformation is often treated as a "project"—a linear path with a start and an end. But in an era of Agentic AI and self-directing systems, transformation is a living organism. It is a system of behaviors, decisions, and capabilities that must evolve in unison. I founded this Lab to move beyond the "hype" and the "tools." My motivation is to provide leaders with the Cohesion, Schema, and Custody required to navigate a world where the boundary between human intuition and machine intelligence is disappearing. People often ask why "6×6." Throughout history, the number six has been the sacred geometry of our existence—from the sexagesimal systems of ancient Sumerian astronomy to the hexagonal patterns found in the farthest reaches of our cosmos. Mathematically, six is the first "perfect number" because its divisors (1, 2, and 3) add up to itself, making it uniquely balanced. In my journey across the U.S., Europe, and Asia, I realized that this ancient symmetry holds the key to modern enterprise. By distilling organizational complexity into six fundamental pillars and viewing them through six transformative lenses, we activate a collective "sixth sense" for the organization—the ability to perceive and act on what others miss. This architecture forms the foundation of our three flagship methodologies—the "*guided navigation system every transformation needs*".

My journey to this blueprint was forged in the engine rooms of the world's most complex organizations. I have led wholesale business transformations for *The Economist*, reinvented Bloomberg's digital presence following the *Businessweek* acquisition, revitalized the iconic *Dummies* brand within Wiley, and optimized the global ratings processes within S&P Global and McGraw-Hill. This strategic work was balanced by deep technical execution: implementing RPA, AI, IoT, and Digital Twins within Raytheon, and architecting CRM, BPM, and digital workplace automations for the global clients of Cognizant and Unisys.

This perspective was further refined through rigorous academic pursuit, drawing on my MBA in Marketing from Jadavpur University, my Master's in Technology Management from Columbia University, and executive certifications from Harvard University and the Indian Institute of Management (IIM-C) to bridge the gap between high-level theory and industrial-scale execution.

Today, through the Ai6x6 Research Lab, I lead the development of the frameworks that are now being adopted by executive teams globally. At the center of this work is three methodologies: 6X6 Blueprint, IGNITE Methodology and REVEAL Methodology our multidimensional navigation system that unifies architecture, execution, and evaluation across the enterprise.

However, my background as a Tech Journalist for *The Telegraph* for over a decade and a columnist for *CIO Magazine* has taught me that true insight requires total independence. I have built this Lab as a space where my team and I can express our opinions freely, without the interference of corporate sponsors. Our goal is not just to report the news, but to "fold" it into the Ai6x6 Blueprints—giving you the critical "so-what" and the "what-next" to make global shifts actionable for your organization.

We are democratizing this knowledge because I believe that the ability to thrive in the AI era should not be reserved only for those who can afford expensive global consulting firms. Our mission is to empower businesses of all sizes with the targeted, practical, and battle-tested information that I often wish I had at my disposal during my own leadership assignments. Through the Ai6x6 Transformation Blueprints and our Executive Mentor Network, we are scaling AI literacy to ensure that as systems become more autonomous, leaders become more intentional. We are not just adopting new tools; we are reimagining the nature of the enterprise itself.

Ai6X6 RESEARCH LAB

NEW YORK | LONDON | KOLKATA

To our esteemed Industry Thought Leaders: Thank you!

Your generous time and profound insights have been the bedrock of this book. Each interview has been converted into sections and one-page cards that have been sprinkled across six books. You have provided a vital perspective, transforming our work from theory into a deeply practical guide rooted in real-world leadership in the AI-Era. Your willingness to share your hard-won wisdom, navigating complex challenges and revealing the future of your industries, enriches every page. You all have successfully navigated the AI transition in your careers, your companies and your industries. This book is a testament to your expertise. We are immensely grateful for your unwavering support and belief in this project. Thanks a lot, from me and my team at AI6x6 and Executive Mentor Network!

TABLE OF CONTENTS

guided navigation for transformation

BLUEPRINT SERIES

SECTION A

Introduction

Ai6x6™ Transformation Blueprint

Ai6x6 RESEARCH LAB – WHO ARE WE?

Ai6x6 Research Lab is a global intellectual institution dedicated to advancing the science and practice of organizational transformation. We develop original frameworks, diagnostic systems, capability playbooks, and research models that help leaders navigate complexity with clarity, precision, and confidence. Our work spans **Leadership, Workforce, Data, Customer, Product, and Operations**—the six pillars that define modern enterprise maturity. We are based in New York, London & Kolkata.

OUR MISSION FOR ORGANIZATIONAL TRANSFORMATION

To build a globally recognized body of knowledge that helps organizations think clearly, act decisively, and transform responsibly in an age of AI acceleration. We essentially play **six roles** for the industry:

THE CATALYST BRING ALIGNMENT TO LEADERSHIP	**THE STRATEGIST** BRING STRUCTURE TO AMBIGUITY	**THE ARCHITECT** BRING CLARITY TO COMPLEXITY
THE ENGINEER BRING DISCIPLINE TO TRANSFORMATION	**THE ETHICIST** BRING INTEGRITY TO OUTCOMES	**THE GUARDIAN** BRING OVERSIGHT TO AUTONOMY

WHAT WE DO?

1. Build Transformation Frameworks: We architect the Ai6x6 Transformation Blueprints— a structured, multi-lens transformation model that helps orgs decode patterns, risks, behaviors, and decision pathways.
2. Develop Diagnostic Systems: Our maturity diagnostics and radar models translate organizational signals into actionable insights, identifying Growth, Risk, Kill, and Shift zones.
3. Publish Research & Methodologies: We codify best practices in enterprise design through multi-volume series and research playbooks, available in bookstores or on our website ai6x6.com.
4. Create Executive Tools: From RACI matrices to transformation roadmaps, we build high-utility models for immediate executive application. We have many accelerated off-the-shelf offerings on our website.
5. Support Global Organizations: Our frameworks empower enterprises, consulting teams, and innovation labs. Our advisory team worldwide including our Executive Mentor Network is here to help.

Meet the Founder & Executive Director : Subrata Mukherjee

ACCOMPLISHMENTS

Subrata Mukherjee has delivered enterprise-scale transformation across the Americas, Europe, and Asia, shaping modernization programs for some of the world's most influential organizations. His work spans AI/ML, automation, digital product strategy, enterprise architecture, and large-scale organizational transformation. He has led digital reinvention in global publishing, driven automation and AI adoption in aerospace and defense, modernized ratings and compliance processes in financial services, and guided product and data transformation across multiple industries. In 2016, he was recognized as one of Folio's Top 100 People in Publishing for his impact on digital innovation.

EDUCATION

Subrata holds a Master's in Technology Management from Columbia University, an MBA in Marketing from Jadavpur University, and a Business Communications certification from Harvard University. He also holds a Management by Values certification from the Indian Institute of Management, Calcutta, deepening his grounding in leadership behavior, organizational culture, and values-driven transformation. He is a Certified Product Owner. His academic foundation spans technology, strategy, communication, and organizational leadership, informing his multidisciplinary approach to AI-era transformation.

INDUSTRIES

His career spans aerospace and defense, financial services, publishing, healthcare, consulting, education, manufacturing, and technology services. He has delivered transformation in highly regulated environments, mission-critical engineering contexts, global media organizations, and fast-moving digital businesses. This cross-industry exposure gives him a multidimensional perspective on how AI reshapes strategy, operations, and organizational identity.

COMPANIES

Across his career, Subrata has delivered transformation programs for organizations including The Economist, Raytheon Technologies, S&P Global, Bloomberg, Wiley, Cognizant, Unisys, and Gujarat Ambuja Cement. His work has ranged from leading The Economist's digital transformation and modernizing Bloomberg's subscription and data infrastructure, to launching Raytheon's enterprise-wide Citizen Development program for Automation and AI, transforming the Dummies brand at Wiley, modernizing S&P's Ratings Process after the sub-prime mortgage crisis, scaling CRM and Automation/AI consulting portfolios at Cognizant and Unisys, and establishing the technology backbone and smart-building systems for the Gujarat Ambuja conglomerate.

EXPERTISE

Subrata's expertise spans the full arc of enterprise transformation, beginning with early work in CAD/CAM for steel plants and smart-city infrastructure, which shaped his systems-thinking foundation. He then became a specialist in enterprise CRM and BPM transformation, leading large-scale modernization programs that re-architected customer operations, workflow orchestration, and enterprise process design. Over time, he transitioned into a strategist, product thinker, and transformation architect, guiding organizations on designing, adopting, and scaling next-generation digital capabilities. Today, he is recognized as a leading voice in AI and Emerging Technologies, with deep expertise across AI/ML, Generative AI, LLMs, RAG architectures, intelligent automation, hyper-automation, conversational AI, chatbots, agentic systems, and decision-automation frameworks. His work is rooted in strategy, product design, operating-model transformation, and enterprise-scale adoption—not coding or engineering execution—and focuses on translating emerging AI capabilities into practical, scalable, enterprise-grade systems that reshape how organizations operate and compete.

THOUGHT LEADERSHIP

Subrata is a prolific speaker who has presented at more than 100 global conferences and events. His writing has appeared in CIO Magazine, and he served as a weekly technology columnist for The Telegraph. Through the Ai6x6 Research Lab, he publishes frameworks, diagnostics, and transformation blueprints that help organizations navigate the AI economy. In 2026–27, he will launch the Executive Mentor Network to develop a global community of AI Strategy–certified coaches and expand AI literacy among senior leaders. Beyond his professional work, he is an avid traveler who has explored more than 40 countries and an accomplished portrait artist.

www.linkedin.com/in/sm360

1. GIVE **VALUE**, REPEATEDLY, BECOME A GOTO BOOK

The 18-book series brings in a lot of content from a practitioner perspective. The advice we are providing have worked for us in our past projects across industries; so we are preaching what we practiced before

2. BE **BRIEF**, PROVIDE CONTENT IN SMALL NUGGETS

I have never read a book cover to cover. I have short attention span, and more and more people are suffering from the same syndrome. Hence each micro-topic we have introduced is covered in two to three pages

3. BE **COMPREHENSIVE** FOR REAL WORLD IMPACT

Transformation is not easy. In the real world there are a lot of inter-connected challenges. We have proposed detailed frameworks that will connect the dots and provide comprehensive solutions to your team/client

4. BE **PRACTICAL**, PROVIDE EXAMPLES TO DEMYSTIFY

Most companies and leaders know they and their companies need to transform, but don't know where to start. I wanted to provide a step-by-step guide or a MOSCOW priority list on getting started

5. PUSH **BOUNDARIES** AND DARE LEADERS TO DREAM

Real transformation is not tinkering around the edges and playing safe; its often about self disrupting and pushing boundaries and reinventing oneself. Applies to individuals, teams and organizations.

6. PUSH LEADERS TO **INTROSPECT** & ASK QUESTIONS

Since the world around us is moving so fast, leaders often have a fear of been exposed about their lack of knowledge and ask "stupid" questions. These book educates you enough, so you can ask intelligent questions

7. BE **EXCITING**, DON'T BE BORING

I have tried to keep the book interesting, small text, lot of structured tables, quite a few images, bold formatting, case studies, Tips and Tricks, Interviews, RACI matrix, buying guides, assignment questions etc.

8. BRING **OTHERS**, HEAR FROM MY COMMUNITY

Tech sector worldwide is very collaborative. We learn from each other every day and we use the knowledge to impact lives positively. This book attempts to bring a few voices from my network and hear their perspective

9. PROMOTE **AI ECONOMY**, USE AI TOOLS YOURSELF

Generative AI has been used extensively for formatting content in a repeatable format & page size control. Without GenAI such an extensive book series would not have been possible in a short time.

10. BE **DIFFERENT**, CHARTER YOUR OWN COURSE

Most management self-help books are pages and pages of long text with guidelines provided by publishers that are a bin antiquated. Modern learning happens byte sized and is multi-modal and multi-channel..

The **Guided Navigation for Transformation** framework is a comprehensive, end-to-end methodology designed to de-risk and accelerate AI adoption within an organization. It moves away from "random acts of digital" by providing a disciplined, three-stage approach that balances human capability with technical execution. The core of the system is the **6x6 Lens Sequence**, which ensures that every strategic move is evaluated through six critical dimensions: **Leadership, Workforce, Data, Customer, Product, and Operations**.

By applying these lenses consistently across the Blueprint, Reveal, and Ignite phases, the framework creates a "Golden Thread" that links high-level strategy to on-the-ground deliverables. This is not just a consulting model, but an educational ecosystem with an 18-book series that ensures that AI transformation is both measurable and sustainable.

"Guided Navigation for Transformation" in the AI Era

We provide guided navigation through three distinct phases: First, we establish your **Transformation Blueprint** to align skills. Second, we run the **Reveal Diagnostic** to assess your current maturity. Finally, we execute the **Consulting Playbook** to turn the 'Art of the Possible' into a 'Science of Deliverables'.

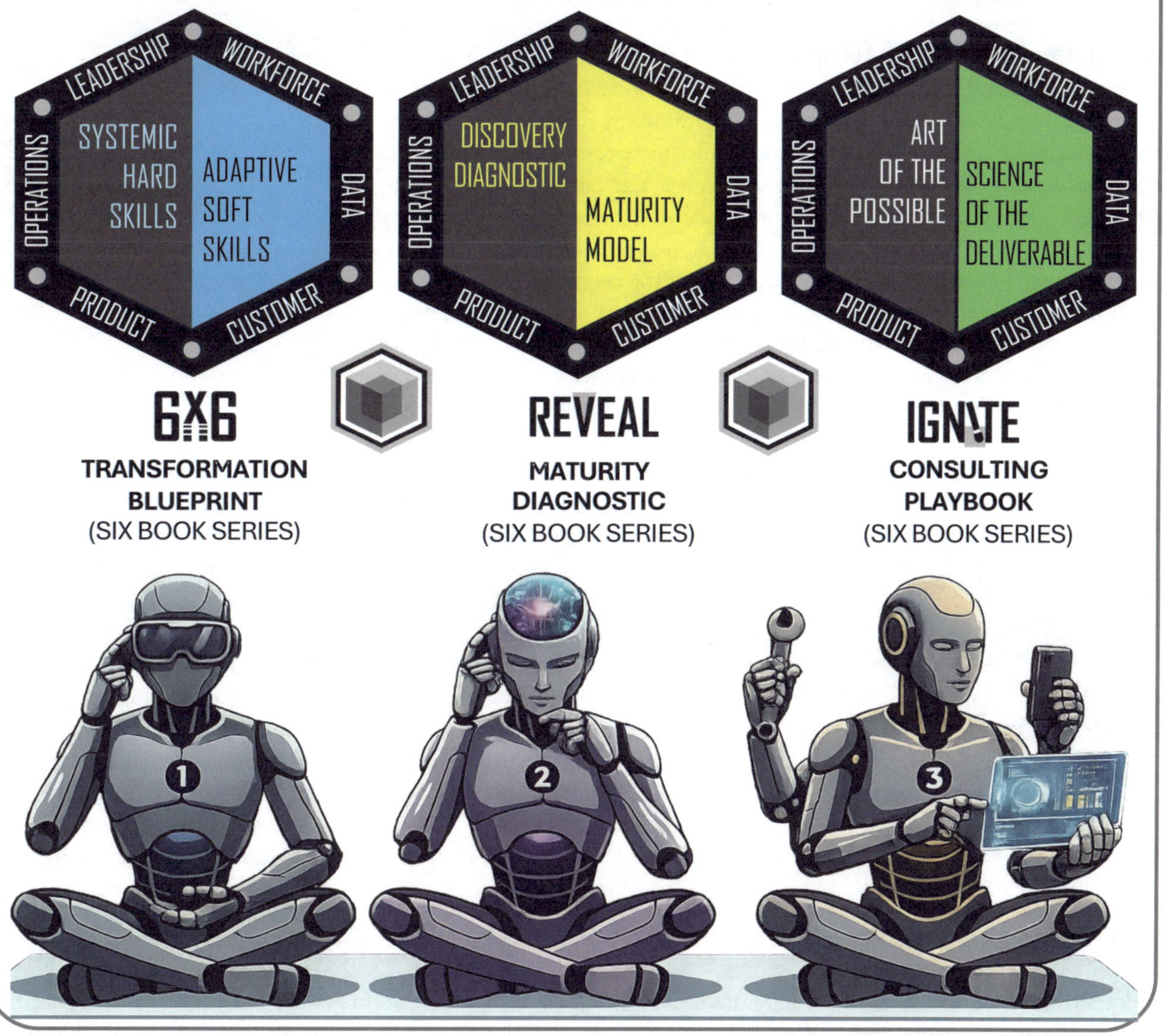

The **AI6x6 Transformation Engine™** is the flagship construct of the AI6x6 Research Lab—a precision-engineered ecosystem that serves as the definitive "Operating System" for the modern enterprise. Built on a proprietary 6×6 geometry, it unifies architecture, execution, and calibration into a single, high-performance unit. The Engine provides leaders with the high-torque machinery and **Guided Navigation** required to convert organizational inertia into scalable momentum in the era of autonomous systems.

The **Transformation Blueprint Series** defines the engine's architecture and establishes the structural logic of the entire system. By introducing the **Six Pillars** and **Six Lenses**, it forms the foundational geometry of enterprise modernization. Written for CEOs, Boards, and Strategy Leaders, this series acts as the **Navigation Deck**, providing the "Architectural North Star." It ensures that every move is strategically calibrated and ethically anchored before the gears are engaged, offering a shared mental model for aligning leadership teams. This intelligence layer removes the guesswork from transformation, providing a rigid yet adaptive framework that allows organizations to visualize their future state with absolute clarity. It is the vital starting point where complex, high-stakes ambition is codified into a disciplined, institutional reality that can withstand the pressures of rapid technological shift.

The **Maturity Diagnostic Series** provides the engine's accountability and calibration system, ensuring peak performance through proprietary **REVEAL ™** methodology. . Acting as the **Sensor-Driven Intake**, it utilizes 36 diagnostic cylinders to quantify organizational readiness and prioritize investments based on evidence rather than intuition. This series allows executive teams to identify hidden gaps and measure progress with scientific clarity, functioning as the **Measuring Layer** of the entire ecosystem. It ensures the engine remains tuned for resilience through a data-driven maturity model that tracks evolution over time. By providing rigorous assessment instruments and scoring logic, it offers the accountability required to sustain a long-term transformation journey. This layer acts as the engine's nervous system, constantly scanning the internal and external environment to calibrate the Blueprint's trajectory and ensure that every unit of energy spent drives the organization toward sustained excellence and institutional stability.

The **Consulting Playbook Series** translates architecture into action by operationalizing the engine through the proprietary **IGNITE™** methodology. It serves as the **Heavy-Duty Gearbox**, providing the repeatable consulting patterns, execution models, and directional force required to activate transformation across diverse domains. Designed for PMOs and delivery leads, these playbooks convert high-level strategy into tangible enterprise-scale outcomes, ensuring that the "torque" of the engine is effectively applied to the road. They function as the essential doing layer, reducing internal friction and shifting the organization from abstract concept to deep-seated conviction. By providing step-by-step pathways and intervention models, the series bridges the gap between executive intent and operational reality. This transmission layer ensures that momentum is not just created, but sustained and scaled, allowing the enterprise to accelerate through complex transitions with precision and measurable impact.

The **Transformation Engine** is built for scale across the global economy. Future **Industry Evolution** volumes will apply this 6×6 engine to specific sectors—such as Healthcare, Banking, and Manufacturing—revealing how the model drives specialized, high-velocity transformation. By adapting the Pillars and Lenses to sector-specific demands, these extensions ensure the Engine remains the definitive standard for navigating the complexities of the age of intelligent systems.

Ai6x6 RESEARCH LAB – OUR FLAGSHIP PUBLICATIONS

SERIES 1 6 BOOKS

SERIES 2 6 BOOKS

SERIES 3 6 BOOKS

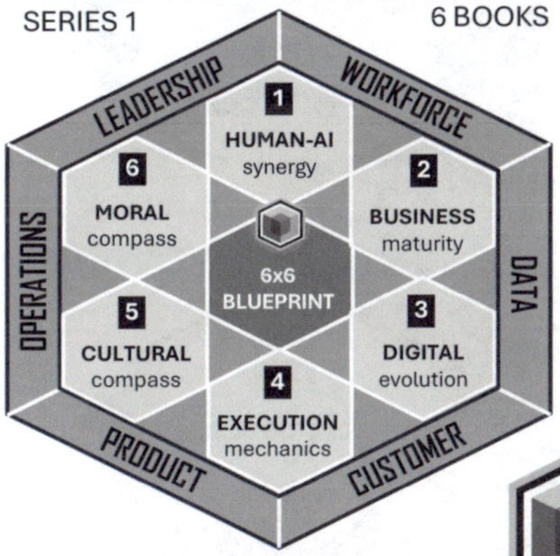

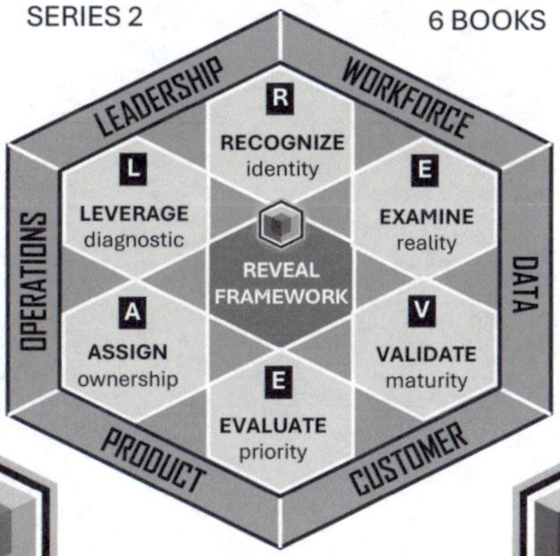

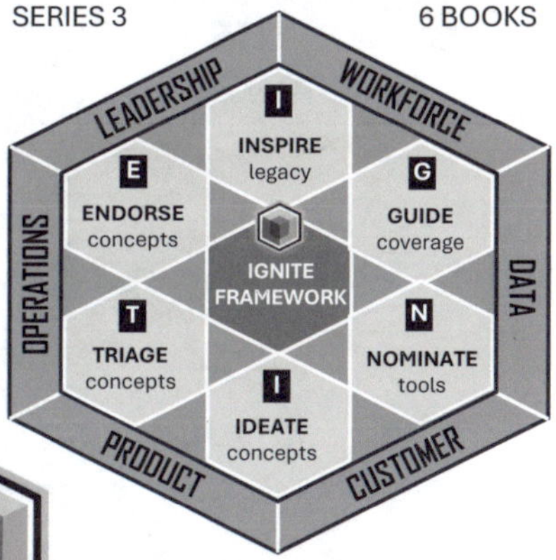

6X6 TRANSFORMATION BLUEPRINT

REVEAL MATURITY DIAGNOSTIC

IGNITE CONSULTING PLAYBOOK

see think act

The Ai6x6 palette is a semantic system, not a decorative choice. Each color encodes the emotional and operational signature of its lens: **pink** captures the human spark that ignites potential; **green** reflects the strategic discipline required for maturity; **purple** signals the depth and computational power of digital evolution; **orange** embodies the kinetic urgency of execution; **blue** anchors the cultural soul in trust and cohesion; and **yellow** illuminates the ethical clarity of the moral audit. Together, these colors form a coherent visual language that guides the reader through the transformation cycle with intuitive, emotionally aligned cues.

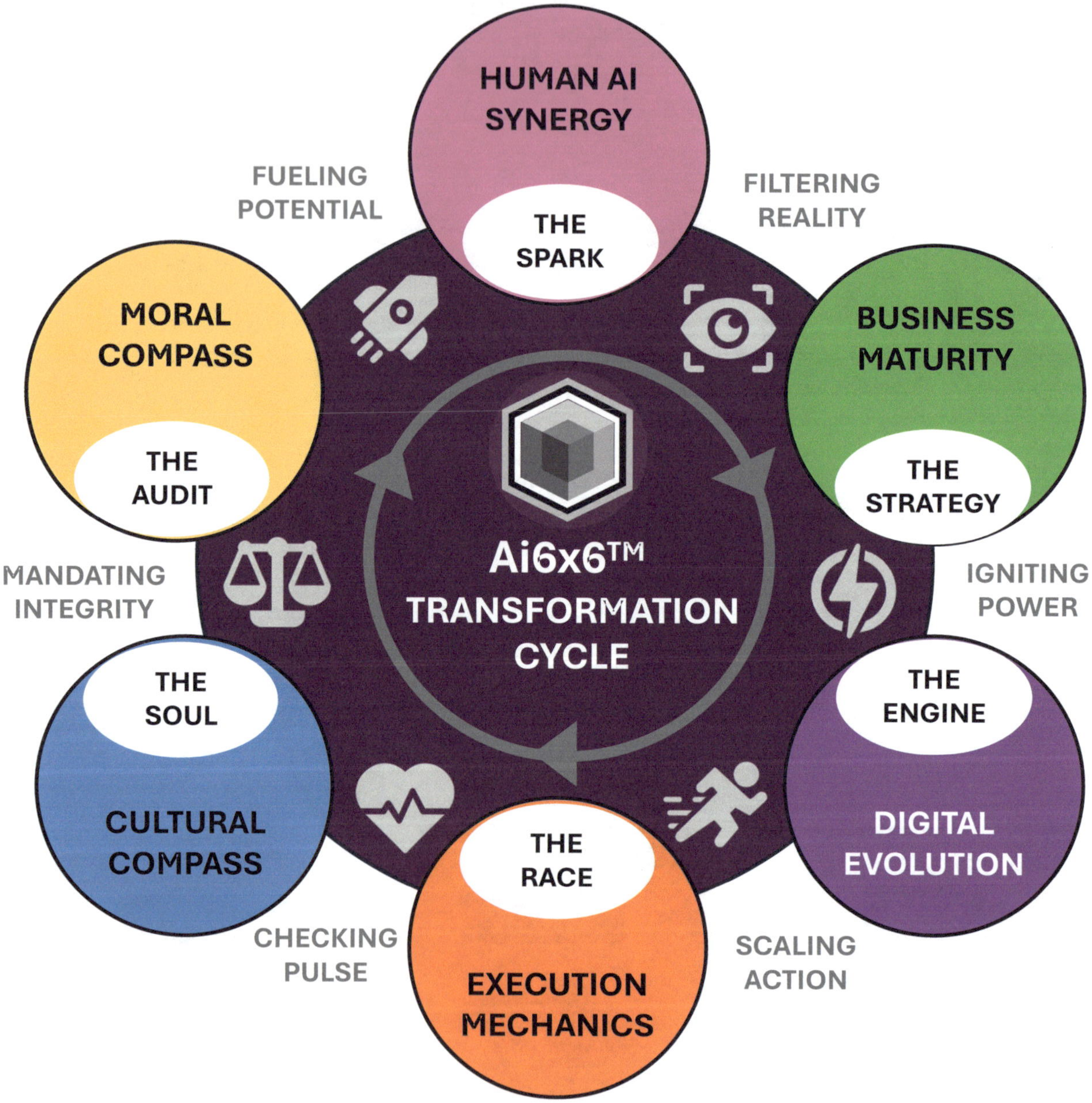

The **COLOR TABS ALONG THE PAGE EDGES IN THE MIDDLE** serve as the physical navigation system of the Ai6x6 series. Each tab mirrors the chromatic identity of its lens, allowing readers to locate themselves instantly within the transformation cycle. This creates a spatial reading experience where orientation is continuous and intuitive. The tabs turn our book into a modular interface—one where Human–AI Synergy, Business Maturity, Digital Evolution, Execution Mechanics, Cultural Compass, and Moral Compass become distinct territories with their own visual signatures. The result is a playbook that reads like a designed system, not a linear text.

6x6™ Transformation Blueprint

The Prologue to a Purpose

"The only way to go forward is to let go of the past."
The Myth of Orpheus and Eurydice

The tragic lesson of this myth is a cautionary tale for business. Orpheus's failure to trust his journey and looking back results in losing everything, underscoring the importance of fully committing to a forward-looking transformation.

The mission of this book series is simple: to transform leaders from passive observers of the AI Era into its active pioneers. We will ignite your thinking with exciting, highly practical content that pushes boundaries and provides the essential tools and frameworks you need to build real-world impact. We won't just tell you what to do; we'll show you how, using a diverse community's experiences and real-world examples that bring the abstract to life. This isn't a passive read; it's an active partnership and a dynamic dialogue designed to turn your natural curiosity into a definitive competitive advantage. This book is the first of a six-part series : Leadership Transformation; which governs your journey on how you should lead from the top and how you should inspire others to take on the AI Era and excel in it.

The **AI6x6 Transformation Engine** is a precision-engineered ecosystem designed to deliver **Guided Navigation** for enterprises facing high-stakes digital evolution. By integrating three proprietary systems—the **Transformation Blueprint**, the **Consulting Playbook**, and the **Maturity Diagnostic**—the Engine provides the robust "Operating System" required to dominate the AI era.

The architecture begins with the **Transformation Blueprint**, the Engine's intelligence layer. Here, the system utilizes **Six Lenses**—including the Cultural and Moral Compasses—to calibrate your trajectory and codify a master architectural plan. This strategic map then engages the **Consulting Playbook**, where the gears of the **IGNITE** methodology translate executive intent into immediate directional force through rapid-response execution cycles. Finally, the **Maturity Diagnostic** acts as a sensor-driven intake, utilizing six diagnostic cylinders to compress raw organizational data into high-pressure strategic insights that validate the transformation's impact. Bolted to a rigid **Chassis of Six Pillars**, the Engine ensures the enterprise remains resilient, balanced, and unshakeable even at peak acceleration. It is more than a strategy; it is a proprietary architecture for sustained institutional excellence.

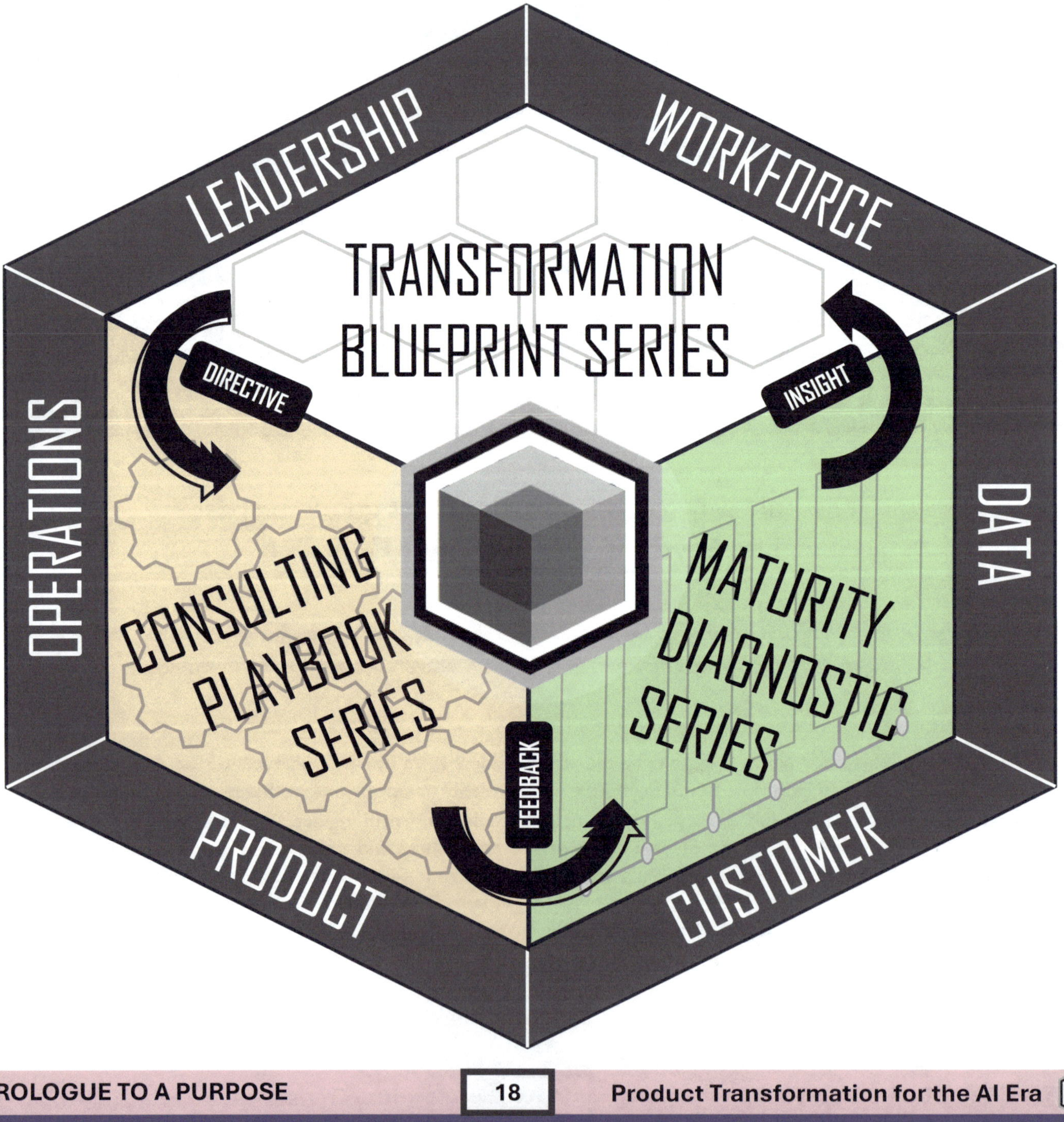

The AI6x6 Transformation Engine is a precision-engineered ecosystem that replaces guesswork with **Guided Navigation**. It is designed to convert organizational inertia into scalable momentum through a disciplined, mechanical approach to digital evolution.

CHASSIS (SIX PILLARS)
The entire system is anchored to a rigid structural frame. This chassis ensures that even as you reach peak acceleration, your enterprise remains resilient and unshakeable.

NAVIGATION DECK (SIX LENSES):
This is the intelligence layer. By applying our **Six Lenses**, we provide the **Guided Navigation** required to calibrate your trajectory, preventing costly drifts in strategy.

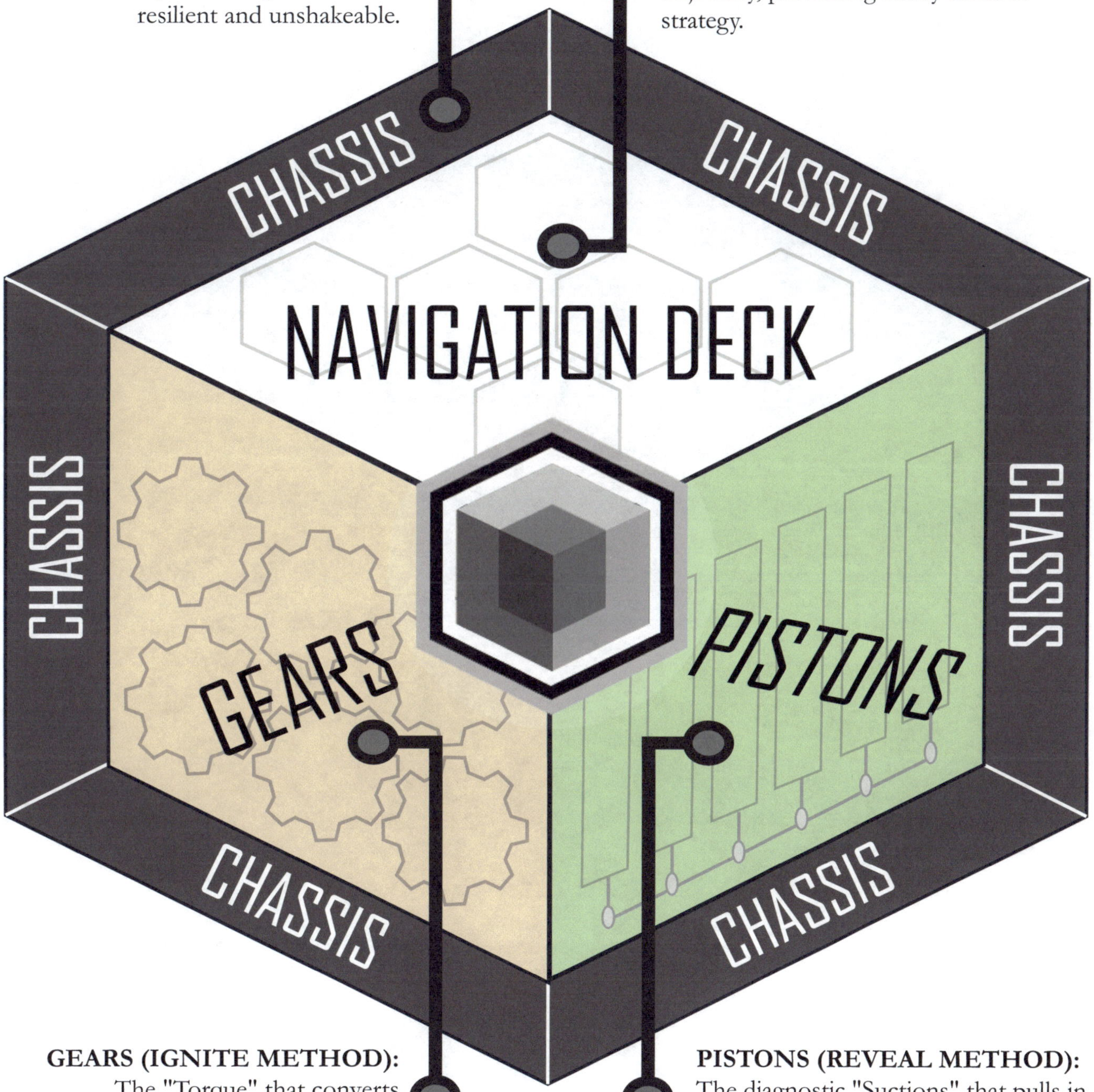

GEARS (IGNITE METHOD):
The "Torque" that converts diagnostic energy into immediate directional force through pilot programs and playbooks.

PISTONS (REVEAL METHOD):
The diagnostic "Suctions" that pulls in raw organizational data and compresses it into high-pressure actionable insights.

A car works best when its actions are coordinated through systems that generate power, transmit motion, and guide direction. Each component plays a precise mechanical role that keeps the vehicle stable, responsive, and moving with purpose.

CHASSIS
Supports vehicle structure, absorbs road stress, anchors mechanical systems, maintains stability, and ensures safe, balanced handling across varying speeds, terrains, and driving conditions.

NAVIGATION DECK
Displays real-time driving information, maps, routes, and alerts, helping the driver maintain direction, anticipate conditions, and make confident, informed decisions.

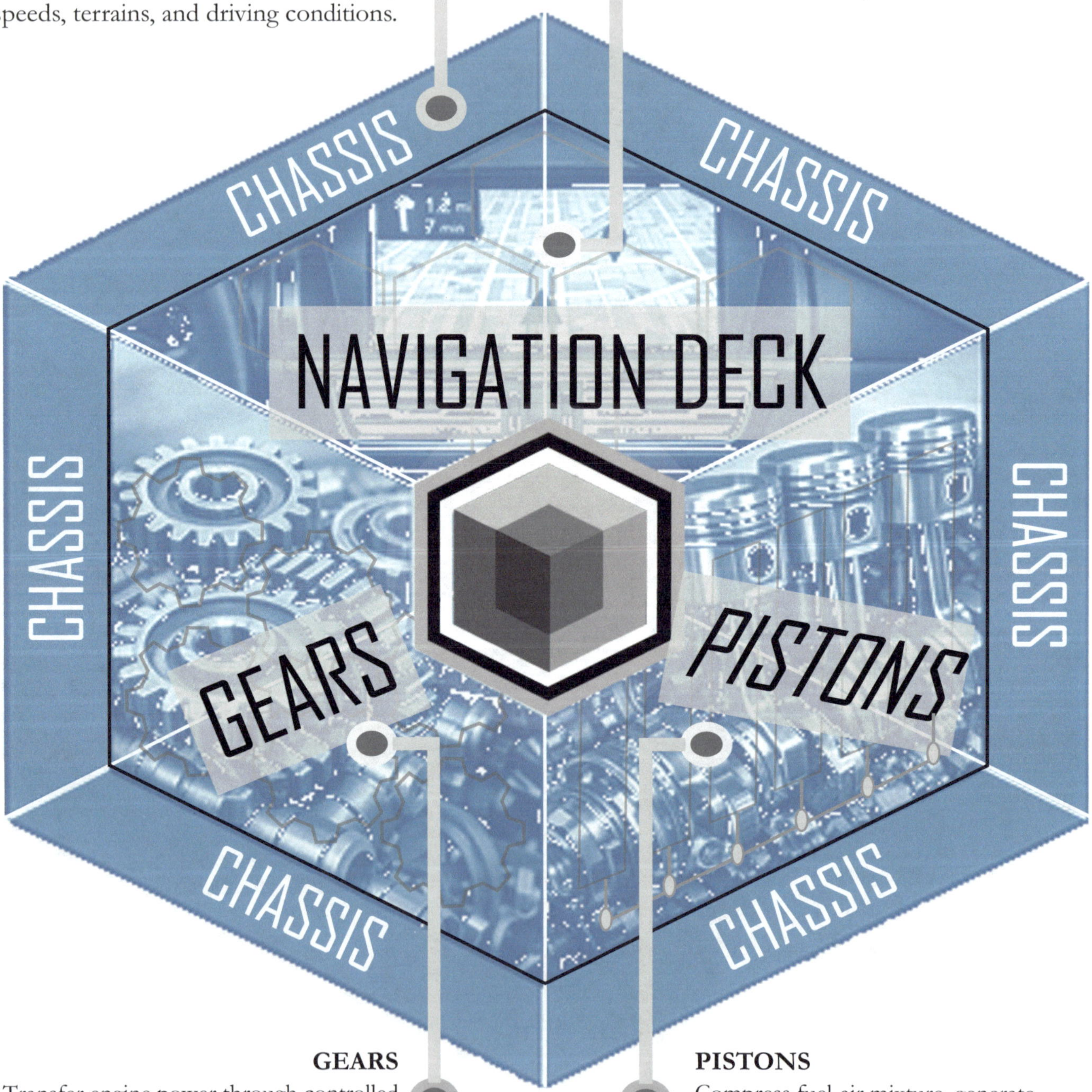

GEARS
Transfer engine power through controlled ratios, regulating speed and torque, ensuring smooth acceleration, efficient performance, and seamless coordination between mechanical components.

PISTONS
Compress fuel-air mixture, generate combustion force, drive the crankshaft, & convert explosive energy into continuous motion powering the vehicle's entire drivetrain system.

The **AI6x6 Transformation Engine** replaces the ambiguity of traditional consulting with mechanical certainty, converting organizational inertia into scalable momentum. By integrating three proprietary systems—the **REVEAL** diagnostic intake, the **IGNITE** consulting transmission, and the **6x6 Blueprint** navigation deck—the Engine provides the robust "Operating System" required to dominate the AI era. The **REVEAL** methodology acts as a sensor-driven intake, utilizing 6x6 diagnostic cylinders to compress raw organizational data into high-pressure strategic insights. This "data fuel" then powers the **IGNITE** phase, where consulting playbooks serve as a heavy-duty gearbox, translating executive intent into immediate directional force through rapid-response execution cycles. This entire power plant is governed by the **6x6 Blueprint**, the intelligence layer that uses **Six Lenses**—including the Cultural and Moral Compasses—to calibrate your trajectory and ensure ethical, operational alignment. Bolted to a rigid **Chassis** of **Six Pillars**, the Engine ensures the enterprise remains resilient, balanced, and unshakeable even at peak acceleration. It is more than a strategy; it is a proprietary architecture for sustained institutional excellence.

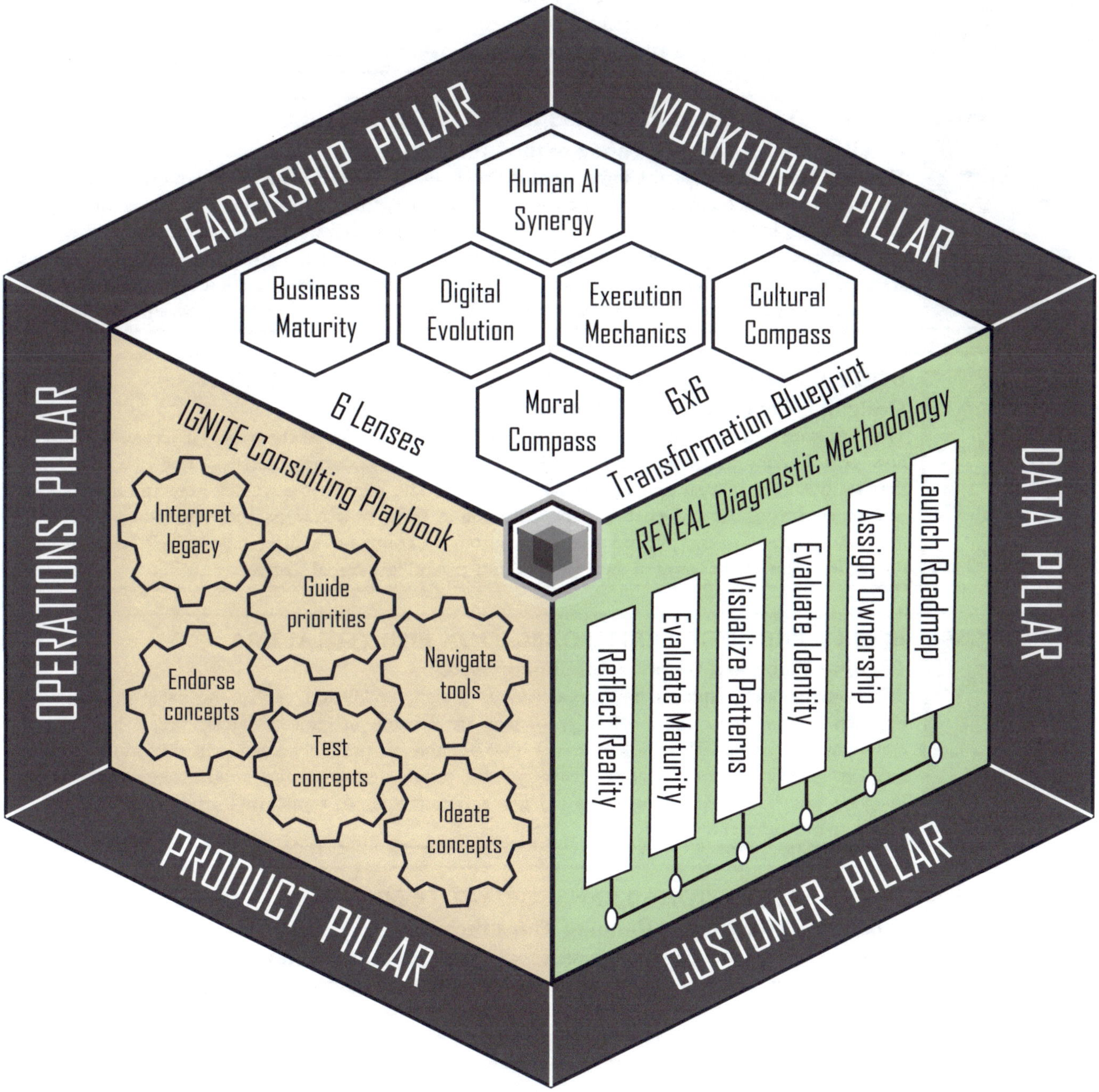

BOOK 1: LEADERSHIP TRANSFORMATION FOR THE AI ERA

Designing the Vision, Not Just Directing It

The Leadership book reframes what it means to lead in an era defined by intelligence, acceleration, and ambiguity. It explores how leaders think, decide, communicate, and shape meaning across the enterprise, emphasizing that leadership is not a role but an architecture. The narrative examines how leaders make decisions systemically, exercise judgment in an AI-augmented world, create alignment through narrative clarity, and uphold ethical stewardship as the foundation of trust. This book sets the tone for the entire series by positioning leadership as the integrator of all transformation forces.

BOOK 2: WORKFORCE TRANSFORMATION FOR THE AI ERA

Elevating Human Capability in a Technology-Driven Era

The Workforce book examines the human system of the enterprise — the skills, roles, mindsets, structures, and experiences that determine how people show up to the work of transformation. It explores how organizations support people as AI reshapes every function, and how leaders build environments where individuals feel empowered rather than displaced. The narrative dives into the evolution of adaptive talent ecosystems, the emergence of AI-augmented roles, the importance of team cohesion and psychological safety, and the recognition of human capability as a strategic asset.

BOOK 3: DATA TRANSFORMATION FOR THE AI ERA

Building the Intelligence Fabric of the Enterprise

The Data book establishes the informational backbone of the organization. It explores how data is collected, governed, interpreted, and transformed into intelligence, emphasizing that data is no longer a technical asset but the raw material of decision-making, personalization, automation, and innovation. The narrative examines how trustworthy data governance is built, how intelligence flows across systems, how data architectures are designed, and how data becomes a source of enterprise-level advantage. This book gives leaders the clarity to build an intelligent, insight-driven organization.

BOOK 4: CUSTOMER TRANSFORMATION FOR THE AI ERA

Designing Experiences That Earn Trust and Loyalty

The Customer book anchors the enterprise in the lived experience of the people it serves. It explores how organizations understand, anticipate, and respond to customer needs — not as a marketing exercise but as a cultural and operational discipline. The narrative examines how emotion shapes experience design, how AI enables personalization at scale, how transparency strengthens customer trust, and how end-to-end journey coherence becomes a competitive differentiator. This book ensures the organization evolves around the customer, not around its internal structures.

BOOK 5: PRODUCT TRANSFORMATION FOR THE AI ERA

Creating Value in a World of Intelligent Systems

The Product book explores how organizations design, build, and evolve the value they deliver. Products are no longer static offerings — they are dynamic systems of intelligence, behavior, and continuous adaptation. This book examines how AI enhances product behavior, how continuous value creation replaces episodic feature delivery, how product strategy and product operations must integrate, and how digital ecosystems redefine differentiation. It transforms product thinking from output to outcome, and from features to value.

BOOK 6: OPERATIONS TRANSFORMATION FOR THE AI ERA

Engineering the Engine Room of Transformation

The Operations book addresses the systems, workflows, and processes that determine whether strategy becomes reality. It explores how organizations orchestrate complexity, manage risk, and maintain resilience in environments defined by acceleration and uncertainty. The narrative examines how intelligent workflows are designed, how operational resilience is built, how AI optimizes processes, and how execution scales with precision. This is the book where transformation becomes tangible.

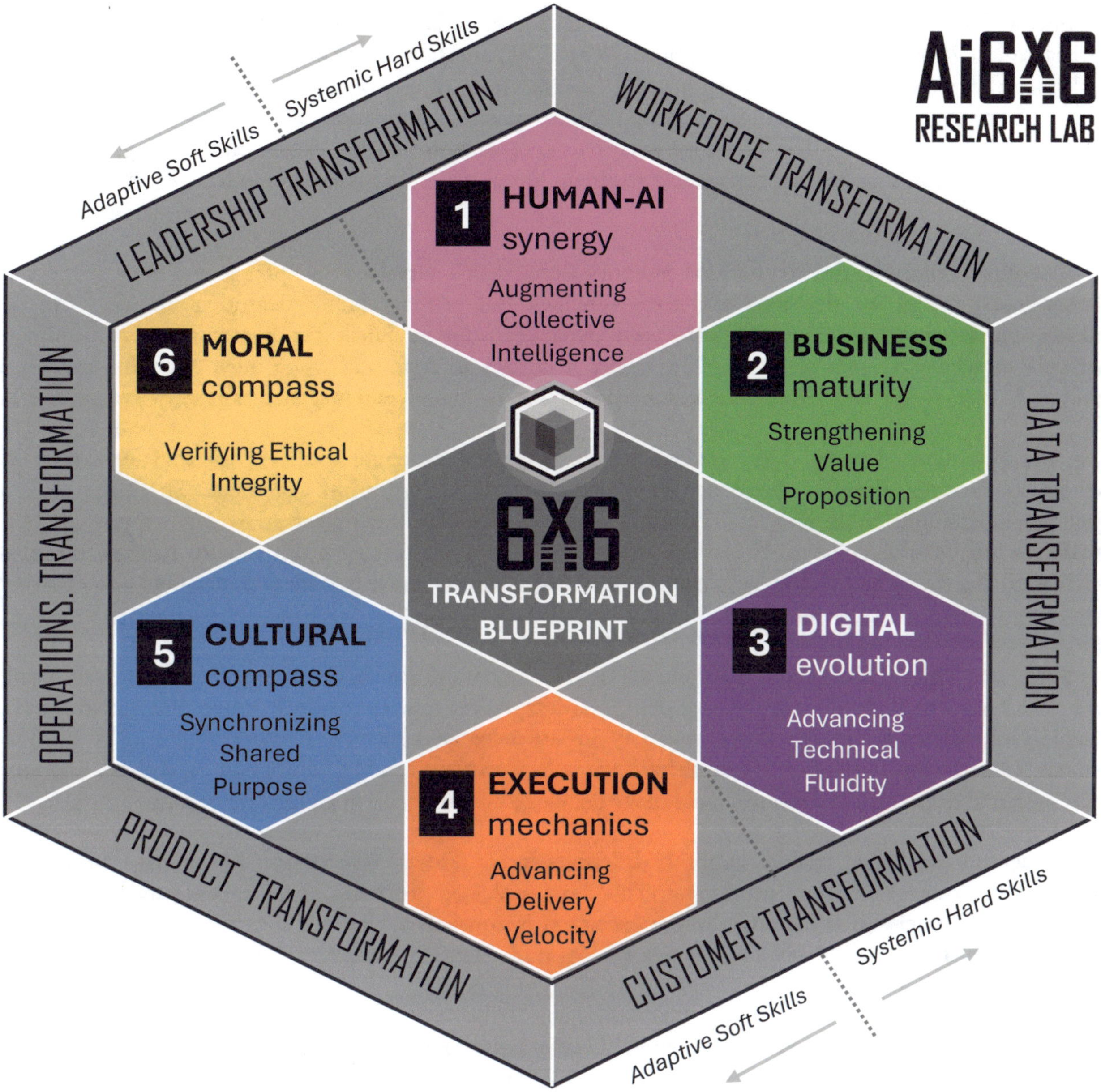

The **Ai6x6 TRANFORMATION BLUEPRINT™** provides the multidimensional structural integrity required for sustainable, high-stakes evolution. Serving as the **Navigation Deck** and **Chassis** of the Transformation Engine, the BLUEPRINT™ moves beyond episodic change to establish a permanent, scalable "Operating System" for the enterprise. The architecture is defined by the intersection of **Six Pillars** and **Six Lenses**. The Pillars—Leadership, Workforce, Data, Operations, Product, and Customer—provide the horizontal breadth and load-bearing strength of the organization. These are bisected vertically by the Six Lenses to reveal the true depth of the enterprise. Leaders apply **Human–AI Synergy** to augment collective intelligence and **Business Maturity** to instill systemic discipline. The **Digital Evolution** lens ensures technical fluidity, while **Execution Mechanics** masters the art of precision delivery. This is anchored by the **Cultural Compass**, which synchronizes shared purpose, and the **Moral Compass**, which verifies ethical integrity.

By aligning these twelve dimensions, the BLUEPRINT™ eliminates strategic drift and structural fragmentation. It ensures that every initiative is not just an isolated project, but a calibrated component of a unified, resilient, and unshakeable corporate architecture. Also remember to follow the Tab colors on the edges to navigate quickly.

The **Ai6x6 Transformation Blueprint™** gives leaders a multidimensional, architecture-grade view of how transformation behaves inside their enterprise. Where REVEAL™ diagnoses the present and IGNITE™ accelerates the future, the Blueprint defines the *structural truth* of the organization. Built on six proprietary Lenses, it converts fragmented observations, cultural noise, and operational ambiguity into a coherent, systems-level understanding. The Blueprint equips leaders with the clarity, depth, and structural intelligence required before designing any transformation roadmap, ensuring decisions are grounded in reality rather than aspiration.

The Blueprint begins with the **Human–AI Synergy Lens**, which examines how people and intelligent systems interact across workflows, decisions, and responsibilities. Rather than simply measuring adoption, this Lens analyzes augmentation, displacement risk, trust dynamics, and readiness for human-machine teaming. It distinguishes where AI enhances capability, where it creates friction, and where new operating models are required. Leaders begin with a grounded understanding of how human and machine roles must evolve together.

Next, the **Business Maturity Lens** evaluates the enterprise's structural readiness for transformation. This includes governance, financial discipline, decision velocity, and leadership alignment. The Lens separates foundational strength from structural fragility, revealing whether the organization can support sustained transformation without collapse. It clarifies which capabilities are stable, which require reinforcement, and which must be redesigned before acceleration. Business Maturity ensures leaders understand the chassis on which transformation will operate.

The third Lens, **Digital Evolution**, maps the depth, coherence, and scalability of the digital ecosystem. Leaders often face fragmented platforms, technical debt, and integration gaps that silently undermine transformation. Digital Evolution cuts through this complexity by assessing architecture, interoperability, data flow, and platform extensibility. It clarifies which digital foundations can scale, which must be modernized, and which create systemic risk. This Lens ensures technology decisions are grounded in architectural truth.

With structural and digital realities established, leaders move into the **Execution Mechanics Lens**, which examines how strategy becomes movement. This Lens evaluates processes, rhythms, accountability systems, and operational precision. It identifies where execution is consistent, where it is brittle, and where it breaks under pressure. Execution Mechanics ensures leaders understand the organization's true capacity to deliver, adapt, and sustain momentum across complex transformation cycles.

The fifth Lens, **Cultural Compass**, explores the beliefs, behaviors, incentives, and informal power structures that shape how people respond to change. Culture is often the silent accelerator or the invisible brake. This Lens reveals alignment, resistance patterns, psychological safety, and leadership influence networks. It distinguishes cultural strengths from cultural liabilities, giving leaders a clear view of the human forces that determine whether transformation will be embraced or rejected.

The final Lens, **Moral Compass**, evaluates ethics, trust, transparency, and responsible AI principles. As organizations adopt intelligent systems, the stakes of ethical decision-making rise dramatically. This Lens ensures transformation is not only effective but also equitable, accountable, and aligned with long-term societal and organizational values. It clarifies where safeguards are strong, where risks exist, and where governance must evolve.

Together, these Six Lenses form a complete, repeatable, consulting-grade architectural framework. The Ai6x6 Transformation Blueprint™ is not a checklist or maturity model; it is a structural intelligence system that moves leaders from assumption to understanding, from understanding to design, and from design to confident transformation planning. It gives leaders the conviction to move forward — not because the future is predictable, but because their decisions are grounded in multidimensional truth, structural clarity, and a deep understanding of how transformation behaves across the enterprise.

Ai6x6™ Transformation Blueprint

Understand Pillars & Lenses

"We don't see things as they are, we see them as we are." — Anaïs Nin

The **Ai6x6 TRANSFORMATION BLUEPRINT** is fundamentally structured around two core concepts: **Pillars and Lenses**. Together, they create a dual-axis architecture that brings both breadth and depth to enterprise transformation. The Pillars provide the WIDTH of the framework, defining the major domains where transformation must take root for the organization to evolve coherently. These domains—Customer, Product, Operations, Data, Workforce, and Leadership—represent the full span of enterprise behavior. Each Pillar receives its own dedicated book, allowing leaders to explore the complexities, tensions, and opportunities of that domain without dilution or compromise.

This ensures that transformation is not treated as a single initiative but as a system of interconnected disciplines that must advance in parallel. Complementing this horizontal structure is the DEPTH created by the Six Lenses, the universal imperatives that cut vertically through every Pillar. These Lenses represent the non-negotiable dimensions of modern transformation—forces that must be activated in every domain, regardless of industry, maturity, or strategic ambition. The six Lenses are: Human–AI Synergy (L1), Business Maturity (L2), Digital Evolution (L3), Execution Mechanics (L4), Cultural Compass (L5), and Moral Compass (L6). Each Lens exposes a different layer of organizational capability, ensuring that no domain is approached superficially, and no critical dimension is overlooked. The goal of the blueprint is to be holistic and very comprehensive so that executives are not blind-sighted by a key aspects being overlooked. Since a lot of ground has to be covered, the books may look very dense, but I have tried my best to keep it brief & engaging.

Transformation has become one of the most OVERUSED words in modern leadership, yet one of the least understood. Organizations declare it, teams attempt it, consultants package it—but few enterprises ever truly achieve it. Not because they lack ambition or intelligence, but because they approach transformation as a sequence of disconnected initiatives rather than a shift in how the organization thinks, behaves, and evolves. I created the **AI6x6 Transformation Blueprint** because I have seen the same pattern repeat across industries, functions, and eras of technology: transformation fails not at the point of strategy, but at the point of structure. It fails when organizations modernize technology but ignore culture. It fails when they redesign processes but overlook trust. It fails when they invest in AI but fail to prepare people. It fails when they build strategy but cannot execute it. It fails when leaders underestimate the complexity of aligning human capability with machine intelligence.

The AI6x6 Transformation Blueprint exists to solve this fragmentation. It brings together **six foundational lenses** and **six transformation pillars** into a single, coherent architecture. Each lens represents a dimension of capability that every organization must master. Each pillar represents a domain— Leadership, Workforce, Data, Customer, Product, Operations —where transformation must take root. Together, they form a 6×6 matrix that reveals not only what must change, but how change becomes sustainable, repeatable, and scalable.

This blueprint is intentionally multidimensional. Transformation is not linear. It is not sequential. It is not a checklist. It is a living system—human, technical, cultural, operational, ethical—moving together or not at all. The AI6x6 model acknowledges this complexity and gives leaders a structure strong enough to hold it.

- It is called a blueprint because it is both **architectural** and **actionable**.
- This blueprint is not a methodology per se. It is not a maturity model in the traditional sense. It is not a checklist as you commonly see it. It is a **HOLISTIC** way of seeing things that are often overlooked.
- It is called AI6x6 because it integrates **AI-driven intelligence** with **six human-centered disciplines**, applied across **six transformation pillars**.
- It is a system designed for organizations that want to evolve with intention rather than react with urgency.

The Six Lenses emerged from a simple but unavoidable truth: "**no single discipline can carry transformation on its own**". Not AI. Not culture. Not governance. Not execution. Not digital modernization. Not ethics. Each is necessary, but none is sufficient. Real transformation requires all six lenses working together—reinforcing one another, correcting one another, and elevating one another.

It reveals the forces that shape organizational behavior, the tensions that undermine progress, and the capabilities that determine whether transformation becomes sustainable or collapses under its own ambition. It provides leaders with a structural map—one that makes complexity navigable and makes evolution repeatable.

Each lens represents a dimension of organizational life:

1. Human–AI Synergy: how intelligence is created, shared, and amplified.
2. Business Maturity: how discipline, readiness, and structure turn ambition into capability.
3. Digital Evolution: how technology becomes fluid, integrated, and trustworthy.
4. Execution Mechanics: how strategy becomes action and action becomes advantage.
5. Cultural Compass: how beliefs, behaviors, and meaning shape the human system.
6. Moral Compass: how ethics, transparency, and responsibility anchor everything else.

Together, they form a complete operating philosophy for modern transformation—one that is as human as it is technical, as strategic as it is operational, as aspirational as it is grounded. The AI6x6 Transformation Blueprint is not about predicting the future. It is about preparing organizations to shape it. It is about giving leaders a language for what they instinctively know but rarely articulate. It is about building enterprises that can evolve continuously, coherently, and confidently. Most of all, it is about honoring the truth that **transformation is not something organizations *do*—it is something they *become*.**

The Pillars of the **AI6x6 Transformation Blueprint** define the WIDTH of enterprise evolution. They represent the six domains where transformation must become real, measurable, and operationally grounded. While the Lenses provide the depth of capability, the Pillars provide the breadth of application—ensuring that transformation is not isolated to a single function but becomes a structural shift across the entire enterprise. Each Pillar receives its own dedicated book, allowing leaders to confront the full complexity of that domain with precision and without compromise.

The Leadership Pillar sits at the top of the architecture because leadership is the force that shapes every other domain. It defines how decisions are made, how priorities are set, how accountability is distributed, and how meaning is communicated. This Pillar examines the evolution of leadership in an era defined by AI, acceleration, and ambiguity. It challenges leaders to move from managing activities to designing systems, from directing work to enabling capability, and from protecting the past to shaping the future. Leadership is not a separate function—it is the integrator of the entire Blueprint.

The Workforce Pillar focuses on the human system that powers the enterprise. It explores how roles evolve, how teams collaborate, how skills are developed, and how people experience their work. In a world where AI augments every function, the Workforce Pillar examines how organizations support human capability, build adaptive talent ecosystems, and create environments where people feel empowered rather than displaced. This Pillar forces leaders to confront the truth that transformation succeeds or fails based on whether people feel equipped, supported, and connected to the mission.

The Data Pillar establishes the informational foundation of the enterprise. It examines how data is collected, governed, interpreted, and transformed into intelligence. In an AI-driven world, data is no longer a byproduct of operations—it is the raw material of decision-making, personalization, automation, and innovation. This Pillar explores how organizations build trustworthy data ecosystems, eliminate silos, and ensure that intelligence flows across the enterprise. It challenges leaders to treat data not as a technical asset but as a strategic capability that shapes every other Pillar.

The Customer Pillar anchors the organization in the lived experience of the people it serves. It forces leaders to confront the reality that customer expectations evolve faster than internal processes, and that loyalty is earned through coherence, trust, and emotional resonance. This Pillar examines how organizations understand, anticipate, and respond to customer needs—not as a marketing exercise, but as a cultural and operational discipline. It challenges leaders to build enterprises where customer experience is not a department, but a shared responsibility embedded in every decision.

The Product Pillar focuses on how organizations design, build, and evolve the value they deliver. Products are no longer static offerings—they are dynamic systems of intelligence, behavior, and continuous adaptation. This Pillar explores how product strategy, product design, and product operations must work together to create experiences that feel alive and responsive. It examines how AI reshapes product behavior, how digital ecosystems redefine differentiation, and how organizations must shift from feature delivery to continuous value creation.

The Operations Pillar addresses the engine room of the enterprise—the systems, workflows, and processes that determine whether strategy becomes reality. It examines how organizations orchestrate complexity, manage risk, and maintain resilience in environments defined by acceleration and uncertainty. This Pillar forces leaders to recognize that operational excellence is not about efficiency alone; it is about adaptability, intelligence, and the ability to respond to change without losing coherence. Operations is where transformation becomes tangible.

Together, these six Pillars form the structural WIDTH of the AI6x6 Transformation Blueprint. They define the terrain on which the Six Lenses operate, ensuring that transformation is comprehensive, coherent, and enterprise-wide.

4
Earned relationships
CUSTOMER
"Who we serve?"
5
Brand Experience
PRODUCT
"What we serve?"
6
Engine room of service
OPERATIONS
"How we serve?"
2
Human Capability
WORKFORCE
"Human Assets"
3
Information foundation
DATA
"Insights"
1
Vision, mission & agenda
LEADERSHIP
"C-Suite"
Each Pillar = "Width" of the Transformation

4
Discerning Diners
CUSTOMER
"The Patrons"
5
Food, Drinks, Ambiance
PRODUCT
"The Feast"
6
Ingredients, Purchase
OPERATIONS
"The Logistics"
2
Cooks, Stewards, Servers
WORKFORCE
"The other staff"
3
Menu, Recipe & Reviews
DATA
"The Menu"
1
Lead Culinary aspect
LEADERSHIP
"The Exec Chef"

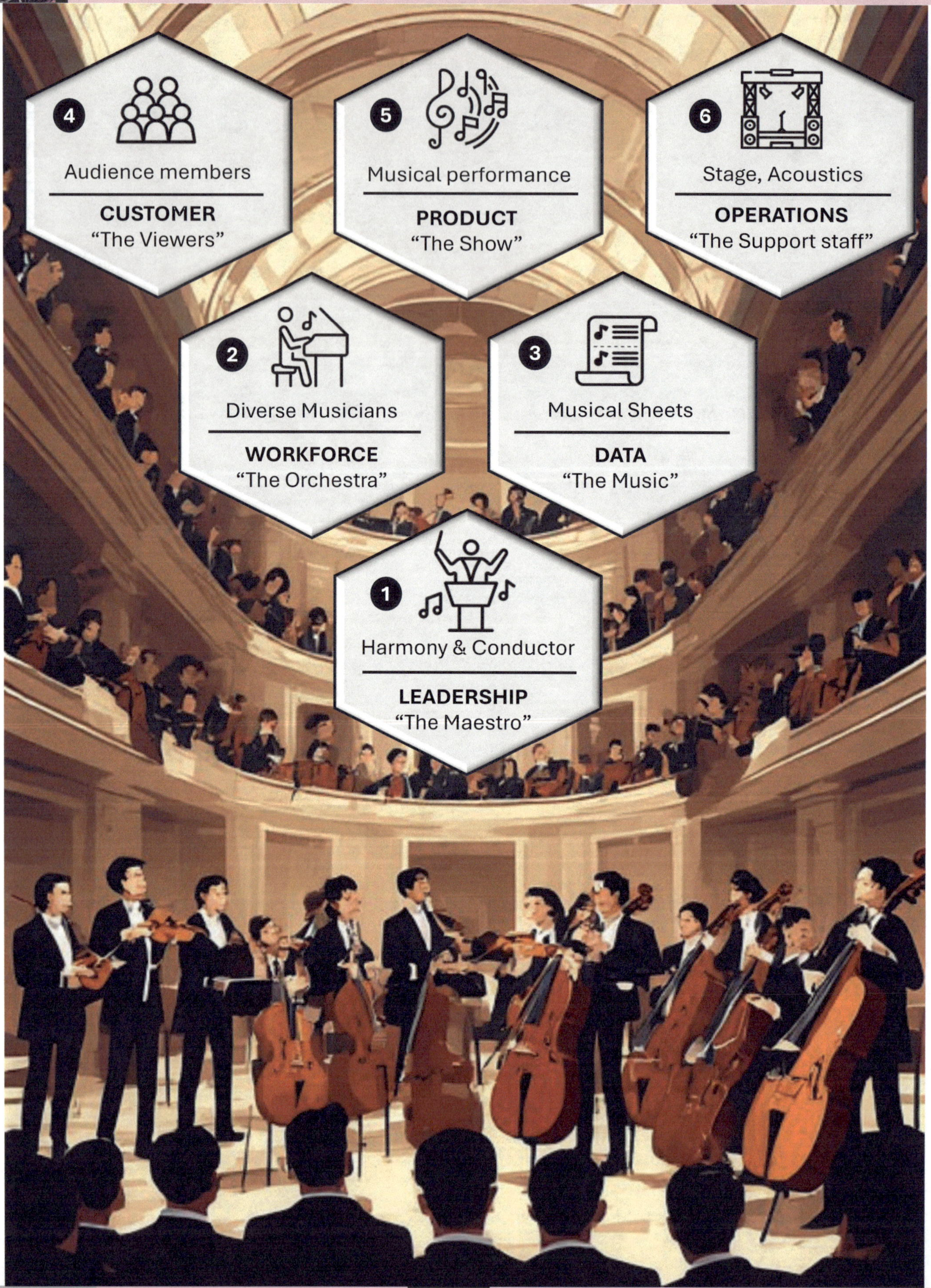
4
Audience members
CUSTOMER
"The Viewers"
5
Musical performance
PRODUCT
"The Show"
6
Stage, Acoustics
OPERATIONS
"The Support staff"
2
Diverse Musicians
WORKFORCE
"The Orchestra"
3
Musical Sheets
DATA
"The Music"
1
Harmony & Conductor
LEADERSHIP
"The Maestro"

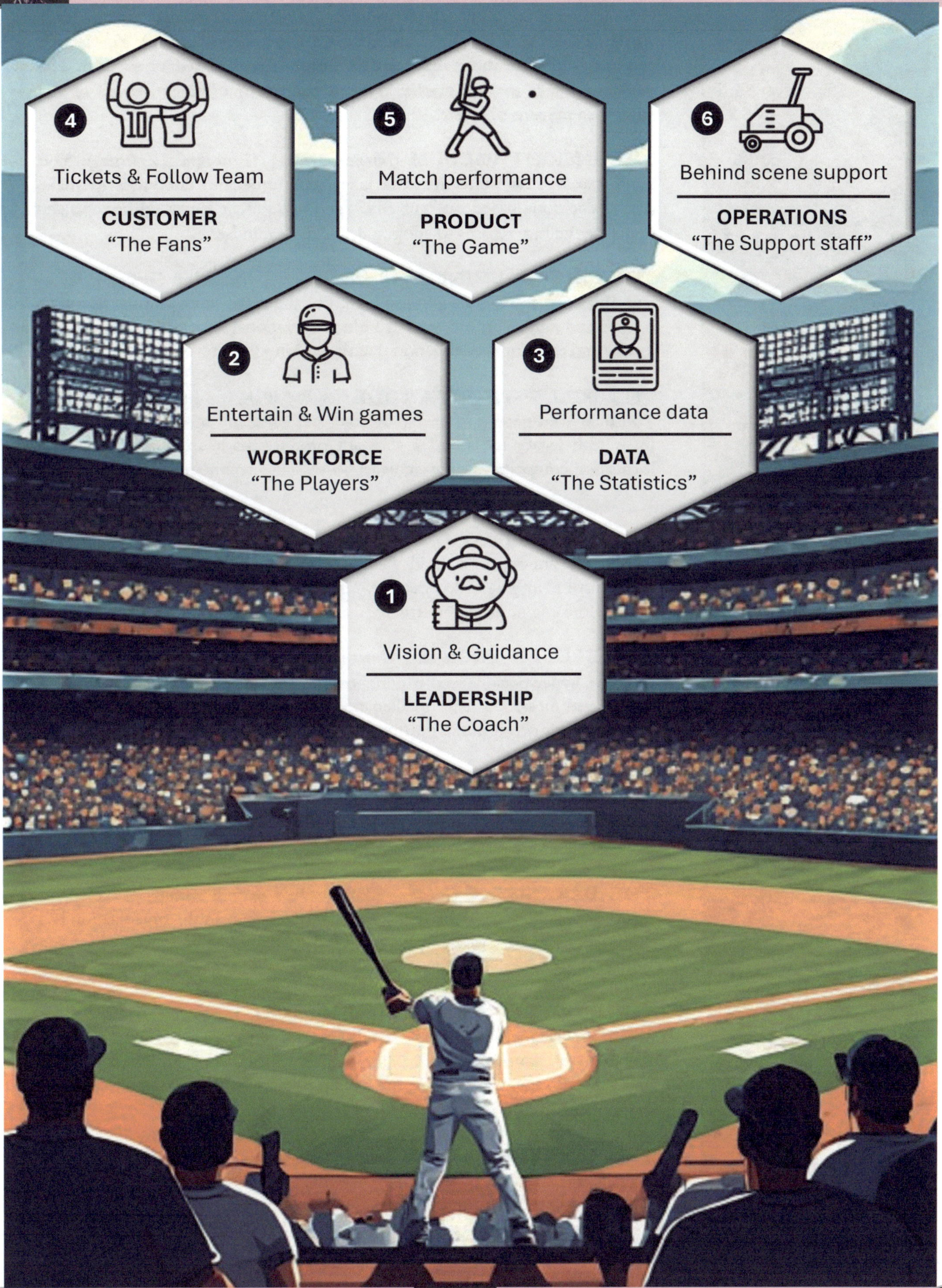
4
Tickets & Follow Team
CUSTOMER
"The Fans"
5
Match performance
PRODUCT
"The Game"
6
Behind scene support
OPERATIONS
"The Support staff"
2
Entertain & Win games
WORKFORCE
"The Players"
3
Performance data
DATA
"The Statistics"
1
Vision & Guidance
LEADERSHIP
"The Coach"

THE LITERACY GAP (*Invested: Product | Overlooked: Workforce*): You launch genius AI-driven features that your staff cannot explain, sell, or support. The technology fails to gain market traction because the "human layer" of the enterprise remains obsolete.

THE INSIGHT VACUUM (*Invested: Data | Overlooked: Leadership*): You build massive, high-integrity data lakes, but leadership continues to make critical decisions based on "gut feel." The investment is wasted because the upper-tier culture refuses to adopt a data-driven mindset.

THE COLD MACHINE (*Invested: Operations | Overlooked: Customer*): Back-end efficiency reaches an all-time high, but the actual user journey becomes robotic and rigid. You successfully reduce operational costs but lose the brand loyalty and emotional connection that drives long-term revenue.

THE SKILL WITHOUT A TOOL (*Invested: Workforce | Overlooked: Product*): You train your people to be "AI-Ready" and culturally agile, but you provide them with legacy tools and a stagnant product roadmap. Your best talent leaves for competitors who actually possess the technology to match their skills.

THE HALLUCINATION (*Invested: Leadership | Overlooked: Operations*): Executives announce a bold "Digital First" vision, but the underlying operational plumbing is decades old. The strategy never leaves the boardroom because the engine lacks the structural integrity to turn the wheels.

THE BROKEN PROMISE (*Invested: Customer | Overlooked: Data*): You market a hyper-personalized experience, but your internal data silos remain a fragmented mess. You end up alienating customers with irrelevant offers and inaccurate information that destroys their trust.

THE PROTOTYPE TRAP (*Invested: Product | Overlooked: Operations*): A brilliant AI-driven tool is developed in a vacuum, but the legacy supply chain and logistics cannot fulfill the demand it creates. The product succeeds in the lab but collapses the moment it hits the real-world operational floor.

THE DARK ASSET (*Invested: Data | Overlooked: Workforce*): You spend millions cleaning and structuring your data, but no one in the organization has the literacy to query or interpret it. The data sits idle and "dark," providing zero ROI because the people aren't equipped to find the value.

THE ROGUE ENGINE (*Invested: Operations | Overlooked: Leadership*): Middle management automates workflows to save time, but leadership has no visibility or governance over these new "black box" processes. The organization moves faster, but the steering wheel is no longer connected to the captain's chair.

THE INTERNAL ECHO (*Invested: Workforce | Overlooked: Customer*): Employees are empowered and "AI-literate," but they become obsessed with internal process improvements rather than user needs. The company becomes a high-performing bubble that is perfectly optimized for a customer that no longer exists.

COLLISION PATHS TO AVOID IN YOUR TRANSFORMATION

THE BLIND CAPTAIN (*Invested: Leadership* | *Overlooked: Data*): A visionary leader attempts to steer the company through a massive pivot based on bold assumptions. Because the data pillar was ignored, they navigate straight into a market reef that the "missing data" would have clearly identified.

THE FANCY WRAPPER (*Invested: Customer* | *Overlooked: Product*): Your marketing and customer service are world-class, but the actual product has not evolved to keep up with competitors. You win the initial sale through great branding, but the churn rate sky-rockets once the customer realizes the product is hollow.

THE ANALYTICS ISLAND (*Invested: Data* | *Overlooked: Operations*): The data team produces perfect, real-time reports on every aspect of the business. However, the operational workflows are so slow and manual that by the time the business acts on the data, the opportunity has already vanished.

THE ORPHAN PROJECT (*Invested: Product* | *Overlooked: Leadership*): Innovation teams build cutting-edge AI tools that solve real problems, but leadership never provides a permanent budget or strategic home for them. These "pilot projects" eventually wither and die because they lack the top-down air cover to survive.

THE FRUSTRATED EXPERT (*Invested: Workforce* | *Overlooked: Operations*): You hire the best AI researchers and data scientists, but you force them to work within manual, "broken" internal workflows. Your experts spend 90% of their time fighting the system rather than building the future.

THE IVORY TOWER (*Invested: Leadership* | *Overlooked: Customer*): The board develops a strategy based purely on shareholder value and financial engineering. By ignoring the Customer pillar, they fail to see a massive shift in user sentiment that makes their entire three-year plan irrelevant.

THE HUMAN BOTTLENECK (*Invested: Operations* | *Overlooked: Workforce*): You automate 80% of your workflows, but the remaining 20% require human intervention that your staff isn't trained for. The entire system grinds to a halt because you optimized the machine but forgot to upskill the operator.

THE REACTIVE PANIC (*Invested: Customer* | *Overlooked: Leadership*): The company chases every new customer trend and social media "vibe" to stay relevant. Without a strategic anchor from Leadership, the organization becomes a fragmented mess of contradictory ideas that stand for nothing.

THE WASTEFUL HARVEST (*Invested: Data* | *Overlooked: Product*): You collect trillions of data points and pay massive cloud storage fees, but none of that data is ever used to iterate or improve the actual product. You are "Data Rich" but "Value Poor," hoarding information with no path to monetization.

THE GRASSROOTS REBELLION (*Invested: Workforce* | *Overlooked: Leadership*): Staff begins using "Shadow AI" and unsanctioned tools because leadership is too slow or fearful to provide guidance. You have an AI-ready workforce that is operating outside of the company's governance, creating a massive security and legal risk.

If the Pillars define the WIDTH of enterprise transformation, the **Six Lenses** define its DEPTH. They are the universal imperatives that cut vertically through every domain—Leadership, Workforce, Data, Customer, Product, and Operations. Each Lens exposes a different dimension of capability, ensuring that transformation is not superficial, fragmented, or episodic, but deeply rooted in the systems that shape how organizations think, behave, and evolve. These Lenses are not optional. They are not enhancements. They are the structural forces that determine whether transformation becomes sustainable or collapses under its own ambition. Each Lens reveals a layer of truth about the enterprise—its intelligence, its discipline, its technology, its execution, its culture, and its integrity.

The Human–AI Synergy Lens (L1) examines how intelligence is created, shared, and amplified across the enterprise. It explores the evolving relationship between human judgment and machine capability, ensuring that AI augments rather than replaces human potential. This Lens forces leaders to confront how decisions are made, how insights flow, and how people interact with intelligent systems. It challenges organizations to build environments where humans and AI operate as partners—each elevating the other. Human–AI Synergy is the starting point because it defines the cognitive foundation of modern transformation.

The Business Maturity Lens (L2) establishes the structural discipline required for transformation to take hold. It examines governance, alignment, prioritization, funding, risk management, and organizational readiness. This Lens reveals whether the enterprise has the scaffolding to support change or whether ambition is outpacing capability. Business Maturity is where leaders confront the operational truth: transformation is not powered by inspiration alone—it is powered by systems, discipline, and repeatable mechanisms that ensure consistency across the enterprise.

The Digital Evolution Lens (L3) focuses on the technological backbone of the organization. It examines how systems integrate, how data flows, how platforms scale, and how digital environments adapt to new demands. This Lens ensures that technology becomes an enabler rather than a bottleneck. It challenges organizations to build digital ecosystems that are fluid, intelligent, secure, and trustworthy. Digital Evolution is not about tools—it is about creating a technological environment capable of supporting continuous change.

The Execution Mechanics Lens (L4) converts strategy into action. It examines the methodologies, operating models, workflows, and performance systems that determine whether the organization can deliver consistently and at scale. This Lens exposes the gap between what leaders intend and what teams experience. It ensures that execution is not left to improvisation but is engineered with precision. Execution Mechanics is where transformation becomes tangible—where ideas become outcomes and outcomes become advantage.

The Cultural Compass Lens (L5) explores the beliefs, behaviors, and shared meaning that shape how people think and collaborate. It examines psychological safety, communication norms, leadership behaviors, and the emotional climate of the enterprise. This Lens reveals whether the culture accelerates transformation or resists it. Cultural Compass ensures that people feel connected, supported, and aligned—because no transformation can succeed in a culture that does not want it.

The **Moral Compass Lens (L6)** anchors the Blueprint in ethics and transparency. Moving beyond a static matrix, it defines the **True North** for AI governance, data privacy, and leadership accountability. This Lens ensures transformation is principled, not just efficient. In the AI era, a Moral Compass is the boundary condition for legitimacy—without it, innovation loses its direction; with it, trust becomes an unbreakable competitive advantage.

Together, these Six Lenses form the DEPTH of the AI6x6 Transformation Blueprint. They ensure that every Pillar—Leadership, Workforce, Data, Customer, Product, and Operations—is approached with rigor, coherence, and integrity. They transform transformation from a collection of projects into a complete operating philosophy.

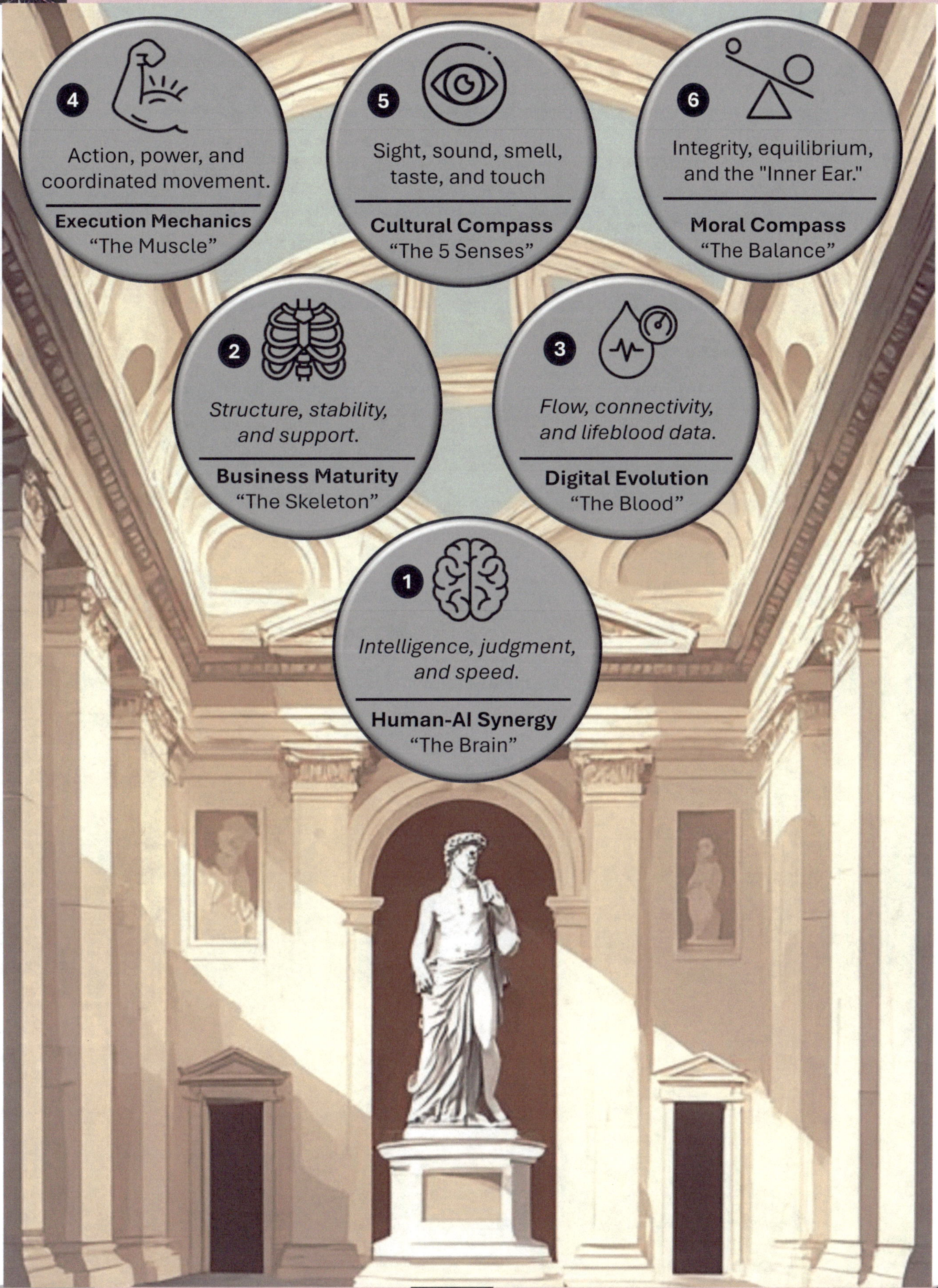
4
Action, power, and coordinated movement.
Execution Mechanics
"The Muscle"
5
Sight, sound, smell, taste, and touch
Cultural Compass
"The 5 Senses"
6
Integrity, equilibrium, and the "Inner Ear."
Moral Compass
"The Balance"
2
Structure, stability, and support.
Business Maturity
"The Skeleton"
3
Flow, connectivity, and lifeblood data.
Digital Evolution
"The Blood"
1
Intelligence, judgment, and speed.
Human-AI Synergy
"The Brain"

4
Gears that Convert energy power to motion
Execution Mechanics
"The Engine"
5
Monitor car health, temperature, audio
Cultural Compass
"The Dashboard"
6
Alignment, stability and stay in 'right lane'
Moral Compass
"The Steering"
2
Structural Frame and body of the car
Business Maturity
"The Chassis"
3
Gasoline, Electric energy & battery
Digital Evolution
"The Fuel"
1
Judgement, GPS, decision-making
Human-AI Synergy
"The Driver"

THE ETHICAL IMPLOSION (*Invested: Digital evolution | Overlooked: Moral compass*): You build a high-speed AI infrastructure with zero ethical guardrails or governance. One biased algorithm creates a public relations nightmare and a massive legal liability that can bankrupt the firm.

THE BUREAUCRATIC FREEZE (*Invested: Business maturity | Overlooked: Cultural compass*): You have established perfect governance and checklists, but you are operating within a terrified culture. People follow the rigid "process" of transformation but stop taking risks, effectively killing all real innovation.

THE THEORETICAL DREAM (*Invested: Human–AI synergy | Overlooked: Execution mechanics*): The organizational "brain" develops brilliant ideas for AI collaboration, but there is no "muscle" to act on them. Innovation stays trapped as a pilot project and never scales to produce enterprise value.

THE AUTOMATED ERROR (*Invested: Execution mechanics | Overlooked: Human–AI synergy*): You build a powerful execution engine but ignore the necessity of human judgment and intervention. The system begins executing "bad" decisions or flawed logic at 100x speed without anyone knowing how to stop it.

THE FUNDED HOBBY (*Invested: Cultural compass | Overlooked: Business maturity*): The internal culture is happy, safe, and collaborative, but there is no structural discipline or funding prioritization. Massive amounts of energy and capital are wasted on uncoordinated "passion projects" that never deliver ROI.

THE PRINCIPLED STANDSTILL (*Invested: Moral compass | Overlooked: Digital evolution*): You have world-class ethics and deep integrity, but you lack the "blood flow" of modern data and systems. You remain a "good" organization, but your technology is too outdated to actually compete with faster rivals.

THE FERRARI IN THE GARAGE (*Invested: Digital evolution | Overlooked: Execution mechanics*): You invest millions in an elite, high-performance tech stack that stays idle because no one has built the workflows to use it. You have purchased the "Engine" but never bothered to build the "Transmission" to turn the wheels.

THE LEGALISTIC LOOPHOLE (*Invested: Business maturity | Overlooked: Moral compass*): You strictly follow every legal rule and regulation while completely violating the spirit of ethics and trust. Your "Governance" is perfect on paper, but your brand is dying because your customers feel exploited by your "technically legal" actions.

THE INTELLIGENT OUTCAST (*Invested: Human–AI synergy | Overlooked: Cultural compass*): You develop AI tools that are technically brilliant and offer high synergy, but the workforce finds them creepy or intrusive. Because you ignored the "Senses" of the culture, the tools are rejected by the very people they were meant to help.

THE EFFICIENT COLLAPSE (*Invested: Execution mechanics | Overlooked: Business maturity*): The organization is moving with incredible power and speed, but it is moving toward a goal that isn't funded or strategically prioritized. You are winning the race, but you've realized too late that you're on the wrong track entirely.

THE NICE LIAR (*Invested: Cultural compass* | *Overlooked: Moral compass*): The culture is polite, collaborative, and highly "engaged," but it is built on a foundation of hidden ethical skeletons. People are "happy" because they are protecting each other from the truth, leading to an eventual systemic implosion.

THE ANALOG SAINT (*Invested: Moral compass* | *Overlooked: Human–AI synergy*): The leadership has high integrity but lacks the "AI-IQ" to understand the risks of the modern digital landscape. They are making ethical decisions based on a world that no longer exists, leaving them vulnerable to new forms of disruption.

THE EXPENSIVE CHAOS (*Invested: Digital evolution* | *Overlooked: Business maturity*): You layer high-tech AI tools on top of disorganized, immature governance and disorganized funding. The tech doesn't solve your problems; it simply makes your existing chaos move much faster and cost much more.

THE STAGNANT FORTRESS (*Invested: Business maturity* | *Overlooked: Execution mechanics*): The organization has perfect defenses, massive funding reserves, and rigid discipline, but zero kinetic movement. You are a perfectly "safe" business that is being slowly surrounded and suffocated by smaller, faster competitors.

THE DR. FRANKENSTEIN (*Invested: Human–AI synergy* | *Overlooked: Moral compass*): You achieve brilliant "Brain-to-Machine" innovation that pushes the boundaries of what is possible. However, because you ignored the "Inner Ear" of balance and ethics, you create a system that causes more social harm than organizational good.

THE BURNOUT FACTORY (*Invested: Execution mechanics* | *Overlooked: Cultural compass*): You optimize your "Muscle" for maximum output but completely ignore your "Senses" and emotional climate. You fail to detect internal fatigue and resentment until your highest-performing talent quits en masse.

THE LUDDITE OASIS (*Invested: Cultural compass* | *Overlooked: Digital evolution*): You have created the best place in the world to work, with high safety and great vibes. However, because you ignored the "Blood Flow" of digital evolution, your company is being disrupted out of existence by competitors you can't even see.

THE TOOTHLESS POLICY (*Invested: Moral compass* | *Overlooked: Execution mechanics*): You have beautiful "Values" posters and world-class ethics policies, but no actual workflows or mechanics to enforce them. Your integrity is a theory that falls apart the moment a team is pressured to hit a quarterly target.

THE GHOST SHIP (*Invested: Digital evolution* | *Overlooked: Human–AI synergy*): You have successfully implemented full automation and "lights out" data systems across the enterprise. However, because there is no human judgment in the loop, the system continues to run on outdated logic that is driving the company into a ditch.

THE PRECISE DINOSAUR (*Invested: Business maturity* | *Overlooked: Human–AI synergy*): You have spent years optimizing and documenting every single process to a level of "Six Sigma" perfection. Unfortunately, you ignored the "Brain" of AI synergy, and those processes are now perfectly optimized for a market that has already moved on.

Human judgment amplified by AI

Human–AI Synergy defines how intelligence is created, shared, and amplified across the enterprise. It examines the evolving partnership between human judgment and machine capability, ensuring that AI elevates rather than replaces human potential. This Lens forces leaders to confront how decisions are made, how insights flow, and how people interact with intelligent systems. It exposes whether the organization is building a truly augmented workforce or simply layering tools on top of old thinking. Ignoring this Lens leads to fragmented automation, poor adoption, and inconsistent decision-making. Human–AI Synergy is essential because it establishes the cognitive foundation of transformation—how the enterprise thinks, learns, and adapts—and without it, AI becomes a disconnected tool rather than a strategic multiplier.

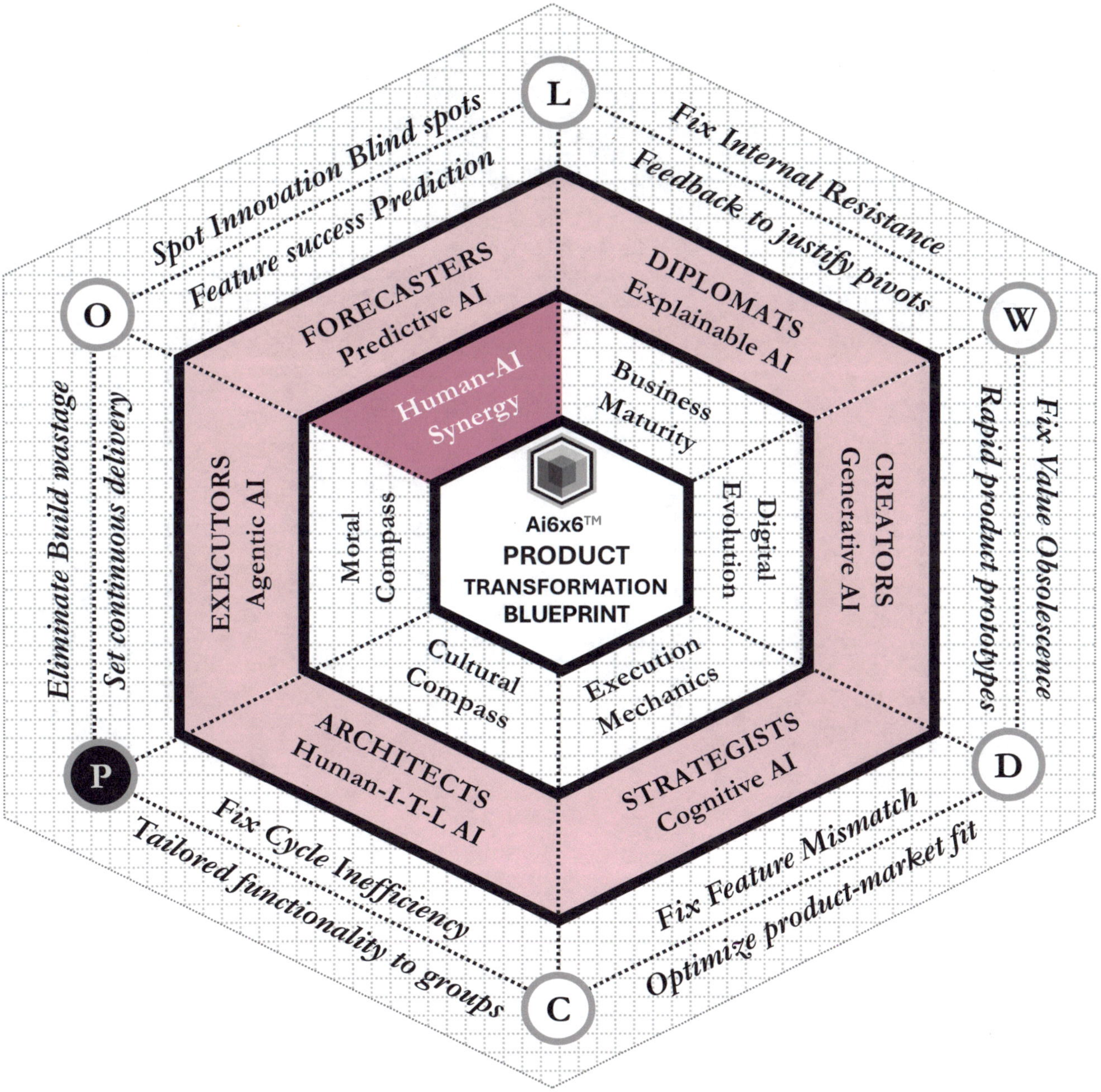

Ai6x6 HUMAN-AI SYNERGY FRAMEWORK FOR **PRODUCT LIFECYCLE**

Discipline that stabilizes transformation

Business Maturity is the structural discipline that transforms raw ambition into a legacy of Authority. It begins at the Foundation, establishing the governance and funding models that anchor an organization's identity before activation begins. As the enterprise moves through Initiation and Cohesion, this Lens evaluates whether the internal scaffolding is robust enough to support the increasing velocity of the transformation engine. This Lens exposes a fundamental truth: sustained success is engineered through systems, not slogans. It ensures that as Momentum builds, the enterprise maintains the Optimization necessary to prevent collapse under the weight of its own initiatives. Mastering this Lens allows an organization to transcend temporary programs and achieve the permanent market Authority that defines industry leadership.

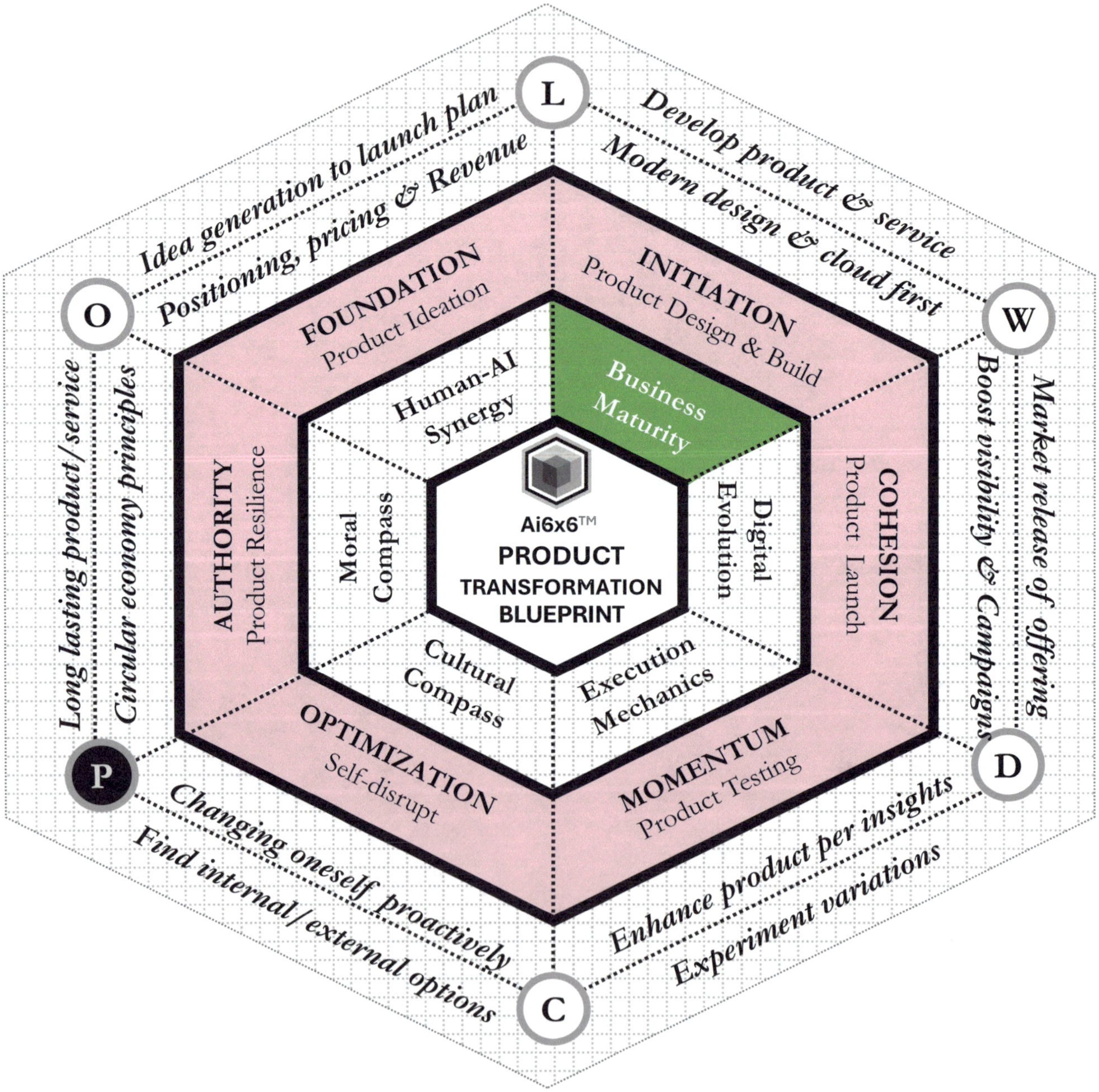

Ai6x6 BUSINESS MATURITY FRAMEWORK FOR **PRODUCT LIFECYCLE**

Infrastructure built for continuous change

Digital Evolution examines the technological backbone that enables modern transformation. It evaluates system integration, data flow, platform scalability, and the adaptability of digital environments. This Lens ensures technology accelerates rather than obstructs progress. It forces organizations to confront technical debt, architectural fragility, and the true readiness of their digital ecosystem. When ignored, enterprises build brittle systems, slow innovation, and create environments where strategy moves faster than infrastructure. Digital Evolution is not about tools—it is about constructing a digital foundation capable of continuous change. It ensures that technology becomes an enabler of intelligence, speed, and resilience. Without this Lens, even the strongest strategies fail because the underlying environment cannot support enterprise-wide evolution.

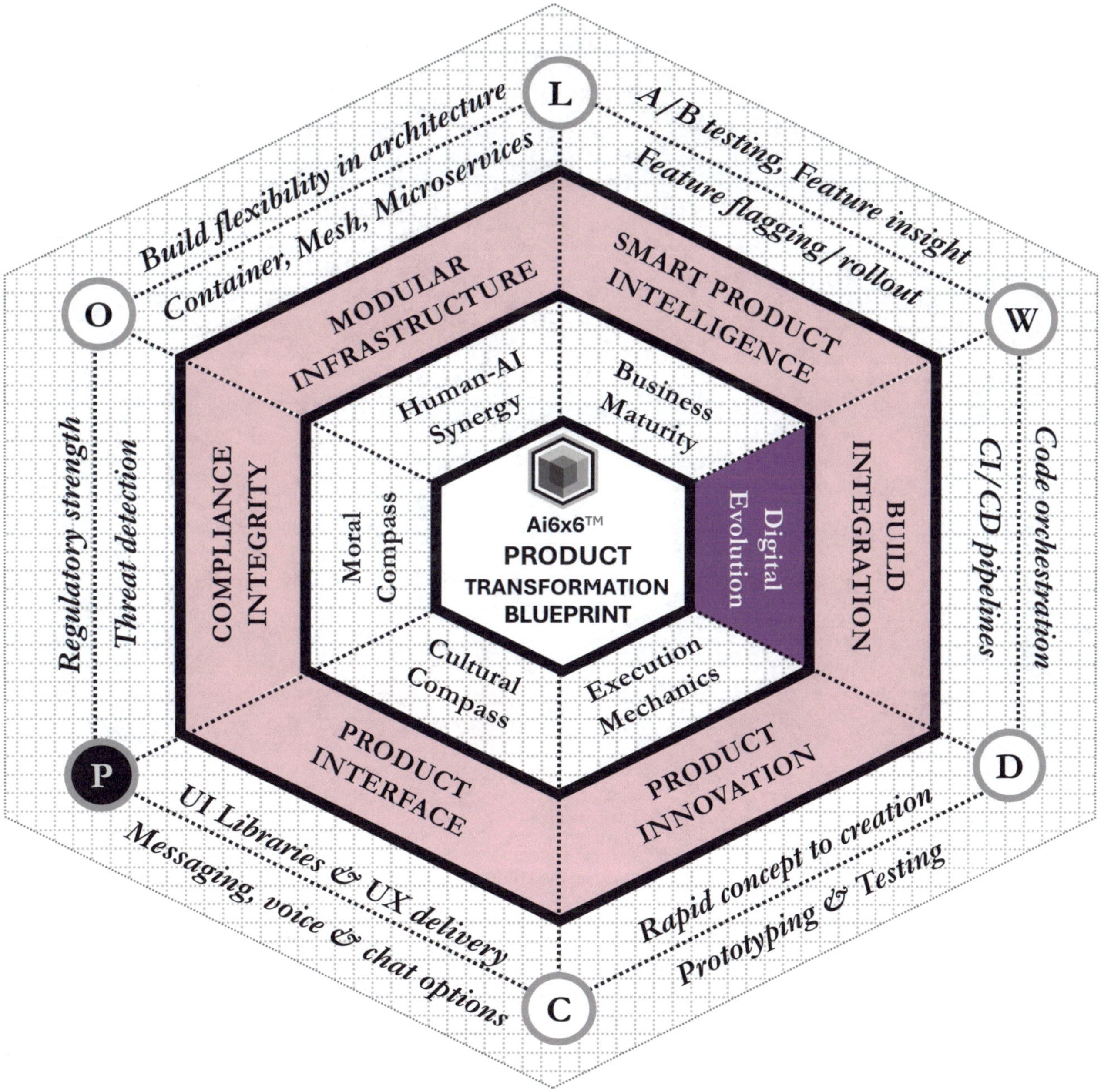

Ai6x6 DIGITAL EVOLUTION FRAMEWORK FOR **PRODUCT LIFECYCLE**

Strategy delivered with operational precision

Execution Mechanics converts strategy into consistent, scalable action. It examines operating models, workflows, methodologies, and performance systems to determine whether the organization can deliver reliably. This Lens exposes the gap between leadership intent and frontline reality, revealing whether execution is engineered or improvised. It forces leaders to confront bottlenecks, handoff failures, and the true velocity of delivery. Ignoring this Lens results in missed deadlines, inconsistent outcomes, and organizational frustration. Execution Mechanics is essential because it transforms ideas into outcomes and outcomes into advantage. It ensures that transformation is not theoretical but operational—a system of delivery that produces measurable impact at scale.

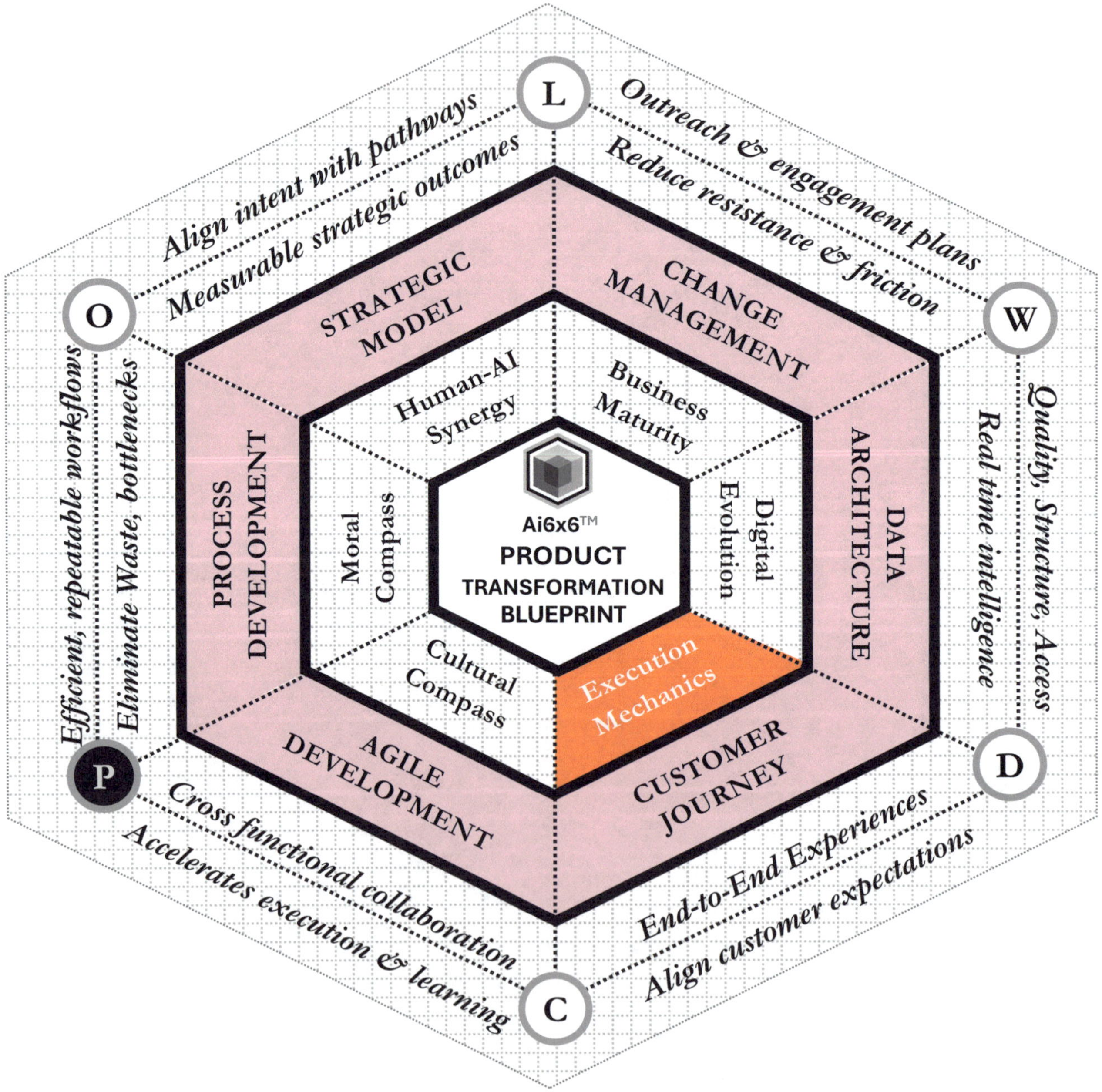

Ai6x6 EXECUTION MECHANICS FRAMEWORK FOR **PRODUCT LIFECYCLE**

Culture aligned to accelerate change

Cultural Compass explores the beliefs, behaviors, and shared meaning that shape how people think, collaborate, and respond to change. It examines psychological safety, communication norms, leadership behaviors, and emotional climate. This Lens reveals whether culture accelerates transformation or quietly resists it. It forces leaders to confront the invisible forces that determine engagement, trust, and momentum. Ignoring this Lens means ignoring the human system that ultimately determines success. Even the best strategies fail in cultures that are misaligned, fearful, or fatigued. Cultural Compass is essential because transformation is a human endeavor—no system or technology can succeed without cultural readiness. It ensures people are not just compliant but committed, creating the conditions for lasting change.

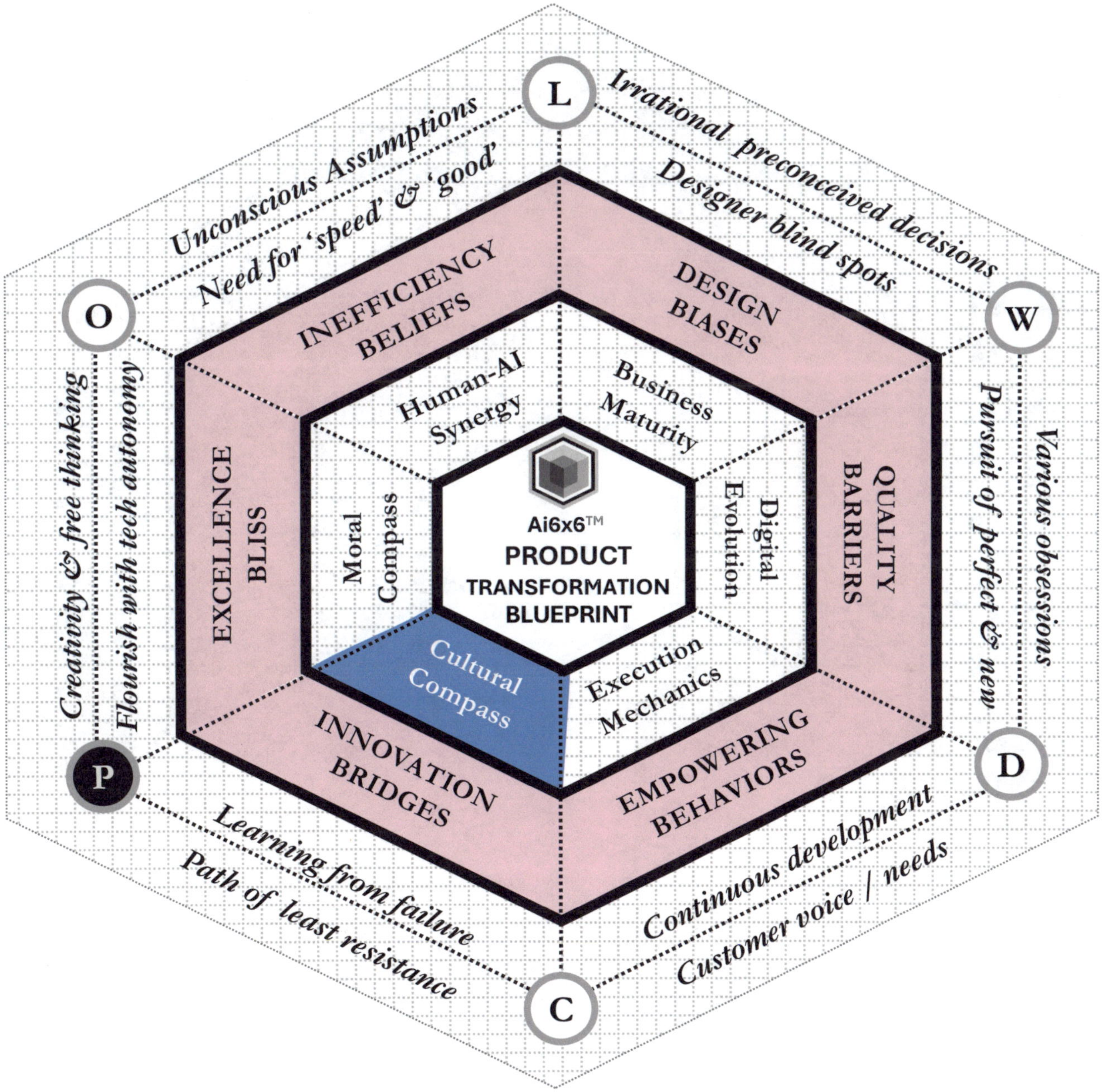

Ai6x6 CULTURAL COMPASS FRAMEWORK FOR **PRODUCT LIFECYCLE**

Ethics governing intelligent enterprise systems

The **Moral Compass** anchors transformation in ethics, transparency, and responsibility. It evaluates data privacy, AI fairness, accountability structures, and the moral obligations of leadership. This Lens ensures that innovation is principled and that stakeholders feel protected, respected, and informed. It forces leaders to confront the ethical implications of their decisions and the long-term consequences of their systems. Ignoring this Lens exposes organizations to reputational damage, regulatory risk, employee resistance, and customer disengagement. Trust is the boundary condition of modern enterprise—without it, nothing else holds. The Moral Compass is essential because it defines the legitimacy of every decision, system, and outcome, ensuring transformation is not only effective but worthy of confidence.

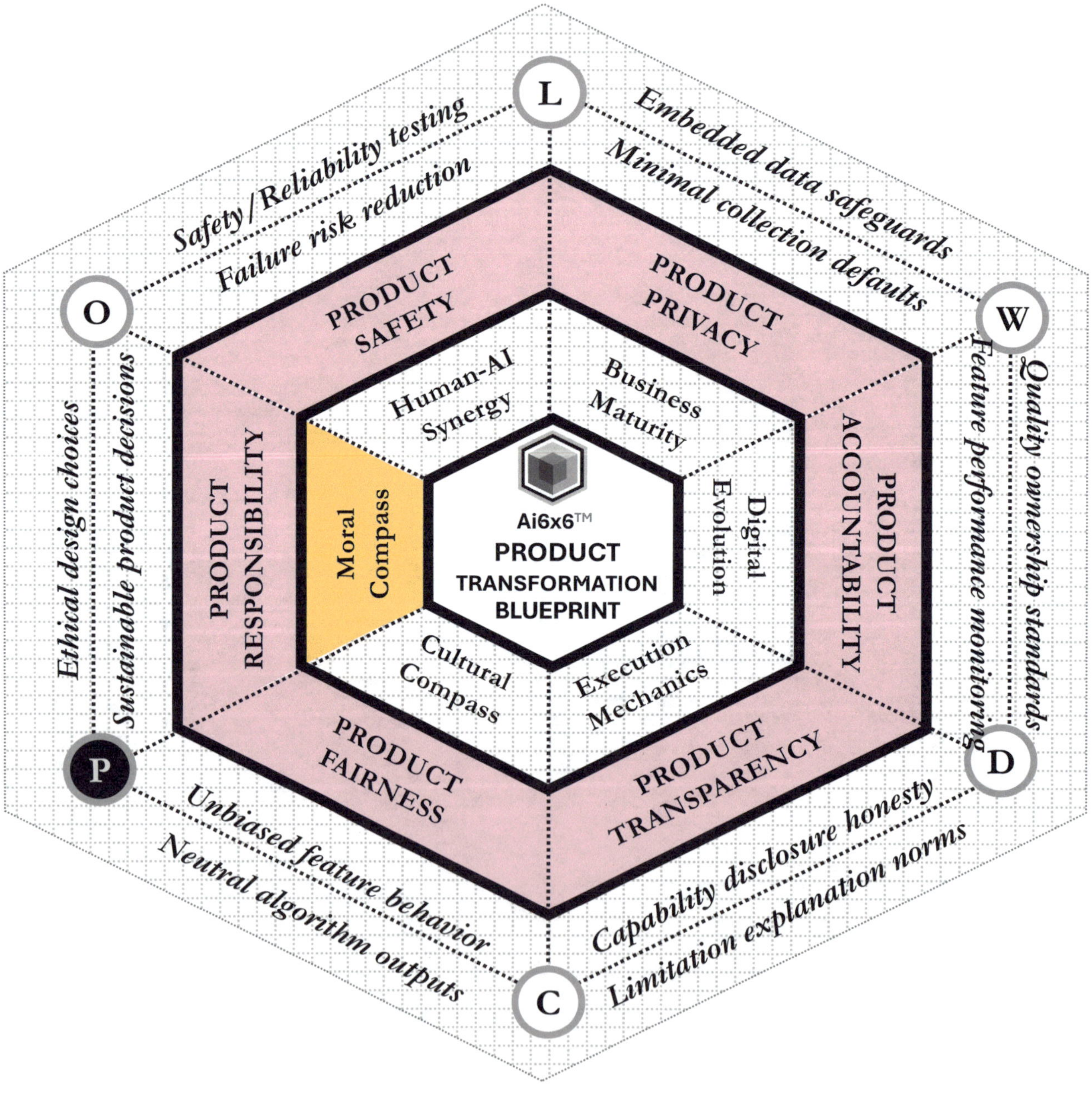

Ai6x6 MORAL COMPASS FRAMEWORK FOR **PRODUCT LIFECYCLE**

Ai6x6™ Transformation Blueprint

Why Transform Product?

"You can have brilliant ideas, but if you can't get them across, your ideas won't get you anywhere." — Lee Iacocca

Lee Iacocca was an American automobile executive and author, best known for his leadership roles at Ford Motor Company and as the CEO of Chrysler Corporation.

In a rapidly changing market, products that fail to evolve become irrelevant. Product transformation is a constant necessity driven by shifting consumer demands and technological advancements. Without it, a business risks being disrupted by more agile competitors who are better at solving emerging problems. This transformation requires a continuous cycle of design, testing, and iteration, prioritizing user experience and feedback above all else. It ensures that a company's offerings remain fresh, valuable, and indispensable. The benefit of this transformation is a product portfolio that consistently meets market demands and a reduced risk of obsolescence.

1. **Faster Time-to-Market (TTM):** Leveraging robust CI/CD pipelines, automated testing, and microservices allows product teams to release new features and updates continuously (often multiple times per day) rather than in slow, monolithic batches. This superior speed allows the organization to respond instantly to competitor moves or rapidly capture emerging market opportunities ahead of the curve. First-mover advantage is maintained across the entire feature set, not just the initial launch.
2. **Continuous Reformation Capability:** Products are built with Structural Optionality—modular, decoupled components that can be independently updated, replaced, or scaled, ensuring no technology decision permanently constrains future innovation. This architectural flexibility eliminates the need for expensive, time-consuming "rip and replace" projects, maintaining a perpetually modern and adaptable technology stack. The system can integrate future technologies with minimal re-engineering.
3. **Tighter Product-Market Fit:** The architecture supports rapid A/B testing, feature flagging, and multivariate analysis directly in the production environment, allowing teams to quickly validate assumptions and optimize features based on user behavior data. By proving value with small user segments before large rollouts, the organization minimizes the risk of launching features that fail to meet market expectations. Investment is concentrated only on features with proven user adoption.
4. **Increased Experimentation Velocity:** Low-cost, isolated testing environments and infrastructure-as-code allow teams to spin up and tear down experimental features rapidly, dramatically accelerating the organizational pace of learning and discovery. A high frequency of safe failure and learning is the single biggest predictor of long-term disruptive innovation success. The speed of testing translates directly into a higher rate of successful commercial innovations.
5. **Higher Software Quality and Stability:** Automated quality gates, shift-left testing practices (testing earlier in the cycle), and smaller, more frequent deployments reduce the likelihood of major bugs and service-impacting incidents. The reduction in major production failures improves customer trust, minimizes SLA penalties, and frees up engineering time from firefighting. Engineering capacity is consistently dedicated to value-adding work.
6. **Reduced Cost of Innovation Failure:** Due to the modular nature of the architecture and early testing, unsuccessful product features can be retired or repurposed quickly and cheaply, minimizing the wasted investment and technical baggage. This low barrier to removal encourages more aggressive, boundary-pushing attempts at innovation without fearing paralyzing organizational cleanup. Failure becomes a cheap learning opportunity, not an existential threat.
7. **Reduced Technical Debt Accumulation:** Standardized development patterns, automated deployment scripts, and mandatory code quality checks prevent the constant accumulation of unmanaged technical debt that typically slows down large organizations. A managed technical debt profile ensures the velocity of future feature development remains high and predictable, preserving long-term profitability. Developers spend less time navigating legacy complexity and more time coding new value.
8. **Optimized Cloud Spend:** Strategic use of serverless components, containerization, and granular resource allocation ensures the organization pays for precisely the compute power required, improving cost efficiency and minimizing resource waste. By dynamically scaling down compute resources when not needed, the transformation provides immediate, measurable savings on utility cloud bills. Infrastructure costs are directly and accurately tied to business activity.
9. **Embedded Security (DevSecOps):** Security policies, vulnerability scanning, and compliance checks are automated and integrated directly into the continuous delivery pipeline, making security a mandatory, non-negotiable part of every release. Proactive security integration means vulnerabilities are caught and fixed in hours by the development team, rather than discovered months later in a costly audit. The security team shifts from gatekeeper to enabler of speed.
10. **Modular Ecosystem Integration:** Products are designed to be exposed externally via clean, well-documented APIs, enabling easy and fast integration with partners, third-party services, and new sales channels, vastly expanding market reach. The ability to quickly participate in new ecosystems transforms the product from a standalone application into a platform for partner value creation.

1. The Feature Factory Trap — Shipping Without Strategy

Every sprint, Priya's team delivered. Tickets closed. Features launched. Stakeholders applauded the pace. But when support tickets surged and engagement plateaued, nobody had answers. A new PM posed an uncomfortable question: "Are we building what matters—or just what fits the sprint?" The silence was louder than any burndown chart. A retrospective revealed the truth: they'd become a feature factory—shipping outputs without outcomes. Roadmaps were filled with pet requests and gut feel. So they reoriented the process. Each initiative required a user pain, goal, or behavioral trigger. Half the backlog was culled. They launched fewer features—but tied to validated needs. One simplification in onboarding boosted conversion by 18%. Another solved a long-standing support issue. Engineers felt reengaged—they now understood the "why," not just the "what." By year-end, usage depth rose. Retention ticked up. Leadership shifted focus from quantity to quality. Product transformation isn't about how fast you build—it's why you build at all. Because volume without value is just motion.

2. Lost in Handoff — When Product, Design & Engineering Drifted Apart

Anna's team had energy, process, and intent. They followed the playbook—research, ideate, wireframe, develop. But when a highly anticipated feature finally launched, users were confused. "This isn't what we wanted," one wrote. Postmortem revealed a subtle but fatal drift: product, design, and engineering had interpreted requirements differently. Research was documented but not discussed. UX nuance got lost between Figma and Jira. Implementation met specs—but missed spirit. The team had checked every box except one: shared understanding. So Anna experimented. She co-located squads. Product, design, and dev sat (or Slack'd) together daily. They replaced documentation with conversation, specs with sketches. Engineers joined user interviews. Designers attended grooming. Suddenly, bugs fell, time-to-clarity dropped, and user sentiment improved. The team wasn't just collaborating—they were co-creating. The best products don't survive handoffs—they dissolve them. Product transformation means teams don't work in phases. They work in flow.

3. The Metrics Mirage — When Vanity Derailed Vision

Every dashboard sparkled: DAUs were up. Click-throughs soared. Feature adoption looked strong. But something wasn't right. Reviews dipped. Churn crept. "Feels overbuilt," one user wrote. Jasper dug in—and saw the trap. His team had over-optimized shallow metrics: activity without impact. Launches targeted engagement spikes, not sustained value. That changed after a tough leadership review. They wiped the slate and rebuilt their measurement model: time-to-value, resolution rates, sentiment by segment. They stopped obsessing over opens and started asking, "Are users better off?" Initially, numbers fell. But insights deepened. Teams learned which friction actually drove drop-off—and what created stickiness. A revised feature with fewer steps improved NPS. A simplified onboarding drove trial-to-paid conversion. The org shifted focus from dashboards to decisions. Jasper smiled as one designer said, "Now I know what success feels like." Product transformation happens when you stop reporting success—and start delivering it.

4. Reinvention from the Edge — When Innovation Came Bottom-Up

At Nuvalent, innovation had always flowed top-down—from R&D to release. But when Priya, a support rep, hacked together an internal tool to automate a customer task, it went viral—internally. Within two weeks, five teams were using it. She'd built something real from a known pain. Leadership paid attention. They launched "Builder Fridays"—where any employee could prototype ideas. Product, design, even legal showed up to jam. Some ideas flopped. But one idea from a success manager—a visual dashboard for onboarding—became a product-line staple. The team realized their best insights weren't always at the top. They were in the margins. Now, 30% of the roadmap came from frontline experiments. Employees didn't just execute—they shaped. Customers noticed. So did retention.

Product transformation isn't just faster sprints—it's wider radars. Your edge knows things your center can't. But only if you listen.

5. Stalled at MVP — When Teams Forgot to Finish

The MVP launched in record time. "We'll improve it later," the PM promised. But later never came. Teams moved on. Customers gave feedback—ignored. Reviews soured. One customer asked: "Is this feature abandoned?" The truth stung. MVP had become MIA. That prompted a rethink. The company introduced "MVRs"—Minimum Viable Roadmaps. Every MVP had a planned next iteration or deprecation. Pilots came with expiration dates. If traction appeared, investment followed. If not, they retired respectfully. Teams began tracking usage before celebrating launch. One feature's second version, informed by feedback, doubled adoption. Another was sunset quickly—with clarity. Confidence rose. Customers saw follow-through. Internally, accountability deepened. Transformation isn't speed—it's stewardship. A product half-launched is a promise half-kept. The new playbook? Ship fast, but finish well.

Because MVP isn't the end—it's the beginning of trust.

6. The Backlog That Ate the Strategy

Jira had 687 open tickets. Grooming meetings felt endless. Nobody could explain why feature X was still on the list—or why bug Y had sat untriaged for months. PMs felt reactive. Designers felt adrift. "We're sprinting in circles," one engineer joked. The problem? The backlog became a to-do list instead of a roadmap. So product leadership issued a bold directive: burn it. They archived everything over six months old. Then rebuilt from scratch—every item had to map to one of three outcomes. The first few weeks were bumpy. But then clarity emerged. Teams aligned on purpose. Developers started asking "why" again. Customers noticed tighter, more coherent releases. Velocity stayed steady—but impact rose. Groomings were shorter. Strategy was visible.

Transformation means treating the backlog as a compass—not a storage bin. Because busy isn't the same as building.

7. Product ≠ Platform — When Scaling Meant Systemization

Every enterprise deal required custom engineering. The core product became a patchwork of forks, toggles, and one-offs. Maintaining it drained velocity. "We're a dev shop, not a platform," one engineer muttered. When the CTO ran the numbers—70% of builds were one-time use—the case was clear. They hit pause. Teams re-architected the product: modular services, APIs, config-based UI. Internal docs treated features like Lego—plug and play. Now, customers self-serve 80% of requests. Sales time halved. Engineering reclaimed roadmap ownership. One customer called it "the most flexible out-of-the-box tool we've ever used." Transformation isn't just technical—it's strategic. Because products scale through systemization, not customization.

When every client request stops breaking your backbone, you know you've built a platform—not just a product.

8. Death by Default — When Design Ignored the Edge Cases

Friday night. Login failed—for hundreds of users. Support flagged a bug affecting people with non-Western characters in their usernames. The fix was simple. The fallout? Not so much. Angry tweets. One-star reviews. And a realization: the team had designed for the 80%, not the full spectrum. QA tests had no global scenarios. Research ignored accessibility. Design systems had "assumed defaults." After the fallout, they audited every edge—cultural formats, screen readers, low-bandwidth. A new mandate emerged: inclusivity is not a bonus. It's baseline. Designers got playbooks. PMs included edge personas in briefs. QA prioritized "infrequent" paths. The results? Fewer surprises. Better sentiment. Broader reach. One user wrote, "You fixed a bug we didn't know we were allowed to report."

Product transformation is realizing the margins are the mission. Because no one should have to fit your product. Your product should flex to fit them.

9. The Launch Illusion — When Release Didn't Mean Delivery

The product launched. Decks were shared. Tweets sent. But users barely clicked. Support tickets climbed. "We told them," said marketing. "They weren't ready," said CS. Dev felt proud. UX felt ignored. The truth? Launch wasn't delivery. The company treated release as finish—not transition. So they reframed everything. Every feature got a pre-launch pilot, co-owned by product, marketing, and enablement. Docs were pre-written. Support got playbooks. Success teams had talk tracks. Usage dashboards launched with the feature. Now, launch meant land. One update tripled adoption due to better onboarding. A customer wrote, "This time, we actually knew what was new."

Product transformation means releasing experiences—not just code. Because features don't matter until someone uses them.

10. Builders Without Borders — When Ethics Came Too Late

An AI tool was supposed to auto-fill job applications. Instead, it labeled a woman's title as "assistant"—even though she was a CFO. It went viral for the wrong reasons. The product team was horrified. "We tested the flows," someone said. "But we didn't test for bias," another whispered. Legal had seen the feature post-release. Ethics hadn't been consulted. That misstep sparked a system-wide shift. The company instituted "Ethics Sprints"—where risk, privacy, and accessibility reviews happened during early-stage discovery, not final QA. Red flags were tracked alongside epics. Designers began mapping potential harms like they mapped features. Engineers received training on bias in training data. Within two quarters, every launch had a "human lens" checkpoint. Velocity didn't slow—intention increased. Customers noticed. One reviewer wrote: "Finally, a product that understands I'm not a persona—I'm a person."

Product transformation means responsibility is upstream, not outsourced. Because if you don't build with care—you'll repair with apology.

11. When Your Product IS Your Brand

The brand had buzz. Their ads oozed cool. Their mission sounded righteous. But when customers finally opened the app, it was glitchy and generic. "Beautiful outside, broken inside," one reviewer wrote. Marketing blamed product. Product said branding was overpromising. Customers didn't care who was right—they just left. Eventually, the CMO sat in on product standups. Designers got trained in brand voice. Product teams started measuring "brand moments"—places where functionality met feeling. Micro-interactions mirrored the brand's warmth. Support copy matched its tone. Even error messages felt on-brand. Slowly, sentiment shifted. One user tweeted: "The product finally matches the promise." NPS and referral traffic rose in tandem.

Product transformation means realizing brand isn't a campaign. It's the experience in every tap, scroll, and click. Because what users remember isn't your logo—it's how you made them feel in motion.

12. Sunsets with Integrity — When Saying Goodbye Built Trust

The chart said it all: less than 2% of users tapped the "collections" feature. Yet a few users loved it—deeply. Product proposed cutting it. "Too niche," said ops. "I use it every day," wrote a teacher in Iowa. That sparked debate. Then a designer proposed something radical: say goodbye with grace. They built a sunsetting plan—notice, context, gratitude, alternatives. A banner popped up: "We're retiring Collections on May 15. Here's why." Users could export data, provide feedback, and even vote on what should come next. Responses poured in—some sad, some appreciative. One wrote, "I've never seen a company retire something with this much care." Internally, the team felt different too. They weren't erasing work—they were closing a chapter mindfully.

Product transformation means not just what you launch—but how you leave. Because endings build reputation. And in a world of churn, grace is your stickiest feature.

Ai6x6™ Transformation Blueprint

Power of Diversified Insight

"The pace of progress in artificial intelligence is incredibly fast... It is growing at a pace close to exponential."

Jeff Bezos, Founder of Amazon. Bezos highlights the velocity of change, signaling that in the AI economy, transformation cannot be a years-long project. Businesses must adopt a mindset of perpetual agility to keep pace with the exponential growth of technological capabilities.

We have deliberately chosen to move beyond the limiting confines of the conventional single-expert **Foreword**, which, by its nature, risks restricting the reader to one singular point of view. In an era where **business transformation** is shaped by a confluence of disruptive forces, relying on a solitary oracle is fundamentally inadequate. Instead, this book opens with a **curated collection of 15 interview snippets** from a diverse array of global thought leaders. This method immediately immerses you in a rich, multi-faceted tapestry of opinions and real-world experiences. By foregrounding this **plurality of perspective**, we offer a wider, more nuanced, and dynamic lens through which to view the challenges and enormous opportunities presented by modern change. Consider these short, powerful excerpts a **Chorus of Counsel**: a comprehensive and inclusive foundation built on collective wisdom that better equips you for your own complex journey of enterprise innovation and transformation. This is a commitment to **diversified insight** from the very first page.

EXPERT VOICES

Gain unprecedented, insider access to the minds of 27 global thought leaders in this six-book series. Sourced from comprehensive 1-on-1 interviews, these experts are recognized worldwide for their work in business transformation and emerging technology. Their insights are drawn from my personal, professional network—including former bosses, colleagues, and business partners — giving you an insider's view into the minds of industry shifters. To maximize focus and utility, we've synthesized these deep-dive conversations into thematic 'interview cards.' Each of the six books offers 15 relevant, concentrated insights on specific topics, allowing you to quickly access and apply the most relevant expertise, no matter where it originated in the original interview.

EXPERT VOICES

Subrata Mukherjee: You served as CIO, CDO, and CISO at Christie's, a role that traditionally focuses on stability, not disruption. You pioneered many "art tech firsts." What was the biggest challenge in getting buy-in from a legacy organization to experiment with new technologies like blockchain and augmented reality?

Richard Entrup: Christie's was a 250-year-old company so change was not easy. I had the support of a new CEO who was digitally savvy and understood the benefits of going digital for certain use cases and functions. At the end of the day, the auction house business is really only about finding great art to sell at the highest price, marketing the sale, and conducting successful auctions and private sales. To that end, I ensured everything we built with technology supported those business objectives, which made it a no-brainer, but using AI was different.

The Specialists (art sales/experts) felt that they could price the art better than AI and even predict what something would sell for, but we used data and AI for all kinds of analysis. We even built something at the request of the CEO which was a huge hit. An internal mobile app that helped track auction results real time. We also implemented **AR** to allow clients to see what a work would look like on their wall at home, which checked the engagement and conversion boxes. So, the punch line is to always align the technology to business outcomes, which usually fall under engagement, conversion, retention, and ultimately, revenue growth, so no one can challenge your efforts.

SM: You also launched an "Emerging Technology Group" at Christie's. What are the key elements of building a small, agile innovation team within a large, established company?

RE: First, develop a plan, overcommunicate, and go slow; Second, address staff and key partners, key man risks, and technical debt with great urgency, Third, measure results and communicate success.

Christie's had purchased a startup called Collectrum before I arrived. This was a true startup, open source, mobile/cloud-first, Agile-first, and they could build fast. The legacy IT organization was not so Agile on the DevOps/Software Engineering front. They had utilized traditional PMO/waterfall processes, maintained a large clunky ERP system on the backend, a legacy and dated web services infrastructure supporting mission-critical auctions, and long tenured staff that were not up to current standards.

The technical debt was great, things failed often, and there were serious cybersecurity concerns. So I integrated the Collectrium startup into the larger IT organization, but maintained it separately, to ensure they would not be negatively affected by the legacy team. The ETG team developed all the modern mobile apps that used AI, AR, etc.

I'm not sure how an IT leader can modernize their DevOps to current standards any other way except to outsource and/or slowly replace legacy resources with newer more modern engineers. The same dynamic applies to cybersecurity and IT infrastructure.

Augmented reality (AR) is a technology that overlays computer-generated digital information—like images, sounds, and text—onto the real-world environment, enhancing the user's perception and interaction with their surroundings in real time.

RICHARD ENTRUP

Managing Director, Head of Emerging Solutions at KPMG US

A multi-award-winning global technology and innovation leader with nearly three decades of experience driving strategic business transformation. He is at the forefront of identifying and incubating disruptive technologies like GenAI, Quantum Computing, and Spatial/XR, pioneering new products and solutions for clients. A two-time recipient of the HMG Global Leadership Institute Award for Exceptional Business Transformation, his distinguished background includes C-suite leadership roles (CIO, CTO, CDO) at globally recognized brands such as Verizon, Disney/ABC, Christie's, Time Warner, Viacom, and Tiffany & Company. He also serves as a strategic advisor and investor to a portfolio of innovative startups and venture capital firms, passionately guiding ventures focused on cutting-edge innovations. He is based out of New York, United States.

EXPERT VOICES

Subrata Mukherjee: In your current role, you are at the cutting edge of using data to solve complex problems in maritime intelligence. What are the unique data challenges in this sector, and how is AI being used to transform this raw data into actionable intelligence?

Saleem Khan: The unique challenge in this sector is transparency. Most competitors in this space will have access to vessel tracking, weather, and port activity from commodity providers. However, what our customers rely on the most is our ability to assess maritime risk and ensure safety of lives at sea which happens with AI to detect anomalies in vessel position, emissions, or other factors but to a larger extent our unique ability to anticipate whether a charterer or fleet may become sanctioned or put on a watchlist is key.

Most people don't realize that when you are 50 miles offshore from any country, you are in international waters which are not governed by any country. Anything can happen at sea from pirates hijacking your vessel to a ship being capsized by bad weather. The unique challenge here is to determine which vessel and operators have a higher chance of engaging in risky practices at sea. This has a tangible impact to shippers of goods, receivers of those goods, the crew that man the vessels, the insurers that insure cargo, and the financiers that loan money for a consignee of goods.

SM: How does Pole Star Global use predictive analytics and machine learning models to identify high-risk events, such as illegal fishing, sanctions violations, or potential piracy threats?

SK: We use custom-developed models, heuristics and first party data to identify risks. Similar to a credit rating we have created a new product called the Maritime Transparency Index™ (MTI) which assesses vessel, voyage, and emission risks to name a few. We do employ models to autogenerate these scores but much of our work is based on anomaly detection. Is a vessel moving at a speed which exceeds the second standard deviation for a VLCC (Very Large Crude Carrier)? Is the ship's draught level exceeding its departure draught? Is the vessel visiting ports that were not on its chartered route? The most important item in identifying risk is understanding the vectors that can change for a vessel, owner, and operator and also understanding that the bad guys are smart and will continue to change the threat vectors they use to avoid detection.

SM: What's the biggest difference between managing data in the financial services industry and managing data for global maritime operations?

SK: With respect to data management, I think the biggest difference is in the cadence and type of data being collected which necessitates a polyglot approach to data architecture. We do have a significant amount of ship position data that can be fed to use in one-minute intervals (not exactly hi-frequency trading numbers) but when you look at a vessel's position over the 70,000-100,000 or so active commercial fleet over time the dataset becomes quite large. We also have the emergence of sensor data being emitted from the vessel. This includes automated draught change detection, emission sensors, and engine condition status. All of this requires a flexible approach to data management.

SALEEM KHAN

Chief Data & Analytics Officer, Pole Star Global

Saleem is a seasoned data and analytics executive with a strong history of creating business value through data-driven innovation and successful company exits. In his current role as Chief Data & Analytics Officer at Pole Star Global, he leverages his deep expertise to lead and execute data strategy. His notable achievements include serving as Chief Digital Product Officer at VettaFi, which was successfully acquired by TMX Group, and as Chief Data & Analytics Officer at Discovery Data, which had a successful exit to Institutional Shareholder Services. He has also held leadership positions in data innovation and advanced analytics at major corporations like Dun & Bradstreet and S&P Global, and was a founder of the fintech startup ESGINE, an AI-enhanced platform for ESG compliance. He studied at Fordham University. He is based out of New York, United States.

EXPERT VOICES

Subrata Mukherjee: Having been at the forefront of digital for decades, you've witnessed the media and entertainment industry transform from the ground up. What is the single biggest change you've seen in the way technology is used to reach and engage consumers?

Abe Cytryn: The challenge was felt most by legacy businesses who could not pivot as quickly to digital models due to existing supply chains and relationships. An example is HBO who was so dependent on Cable Companies. This allowed Netflix to rise and surpass them with a business model that was built from the ground up to be direct to consumer. Large companies in almost every industry were slow and disrupted by start-ups. Consumer adoption also took time.

SM: In your view, what are the two or three most critical innovations that defined the digital evolution of the industry?

AC: Digital by definition means real time. Our **Analytics** can tell us very quickly what is working and what is not. **Mobile devices** have connected the global population like nothing else. **ON Demand** delivery of content, goods and self service for just about everything.

SM: How has the role of content itself changed in a world of endless digital delivery & user-generated media?

AC: In mostly good ways and some cautionary ways. In this booming era of AI, delivering relevant content to support our inquiries is more essential than ever. Our decisions and opinions are being shaped daily by content that is made available in so many instantaneous ways. The idea of trusted sources has waned over the years hence the caution. I think our ability to weed out incorrect or false information will be a key success factor. Users need to be more empowered to flag and correct information. I still can't get Google to correct the image of my street address!

SM: As a leader of product roadmaps, what was your playbook for moving a product from an idea to a successful launch?

AC: Our motto was show don't tell. It is easier and cheaper than ever to create a strong representation of an idea with a compelling user interface. Keep it simple and evolve it with functionality that is based on your learning from the users. You no longer have to debate if a feature is important. The analytics will tell you quickly.

SM: You've overseen massive efforts in Content Management and data strategy. What were the biggest technical and operational challenges you faced in managing large-scale digital assets and content?

AC: Third party tools in this space were just evolving when we started out. We tried a mix of home grown and open source solutions with mixed results. We also had unique requirements. The ability to scale and also publish very quickly. Think Time and People Magazine. Workflow was essential. A meta-data strategy that supported search and reuse. We were also an early user of Watson AI to automate tagging to a taxonomy we created.

ABE CYTRYN

Former Chief Technology Officer – Time Inc and Magster

Abe is an accomplished digital leader with over two decades of experience guiding digital product development and technology strategy in the media and entertainment industry. His expertise spans a wide range of critical areas, including eCommerce, mobile, big data, video, and content management. As a proactive and business-oriented leader, he has successfully developed technology roadmaps, built high-performance global teams, and led agile transformations to drive consumer growth and engagement. His extensive career includes pivotal leadership roles at Time Inc., Magzter, and HBO, where he consistently translated business objectives into innovative and effective technology solutions. Known for inspiring high-energy teams and fostering strong staff retention, Abe is an executive who empowers his people and delivers results. He is based out of New York, United States.

EXPERT VOICES

Subrata Mukherjee: The AI economy is rapidly changing the landscape for IT services. From your perspective, how is AI directly impacting the day-to-day operations of field services and the digital workplace,

Rod Rinker: I believe AI can play a vital role in diagnosing customer issues and providing known solutions. In addition, AI can provide the remote "diagnose-before-dispatch" function before dispatching a field engineer with a part or a recommended process to follow when the engineer shows up onsite.

SM: What new skills are becoming essential for the next generation of field engineers and service professionals to thrive in an era of intelligent automation?

RR: Know how to utilize virtual reality to improve remote tech assistance for field engineers. Know how to utilize Just-in-Time Training Refreshers prior to arriving onsite to keep technical skills fresh and relevant for the task at hand.

SM: How do you balance the drive for automation and efficiency with the need to maintain a human-centric, high-touch customer experience?

RR: Repetitive tasks should always be evaluated to determine how automation can streamline a process for greater efficiency. However, the human-centric, and high-touch customer experience is a key differentiator for competitiveness. Having a human involved to "guide" the customer experience requires the right culture and the right performance management system that rewards the behaviors you want. Empowering your team is key, but requires the right reward and recognition systems in place to motivate your team.

SM: You have deep expertise in Managed Services and XaaS.. What the managed services landscape will look like in the next 5 to 10 years, and how AI-powered capabilities will become a new form of currency in the market?

RR: The managed services landscape will become less about hiring an "IT staff" and more about a symbiotic partnership between human experts and intelligent AI systems. AI-powered capabilities will be the new currency by enabling predictive, proactive, and hyper-personalized service delivery. For example, instead of waiting for a server to fail, an AI-powered managed service can use a digital twin of a client's IT environment to simulate load and identify potential hardware failure points before they impact operations. The AI can then automatically order replacement parts and schedule a technician, or in many cases, resolve the issue virtually. Another example is how AI will become a currency by providing clients with intelligent cost optimization.

SM: What is the biggest strategic opportunity for a service delivery organization in the age of AI?

RR: The greatest strategic opportunity is transforming from a reactive support model to a proactive, automated, and self-healing service framework. The goal is to solve problems once and let the AI system learn from it, freeing up human talent to focus on innovation and high-value customer engagements. For example, when a network outage occurs, an AI system doesn't just log the ticket. It automatically analyzes logs, identifies the root cause using pattern recognition from millions of past incidents, and applies a known fix. If it's a new problem, a human engineer intervenes, and as they document the solution, the AI "learns" and adds this to its knowledge base for future automation. For complex issues, an AI can assemble a "virtual swarm" of relevant experts and an AI assistant. The AI can pre-populate the incident chat with diagnostic data and suggested solutions, allowing the human team to start problem-solving at a higher level of insight.

ROD RINKER

Senior Director, Global Field Services, Digital Workplace, Unisys

Rod is a results-driven executive with extensive expertise in global and regional IT service delivery and customer success. He has a proven track record of leading large, complex organizations (over 4,000 employees) and consistently exceeding service level agreements (SLAs) and contractual commitments. At Unisys, he is responsible for the global field services operation, focusing on modernization and transformation. His career is marked by key roles at NCR and Hewlett Packard Enterprise, where he drove significant revenue growth, improved customer satisfaction (as evidenced by high Net Promoter Scores), and successfully managed multi-million-dollar services contracts. An expert in leadership, he emphasizes building a "trusted advisor" status with clients to foster loyalty and expand business. He holds an MBA from Georgia State University & B.A. in Computer Science - Anderson University. He is based out of Atlanta, in the United States

EXPERT VOICES

Subrata Mukherjee: What are the signs that a platform is no longer fit for purpose, and what is the most challenging part of a migration?

Karen A. Brophy: The biggest sign is overall velocity of the platform. Publishing speed, the ability to roll out features quickly and then the performance of the products are all key indicators. If you are struggling in any of those areas, especially for news which is urgent, the platform is not acceptable. The hardest part of migration is business continuity, ensuring no dip in traffic or revenue during transition. We solved this with building the platform in parallel to current operations. Migrations were done in groups with the highest risk going last to minimize on any traffic or revenue impacts. And we worked closely with local teams on a train the trainer model to expedite training across such a large group.

SM: What is your strategy for winning in OTT and FAST, especially in local news?

KAB: Local news has habit and trust advantages. Our strategy was to lean into those with utility like breaking alerts, weather, and lifestyle, then add entertainment to drive time spent. Local stations would promote their OTT and FAST on air which drove awareness and eventual adoption. Distribution was important as well, so we secured placement on Roku, Amazon, Samsung, and Apple. Monetization came from blending programmatic video, sponsorship, and local ad packages. In broadcast especially, OTT inventory can be sold as premium and is highly valued by local advertisers.

SM: What key roles do you prioritize when building a new digital/product team, and how do you align them with revenue?

KAB: I prioritize product management to set the roadmap, engineering to build, design to ensure usability, data and analytics to measure, and revenue operations to ensure investment in resources generates a profitable return. I use performance incentive and shared OKRs to drive alignment. I am transparent with the team with the goals on revenue and EBITDA. Every role, from engineer to editor, can see how their work drives revenue and engagement.

SM: What challenges are companies bringing you today?

KAB: Common themes include aligning product strategy with revenue, figuring out AI use cases with integrity, and scaling from founder-led to process-driven organizations. I also advise a variety of startup companies in the AI automation, ad tech and AI tools for media companies. There is a lot of activity in that area and I am helping them consider their go to market, roll out and scaling.

SM: How will AI change hyperlocal journalism?

KAB: It's a big question and results will vary depending on the outlet. Some companies are using it primarily for automation, rewriting headlines for SEO, or article summaries. There's a thought that stories could be drafted from an outline and then editors will add the human touch. It is already changing, one company advised has built AI aggregated newsletters for hyperlocal communities. The AI isn't writing content, but it is aggregating content relevant to the community, and it can be autogenerated after a user puts in a new zip code. I think communities will be wary of AI created journalism and it can erode trust. So I see it's used more in automation than in news coverage.

SM:: You are now a Fractional Product Leader. What attracted you to the fractional model?

KAB: After scaling billion-dollar media divisions, I wanted to apply my playbook across multiple companies. Fractional leadership lets me parachute into high-stakes challenges, deliver impact quickly, and help founders or CEOs avoid costly missteps without them needing to build a full-time C-suite. It's something I've been doing as I consider my next full-time role. I've had some amazing clients and enjoy the work, but as an operator, I eventually want to build something new again.

KAREN A. BROPHY

General Manager, President, COO/CPO | Ex Hearst, Yahoo!

Karen is a dynamic executive specializing in P&L Management, Digital Transformation, and Revenue-Driving AI and technology business strategy for national media brands and consumer platforms. As the founder and President, Digital Division at Nexstar Media Group, she oversaw all digital business, product, and monetization strategies for over 200 local stations, The CW, and The Hill, delivering the highest digital revenue in company history with 70% growth. She is an expert in launching scalable platforms, notably developing the AI-powered advertising platform, Stellar, and spearheading the rollout of OTT Video and FAST channels. Karen held senior Product Leadership roles at major companies like Hearst and Yahoo!, in addition to her current advisory work as a Fractional COO/CXO, focusing on operational excellence and Artificial Intelligence. She is from Los Angeles, California, USA.

EXPERT VOICES

Subrata Mukherjee: During your time as a founding partner at Kaliop Canada, you worked on the digital transformation of major companies like The Economist. What were the unique challenges and opportunities in helping a legacy media brand like The Economist navigate its digital shift?

Gauthier Garnier: The digital transformation of a centuries-old brand like The Economist was a necessity to survive when information became universally accessible on the internet. The challenge was preserving its reputation while innovating its core processes. Our work focused on their e-commerce funnels—buying, gifting, and renewing subscriptions—and managing global channels with partners like airlines and credit card companies, navigating the complexities of taxes, currencies and logistics in over 55 countries. The biggest challenge was replacing an end-of-life system without carrying over its painful processes, while other systems were also being replaced. Our ability to listen, synthesize, and iterate allowed us to deliver a complete solution in under 18 months.

SM: How did you approach defining the technology roadmap for a company with such a valuable brand and a global audience?

GG: A key decision was balancing custom code versus productized software. It was clear no software on the market could deliver the expected features, so we built a custom solution. To limit technical debt and accelerate development, we used the Symfony framework coupled with an Enterprise Open-source DXP: Ibexa. This gave us the freedom to implement specific requirements while using Ibexa's out-of-the-box admin UI for content management. Long-term cost, flexibility, and performance were central to our decision-making.

SM: What is the most critical lesson you learned from that project about the relationship between technology and brand integrity?

GG: The most critical lesson is that technology must always be in service of the brand promise, never the other way around. For The Economist, the brand promise is intellectual rigor, clarity, and trust. Any technology we implemented had to reinforce those pillars. A slow-loading page, a confusing user interface, or a data breach would do more damage to their brand than a poorly written article. Brand integrity in the digital world is inextricably linked to the quality and reliability of the user's technical experience.

SM: How has the rise of AI changed the way you approach strategic planning for digital transformation and what are the most common challenges and fears companies face when trying to integrate AI into their existing business models?

GG: The rise of AI acts as a powerful accelerator, forcing us to discover, test, and iterate faster than ever to stay relevant. That said, the fundamentals of strategic planning remain the same: our core focus is always on the long-term vision and solving customer pain points. AI is not a strategy, but rather a formidable new tool in the developer's arsenal. It provides an unprecedented ability to solve complex business issues and enhance customer experiences. AI reinforces our core mission by providing new ways to address old problems, rather than fundamentally changing our strategic direction.

A lot of companies are pushed to jump into an AI transformation they are not ready for. This requires an advanced digital transformation, including access to enterprise data and a governance structure, but also the right mindset. Then there is the common "black box" problem—a lack of understanding of how AI models arrive at their conclusions, which creates concerns around control, bias, and accountability. Data is another significant challenge, as many companies struggle with data silos, poor data quality, and ensuring data privacy, which are prerequisites for effective AI.

GAUTHIER GARNIER

General Manager, DACH & Channels at Platform.sh

Gauthier Garnier is a highly experienced General Manager and business leader with a career focused on building strategic partnerships and long-term, win-win relationships with clients. As General Manager for the DACH (Germany, Austria, Switzerland) region and Channels at Platform.sh, he is responsible for driving growth and managing strategic alliances. His background is deeply rooted in digital transformation and web agency leadership. Previously, as a partner and General Manager at Kaliop, he successfully launched and scaled a web agency in the North American market, overseeing production, human resources, and finances. He has a strong expertise in strategic planning for complex digital projects, particularly within the retail and media industries, and has taught and coached on these topics. He has been recognized for his projects, including an "Innovation of the year" award for an e-commerce platform for The Economist. He is based out of Montreal, Canada.

EXPERT VOICES

Subrata Mukherjee: Having been at the forefront of content management for decades, you now see a new frontier with artificial intelligence. How do you believe traditional, monolithic CMS platforms are poorly suited to meet the needs of an AI-driven economy?

Bard Farstad: Traditional monolithic CMS platforms are, by design, rooted in the previous generation of digital publishing. They are not built with modern, API-first principles in mind, which makes them poorly suited for machine-to-machine communication and seamless integration with AI services. In an AI-driven economy, data must flow freely across systems to power automation, personalization, and predictive analytics. Monolithic systems tend to lock content and functionality into rigid silos, making it harder to extract, enrich, and repurpose information. This lack of flexibility prevents businesses from leveraging real-time insights or dynamically adapting content for different audiences and channels.

SM: In what ways is a headless architecture uniquely positioned to serve as the ideal content hub for Large Language Models (LLMs) and other AI applications?

BF: Headless architecture is uniquely positioned as the foundation for AI-driven applications because it is inherently API-first. In today's API economy, every modern service communicates through APIs, making real-time access to content and data seamless. A headless CMS exposes structured content directly through APIs, which allows Large Language Models and other AI systems to consume, enrich, and repurpose that data without the friction of presentation layers. This design supports true machine-to-machine communication, where AI agents can request, transform, and deliver content instantly across channels. By serving as a clean and structured content hub, a headless system enables advanced use cases like dynamic personalization, automated translations, and predictive recommendations. With the emergence of Model Context Protocol (MCP) servers and agentic AI interactions, headless platforms become not just content repositories but active participants in intelligent workflows.

SM: As AI becomes a co-pilot for content creation and a driver for personalized experiences, how will the role of a CMS need to evolve in the next 3-5 years?

BF: AI has already proven valuable for tasks like translations, summarization, and language refinement, and these capabilities are now standard in many workflows. Looking ahead, the role of a CMS will evolve from being a static repository of content to an active co-pilot in the entire content lifecycle. We will see far more automation embedded directly into the CMS or PIM, from idea generation to real-time personalization across channels. Over the next 3–5 years, CMS platforms must act as orchestrators that connect human creativity with AI-driven optimization. This means enabling agentic workflows, where AI autonomously executes tasks such as updating product data, tailoring experiences, or triggering campaigns. It is important to note that 3–5 years is an incredibly long horizon in AI terms. Considering that systems like ChatGPT have existed for less than three years, the pace of change will likely redefine what we even mean by "content management."

SM: The term "headless" is central to your business. Let's dig into the technical side. For a developer, what is the single biggest benefit of working with a headless commerce platform?

BF: For a developer, the single biggest benefit of headless commerce is simple: APIs. With a headless platform, every capability is exposed through APIs, which means developers have full control. Whether it is building a unique storefront, extending core functionality, or integrating with other services, everything becomes possible without being constrained by a rigid backend. Of course, the quality of those APIs matters.

BARD FARSTAD

CEO, Crystallize Headless Commerce

Bård is a serial entrepreneur and the CEO of Crystallize, a modern headless e-commerce service built on his vision to reimagine e-commerce platforms. His extensive career is defined by a history of founding and leading multiple companies, demonstrating a broad range of expertise that spans product, engineering, marketing, and digital strategy. In addition to Crystallize, he founded Snowball Digital, a product design and engineering firm, and Skiens Cykkelfabrik, a bicycle design company. Notably, he spent nearly two decades as the founder and chief strategist of eZ Systems, a content management software company, where he was a key driver of its engineering and product direction. With long-standing experience working with organizations across Europe and the USA, showcasing his commitment to strategic guidance and business development. He is based out of Norway.

EXPERT VOICES

Subrata Mukherjee: Your deep background as an Enterprise Architect for the BBC, the Telegraph, and the UK Border Agency and your work with AI Labs at Intive highlight your technical prowess. How do you define a "world-class" digital architecture that is both scalable and resilient, and what is the most critical first step a company should take when planning a re-engineering project?

Mark Brincat: For me, world-class architecture is defined by scalability, resilience, and adaptability. At the BBC and the Telegraph, scale meant millions of concurrent users accessing content in real time. It also meant working across an enterprise ecosystem that has evolved over several years. At the UK Border Agency, resilience meant 24/7 mission-critical systems with zero tolerance for downtime. Across all of these, the key was architecting for change—building modular, API-driven systems with governance that allowed new capabilities to be integrated quickly without destabilizing the core. The most critical first step is clarity of business objectives and constraints. Too often, re-engineering projects fail because teams focus on the "technology refresh" instead of the outcomes. The architecture must be driven by the business strategy—whether that's speed, compliance, scalability, or innovation.

SM: What are the specific architectural considerations and technical challenges involved in integrating new AI solutions into a legacy enterprise system?

MB: Integrating AI into legacy environments raises three big considerations: data readiness, integration layer, and governance and compliance. Legacy systems rarely provide clean, structured, or governed data. Without addressing that, AI underdelivers. Legacy platforms need modern APIs or middleware to connect with AI services, whether on-prem or in the cloud. And finally, explainability, bias, and regulatory reporting need to be built into the design from the start. These challenges are significant and require a holistic approach to data, technology, and governance.

SM: How critical is a strong cloud stategy for enterprise architecture today, especially with the growth of AI models?

MB: A strong cloud strategy is now non-negotiable. AI models are data- and compute-hungry, and trying to run them in siloed on-prem environments limits innovation. The cloud enables elastic scaling, global availability, and integration with AI services. But it's not just about "lift and shift"—it's about designing hybrid or multi-cloud architectures that balance sovereignty, cost, and performance, while keeping an eye on edge computing for latency-sensitive use cases. A well-defined cloud strategy is the very foundation of an AI-ready enterprise, as it provides the necessary infrastructure and tools to build, train, and deploy AI solutions at scale.

SM: Looking ahead, what is one technology trend, outside of AI, that you believe will have the most impact on the business world in the next few years?

MB: Outside of AI, I believe the most significant trend will be the evolution of data privacy and sovereignty. Regulations around data residency, ethical usage, and privacy are rapidly reshaping how global businesses operate. The convergence of cloud, blockchain, and advanced cybersecurity is creating an era where trust, transparency, and compliance are as valuable as raw computing power. Companies that build products with embedded trust and verifiable data handling will gain a significant competitive advantage, much like early adopters of mobile and cloud did a decade ago. It's a new frontier where reputation is directly tied to a company's ability to be a trusted steward of data.

MARK BRINCAT

Ex-CTO @ SHL, Economist | Architect & BBC, Telegraph, Home Office

Mark Brincat is a technology executive with deep expertise in digital transformation, AI, and data analytics. In his current role as the Global Head of Presales and Technology Partnerships at SHL, he leads go-to-market strategy and advises clients on their transformation journeys. Concurrently, he serves as a Fractional CTO for Castille Resourcing, where he oversees the engineering and scalability of their digital platform. He previously held key Chief Technology Officer (CTO) roles at both SHL and Intive, and senior architectural positions at major media organizations like The Economist, Telegraph Media Group, and BBC Worldwide. His career highlights include implementing large-scale digital and cloud solutions and forging strategic partnerships. He is based out of London, United Kingdom.

EXPERT VOICES

Subrata Mukherjee: As the Chief Product Officer and Head of Client Engagements at AInspire, you hold a dual role in shaping both your firm's offerings and your clients' strategies. How does your dual responsibility as both a product leader and a client engagement lead influence your approach to building and deploying Generative AI solutions?

Brian Glassman: Holding a dual position as Chief Product Officer and Head of Client Engagements at AInspire provides me with a unique advantage that integrates technological insight with client-centered strategy. From the product side, I focus on the evolving capabilities of Generative AI in areas such as text-based analysis, finance, and software engineering. My priority is to build fluency in the strengths and limitations of these tools so they can be applied precisely to real-world use cases. From the client engagement side, I begin with a deep understanding of each client's strategic priorities. This reveals where Generative AI can fill capability gaps or drive efficiency through automation. I only recommend deployment when the technology exceeds human benchmarks, ensuring results are not only innovative but reliable. This dual lens helps create solution architectures that are technically sound, operationally viable, and deeply aligned with business goals.

SM: You are developing a proprietary AI-driven solution for call centers. What are the key challenges and opportunities you see in transforming a traditional industry with a technology like AI?

BG: Transforming the call center industry with AI presents both significant opportunity and complex challenges. A key opportunity is unlocking underutilized data from recorded customer interactions. Most call centers record calls, but these are rarely analyzed beyond compliance or dispute resolution. Generative AI enables transcription and large-scale analysis of every conversation, offering a full view of agent performance, issue resolution, product trends, and customer sentiment. This allows call centers to shift from being cost centers to sources of actionable business intelligence. However, AI limitations must be considered. Text transcripts often miss emotional nuance. A phrase like "I'm very happy with your service" may be sincere or sarcastic, and without tone detection, intent can be misread. Advancements in emotional and tonal analysis will be essential to improving accuracy.

SM: In your view, what is the most critical first step a company should take when beginning to build its first Generative AI product?

BG: The most critical first step in building a Generative AI product is adopting a new mindset toward software. Traditional systems are deterministic, producing the same output every time. Generative AI, by contrast, is probabilistic. It generates multiple plausible outcomes, some of which may be ineffective or even harmful without appropriate oversight. Embracing this shift means accepting uncertainty and actively designing for it. Teams must invest in rigorous testing to understand where the model performs well, where it breaks down, and what kinds of edge cases require safeguards. One of the most common mistakes is assuming that AI can simply replace existing systems. It cannot. Success starts with careful evaluation, risk management, and human escalation protocols. Equally important is choosing the right entry point. A practical method is to use a "risk spectrum" approach: high risk tasks stay with humans, moderate risk tasks pair AI with human review, and low risk tasks such as detecting offensive language can be safely automated. Selecting a low complexity, low risk use case enables teams to gain experience without overexposing the business. The goal is not just to ship a product but to build internal capability.

DR. BRIAN SCOTT GLASSMAN

Chief Product Officer & Head of Client Engagements at AInspire.ai

Dr. Glassman is a seasoned product management executive, engineer, and product strategist specializing in generative AI. As the co-founder and Chief Product Officer at AInspire.ai, he leads client engagements and product strategies, collaborating with C-level executives to develop and deploy AI solutions for large enterprises. With a career dedicated to leading technology innovation, he previously served as Director of Product Management at Electra Vehicles, where he built the product function for AI-powered solutions in the automotive sector. His extensive experience includes a seven-year tenure at Rave Digital, where he led product management and customer engagements for Fortune 1000 clients. Furthermore, his background as an engineer and product leader at Cameron, a Schlumberger company, highlights a unique expertise in technical product development. He is based in New York.

EXPERT VOICES

Subrata Mukherjee: Your experience at EY and Razorfish involved advising Fortune 500 clients. How do you advise a technology leader to build strong, strategic relationships with key business stakeholders like CMOs and revenue leaders?

Chris Pena: The foundation is empathy, listening deeply to their priorities and consistently framing technology as an enabler of their specific goals. CIOs who show up as collaborative partners rather than gatekeepers build genuine credibility over time. Practical approaches include joint planning sessions, shared KPIs, and co-ownership of critical initiatives to break down organizational silos. Strong relationships develop when business leaders feel that IT is genuinely invested in their success, not just technical excellence.

SM: You helped build partnerships with major tech providers like Adobe, SAP, and Salesforce. What's the key to selecting the right technology partner for a company's specific needs?

CP: The critical factor is strategic alignment. Does the partner's product roadmap support where you want to be in three to five years? Beyond feature functionality, evaluate trust, cultural fit, and their genuine willingness to co-innovate with your organization. The best technology partners function as extensions of your internal team, deeply invested in your outcomes rather than just license revenue. Seek partners who can adapt and evolve with your changing business model and strategic priorities.

SM: What is one critical insight you gained from your time as a consultant that has influenced your approach as an in-house technology leader?

CP: Consulting taught me that transformation fails without executive sponsorship and compelling storytelling. Even the most technically superior solution will stall if leaders and employees do not believe in its value and necessity. As an in-house leader, I place much greater emphasis on change management, strategic communication, and aligning technology initiatives with existing business narratives. Technology represents about 30% of successful transformation, the other 70% is people, culture, and belief.

SM: At Aspire Lifestyles, you successfully transitioned operations to a cloud-based infrastructure. What were the primary benefits of this move beyond cost reduction, and what were the biggest challenges?

CP: Beyond cost optimization, the transformative benefit was organizational agility. Cloud infrastructure allowed us to scale capacity dynamically with demand, improve system resilience, and deliver customer-facing features significantly faster. It also created a foundation for integrating emerging technologies like RPA and AI. The primary challenge was change management, convincing stakeholders that security and operational control could actually improve in the cloud environment. We invested heavily in communication, comprehensive training programs, and robust governance frameworks to build trust and confidence in the transition.

SM: You also developed a modular B2B2C platform that achieved 60% efficiency gains. What is the most important lesson you learned about building scalable and reusable technology platforms?

CP: Design for flexibility from day one. A truly modular platform enables you to integrate new capabilities without re-architecting core systems. This approach reduced our delivery timelines from months to weeks, but more importantly, it gave business leaders confidence that technology could keep pace with rapidly shifting market demands. Building with reusability as a core principle is not just an IT best practice, it is a business enabler that directly accelerates growth and competitive responsiveness.

CHRIS PEÑA

VP, CIO Go To Market Industry Lead at Salesforce | Ex-CIO Aspire

Chris is a strategic technology leader and CIO advisor with a deep expertise in AI, digital transformation, and e-commerce. In his current role as VP, CIO Go To Market Industry Lead at Salesforce, he is responsible for developing and executing the company's value proposition for Chief Information Officers across various industries. As Global VP of Consumer Technologies at PVH Corp., he managed the global Salesforce platform. Prior to that, as General Manager CIO at Aspire Lifestyles, he built a global technology team from the ground up, modernizing operations with cloud infrastructure and automation, and successfully managing massive budgets. His extensive experience also includes senior leadership roles at EY, Razorfish, and Pearson, where he led digital strategy, platform implementations, and engineering. He is based out New York in United States.

EXPERT VOICES

Subrata Mukherjee: How do you use technology, particularly AI and digital solutions, to create a more personalized and effective customer experience in e-commerce?

Françoise Nauton-Inglis: Personalization should feel like assisting someone in their unique way of describing their problem, not remote guiding users based on what we want to sell them. We combine three signals: live intent (what the shopper asks for), product understanding (attributes and compatibility), and contextual rules (availability, delivery promises, brand priorities). Juliia fuses these in real time to propose "why-this-product" explanations customers can trust. For clients, we plug into Shopify, Adobe Commerce, Magento, Big Commerce and others platforms without the need to replatform or amend the existing site, so the personalization layer rides on existing stacks. Success is measured in fewer dead-ends, faster paths to suitable products, and a reduction in unnecessary returns—outcomes shoppers feel immediately.

SM: You have a deep understanding of e-commerce platforms like Shopify. How can businesses leverage platforms and custom solutions to balance the need for speed with the desire for a unique brand experience?

FNI: Adopt a "platform for 80%, product for 20%" mindset. Use the platform's native strengths—catalog, checkout, ops—to move fast and stay maintainable. Invest your custom layer where your brand wins: discovery, configuration, and service moments. With Juliia at Spiriit, we ship features that respect platform conventions but expose a flexible UX and reasoning layer you can truly brand. Governance matters: integrate design systems and content ops so every custom touch is repeatable. This balance lets you iterate weekly on high-leverage experiences while relying on the platform to carry the operational load.

SM: How do you measure the success of a digital transformation project for a client, and what metrics matter most beyond simple revenue?

FNI: We look at two things mainly. From a business perspective: conversion lift, average order value, return rate, support deflection. From a client perspective: task completion time, first-fit rate (first suggestion that meets criteria) after assisted discovery. We also use the content of past conversations for semantic search that provides unique and actionable customer insight. Finally we aim to track "digital sobriety" indicators—reduction of the number of calls to servers to generate a page, redundant calls avoided—because speed and sustainability often align. The narrative we want is "fewer clicks, clearer choices, happier customers," with revenue as the downstream proof.

SM: Beyond your work with AI for e-commerce, what emerging technology will have the most significant impact on the business world in the next five years?

FNI: I'm watching three trends that need to converge: agentic AI with verifiable context (retrieval + actions), privacy-preserving analytics at the edge, and sustainable IT. Together, they make digital services more proactive and accountable—able to act for customers while proving why those actions were safe, compliant, and beneficial.

SM: How can businesses stay ahead of the curve and prepare for the next wave of technological disruption?

FNI: Build muscles, not roadmaps: continuous discovery interviews, weekly instrumentation reviews, small dual-track teams that can ship and learn. Invest in integration layers—APIs, event buses, observability—so you can plug in new capabilities without tearing down the house. And cultivate external perspective: clusters like Digital 113, academic partners, and customer councils give you early signals you won't find in dashboards.

FRANÇOISE NAUTON-INGLIS

CEO @ Spiriit / Juliia, Ex GM/CFO of Kaliop Interactive Media

Françoise Nauton-Inglis is an accomplished business leader with over 20 years of experience in the digital and technology sectors. She is the CEO of Spiriit, an agency that designs and builds custom digital and AI solutions, and its AI-first product, Juliia, which focuses on intelligent search and product recommendations for e-commerce. Françoise's expertise lies in growth management, international development, and strategic business planning. Prior to Spiriit, she spent a decade at Kaliop Interactive Media, where she served as the Group General Manager, overseeing M&A, international expansion, and financial operations. She has a deep passion for structuring and growing innovative, profitable, and human-centered businesses, and she holds leadership positions in several organizations, including as President of Digital 113 and CI Montpellier Management. She is based out of Montpellier, France.

EXPERT VOICES

Subrata Mukherjee: A core part of your work is facilitating Design Sprints and AI Design Sprints to uncover product value. What is the single most important outcome a company should aim for at the end of a successful Design Sprint?

Sandy McCarron : A clear decision. At the end of a sprint, you should know definitively whether customers want what you're building and whether they would buy it. The core question is: Does this solve a real problem? A successful sprint provides that clarity—yes or no. It's about moving from a hypothesis to a validated, data-backed conclusion in a matter of days. You either get a clear signal to proceed with confidence, or you get a definitive signal to kill an idea before investing millions in a product nobody wants. That's invaluable.

SM: Can you walk us through how you've adapted a traditional Design Sprint to the unique challenges of building AI products, particularly in a data-rich environment like FinTech?

SMC: With AI, especially in data-heavy sectors like FinTech, we put extra emphasis on grounding the sprint in the problem rather than the technology. We prioritize testing whether the product direction resonates with customers and whether the data is reliable enough to support it. That helps prevent building solutions that look great on paper but can't deliver in production. We have also included Quants and Data Scientists in the Design Sprint as participants so they can share the deep understanding of the product the team is trying to build and for whom. We also dedicate more time to data exploration and validation in the early stages, essentially running a mini-sprint on the data itself to understand its limitations and capabilities before we even start prototyping.

SM: What is your philosophy for bridging the gap between product strategy and user experience to ensure a product delivers real value?

SMC: Product strategy is about identifying the next ten things to build—the roadmap. The "what" that will delight the customer. A well-designed user experience should feel natural and seamless, with the product's branding shining through but without calling attention to the mechanics of the interaction. The "how" features are displayed that is most impactful for users. My philosophy is that these two things are two sides of the same coin. Product strategy defines the north star, and UX design is the navigational tool that guides the user there. You can have the best strategy in the world, but without a compelling user experience, your users won't care.

SM: What are the unique challenges of implementing agile methodologies and running a PMO within the highly regulated and risk-averse financial services industry?

SMC: The biggest challenge is coordination—releases have to cross teams with varying access levels, which creates bottlenecks. However, one mistake can cost millions so it's wise to be cautious. On the positive side, agile provides transparency that's invaluable in such an environment. The daily stand-ups and regular sprint reviews create a public, real-time view of progress, which is a powerful tool for a PMO trying to track progress and identify risks. The key is to adapt the methodology to the environment, not the other way around. The goal is to make a process that enables teams, not one that burdens them.

SM: From your experience, what is the key difference between a successful PMO that enables transformation and one that is seen as a bureaucratic roadblock?

SMC: A successful PMO listens to the team. Forcing the team to use a tool or process that adds no value makes the PMO a bottleneck. A PMO becomes a roadblock when it demands busywork—like filling out spreadsheets without saving time, cutting costs, or moving features to market.

SANDY MCCARRON

Co-Founder & COO at Sapient Advisors | Ex- Moody, Ex-S&P

Sandy McCarron is a co-founder and Chief Operating Officer at Sapient Advisors, a lean, AI-enabled agency specializing in product and data strategy. He focuses on guiding companies through AI adoption and achieving operational efficiencies through rapid AI Design Sprints. With a career spanning both media and fintech, his expertise is built on a strong foundation of innovation and program management. At Moody's Analytics, he was a Senior Director of the Innovation Unit, where he introduced Google Ventures Design Sprints and led a major program to upgrade a flagship product, generating a $60M increase in profit. He also held a senior technology director role at S&P Global. His early career includes positions as Director of Online Content at Penguin Publishing Group and Director of Online Production at Penguin Random House. She is based out of New York.

EXPERT VOICES

Subrata Mukherjee: Can you describe a project that won an award and walk us through the journey from initial client need to final delivery?

Preeti Sharma: One of my favorite award-winning projects was the Insta-Challenge Social Media Campaign our creatives developed for John Jay College. The challenge was clear: keep college top of mind for incoming students during the summer, a time when many tend to disconnect and risk not showing up for orientation. The client initially asked for a print + experience approach, but we saw an opportunity to reimagine it digitally. Our team designed a "Choose Your Own Adventure" Instagram game where students followed a story and discovered hidden codes along the way. It transformed recruitment into a social, playful experience that met students where they already were—on their phones. The campaign went on to win both an AVA Platinum Award and a Shorty Award, but what mattered most was that it helped John Jay connect with students in a way that was authentic & innovative.

SM: What is the key to managing client expectations while delivering innovative solutions, especially when dealing with tight deadlines?

PS: It starts with understanding the client and how they work. We tackle this during discovery and spend a lot of time validating the problem before we ever recommend—or validate—a solution. From there, clear communication and straight talk are essential. I believe in painting a realistic picture: what's possible now, what might require trade-offs, and where innovation can push boundaries. That upfront clarity builds trust, even under tight deadlines, because clients know we're aligned on the problem we're solving and the path forward.

SM: Your work has been recognized with multiple AVA Digital Awards. Looking ahead, what is the most significant challenge you anticipate for Chief Information and Digital Officers in the next five years?

PS: Keeping pace with exponential change while ensuring inclusivity and security. Technology is moving faster than governance, workforce adaptation, or even culture.

Leaders will need to bridge that gap thoughtfully. Another major challenge will be finding and retaining talent—because no matter how fast technology evolves, your people are the most important cornerstone of your organization. Technology will always change, but it is the strength, creativity, and resilience of your teams that ultimately determines success.

SM: As a leader who believes in "possibility," what is one major innovation or change you hope to see in the technology services industry?

PS: I'd love to see today's smart note-takers evolve into real brainstorming buddies at the whiteboard—tools that don't just capture information, but actively contribute ideas, connections, and insights to spark creativity. Imagine AI not just recording meetings, but helping teams connect the dots in real time, so colleagues and AI tools can collaborate side by side.

SM: What is the most crucial piece of advice you would give to a woman aspiring to a C-suite role in technology today?

PS: First—know your stuff. Build your expertise and confidence in your craft, because credibility is the foundation. Then, build your network. Surround yourself with mentors, allies, and peers who will both support and challenge you.

Own your voice and your story—you don't need to "fit" a mold. Bring your perspective, empathy, and passion unapologetically. The tech industry needs more women who lead authentically, not by imitation.

PREETI SHARMA

Chief Digital Officer & CIO at VGD NYC

Preeti is a distinguished technology executive and a "Notable Woman in Tech" with a proven history of driving digital strategy and innovation. As the Chief Digital Officer and CIO at VGD NYC for over 18 years, she leads the delivery of digital services, technology strategy, and solutions for a diverse range of clients. Preeti's expertise is deeply rooted in her long-standing focus on AI and technology transformation, backed by recent certifications in AI Leadership and Digital Innovation from Carnegie Mellon University. Her foundational experience includes nearly a decade at ISI Emerging Markets, where she served as Global Operations Manager, playing a key role in scaling the business from a small startup into a profitable enterprise. A member of Chief and the Forbes Technology Council, she has been recognized with multiple AVA Digital Awards for her work. She is based out of New York.

EXPERT VOICES

Subrata Mukherjee: What are the key differences between leading data strategy as an in-house expert (e.g., at Bristol Myers Squibb) versus as a consultant at a major firm like Accenture?

Mousumi Sengupta: As an in-house leader at Bristol Myers Squibb, my focus was on deep, long-term stewardship. My role was to understand a single, complex data ecosystem intimately, reduce data siloing, and build a lasting capability within the organization. This involved extensive stakeholder engagement and organizational change management. At Accenture, the role was one of breadth and short-term problem-solving. My focus was on diagnosing diverse data challenges across multiple clients and developing scalable frameworks. I was an agent of change, delivering high-impact solutions in a compressed timeframe to provide immediate value.

SM: You worked on frameworks for Generative AI at Accenture. What does it take for a large pharmaceutical or life sciences company to successfully integrate these powerful new tools into their existing data ecosystem?

MS: Successful integration of Generative AI requires a multi-faceted strategy. First, it requires engagement and buy-in from a breadth of senior leaders across the organization. A robust data governance framework must be developed to ensure patient and proprietary data is handled with the utmost security and privacy. Finally, companies must move beyond novelty and identify high-impact use cases—such as R&D knowledge synthesis or clinical trial design—and build a pilot program that can demonstrate tangible ROI and pave the way for broader adoption.

SM: How did your work at Accenture inform your decision to found your own company?

MS: It was my experience transforming an NLP engine at Accenture that was the catalyst for starting my own company. I saw a recurring, fundamental pain point across my many roles: the massive, unsolved problem of unstructured data. We were building custom solutions for each client, but the core issue remained the same. I realized there was a market gap for a platform that could provide a repeatable, automated way to create a contextual data layer, enabling multiple meanings. This insight and the conviction that this problem was ripe for a product-based solution fueled my entrepreneurial journey.

SM: What is the most significant technological challenge that the healthcare industry must overcome in the next five years?

MS: I believe the most significant technological challenge that must be overcome is the persistent problem of data sharing and interoperability, as well as data siloing. Even with advances like HL7 and FHIR, systems are still not geared to talk to each other. Overcoming this requires more than just technology; it requires a concerted effort to establish a shared ontology and governance model that can handle the massive scale and complexity of clinical and research data. This foundational work is essential to unlock the full potential of AI and truly realize the promise of personalized medicine.

SM: What advice would you give to a young professional looking to specialize in bioinformatics or data strategy within digital health?

MS: I would encourage them to pursue this area for a career, as it is very rewarding and offers many opportunities to fix broken workflows and introduce real AI solutions. My advice is to gain a deep, specialized expertise in a technical area like bioinformatics, ontology engineering, or data science. At the same time, cultivate a broad understanding of the entire healthcare value chain and system. Learn the language of clinicians, researchers, and drug developers. Understand the complex regulatory landscape and business models. The most impactful leaders in this field are not just technologists; they are domain experts who can connect a technical solution to a real-world, high-impact clinical problem.

MOUSUMI SENGUPTA

Chief Data Strategist at Condext, Ex-Accenture, Ex- Mount Sinai Hospital

Mousumi Sengupta is an accomplished leader in digital health, currently serving as the Founder and Chief Data Strategist at Condext, a startup automating the contextualization of healthcare data. Her expertise in Data Strategy, Ontology Engineering, and AI is built on a career at prominent organizations like Accenture, where she helped develop a Life Sciences practice with a focus on Knowledge Graphs and Generative AI. Her prior roles include leading a FAIR Data Management function at Bristol Myers Squibb and driving semantic interoperability at Mount Sinai Health System. She was also a Director at organizations like Nuance Communications and the Multiple Myeloma Research Foundation (MMRF). She is a proven innovator at the intersection of technology and clinical data. She has a Masters in Bioinformatics from Columbia University. She is based out of New York, United States.

EXPERT VOICES

Subrata Mukherjee: Your role as the initial executive spearheading the HBO GO initiative was a landmark moment in media. Can you walk us through the biggest strategic challenge you faced getting it off the ground?

Michael Gabriel: The biggest challenge was in understanding the amount of resistance some of the executives would have based on their job responsibilities and personal ambitions. Before I even started at HBO, and due to my being in the music industry when Napster almost destroyed their business, I asked the CEO and CFO whether they supported trying to create an Internet streaming product for HBO. They both said they did, and that it needed to be done cautiously and slowly, to neither create concern with our existing distributors nor spend a lot of funds too early, since technology products need to continuously be updated. The goal was to build some of the key components to prove out the technology and then determine how we could best introduce it in partnership with our distributors. But, even with the support of the CEO and CFO, some key executives were hesitant to do anything to disrupt our extremely successful business model. The General Counsel actually pointed at me during a CEO monthly meeting and stated, "You're going to destroy a 3-billion-dollar business!". A few other business executives had similar concerns, either because their role was working with our existing affiliate distributors, or they were responsible for protecting our current business profitability. And, of course, if their worst-case fears came to fruition, it would impact their compensation, or their roles and influence would be adversely impacted. Anyone who has read the Harvard Business School book "The Innovator's Dilemma" would understand the conflict this creates, and that it's real. That makes it incredibly hard at most companies to introduce new products that compete with the existing profitable products. By working with the entire executive team, those who embraced this new direction, as well as mitigating the fears of those who were hesitant, we were able to move forward and successfully introduce the media industry-changing HBO GO. We also had to spend a lot of time modifying our operational business and digital transformational flows and focus on change management with respect to our employees, third parties, and customers. The impact of that company-wide effort forever changed the media & entertainment industry.

SM: As an inventor on six patents, how do you see the role of intellectual property within technology development in the media space, particularly streaming?

MG: Since we were doing things on a scale that had never been done before, we were in an excellent position to consider creating utility patents on processes and products. We created a newly formed HBO Patent Committee, co-chaired by both me and the CTO, to guide those intellectual property efforts. Creating patents allowed us to either impede our competitors from pursuing that direction, license the patents to create a revenue stream, or hold the patents defensively in case a competitor wanted to block us from infringing on their patents. The streaming industry continues to undergo change at a rapid and diverse pace for live events and recorded shows, for processes, and for software and hardware-related products, including the impact of AI. I believe that media companies must continue to innovate, no matter which way they want to strategically utilize the resulting patents.

SM: Looking back, what was the most valuable lesson you learned from that project?

MG: Innovation is not a switch that can be turned on and off. You either continue to innovate, or you risk being replaced by those who do. And the only constant is that things will continue to change, so a focus on how that impacts your company, business partners, and customers is crucial to remaining successful.

MICHAEL GABRIEL

Ex CIO at HBO, NBA | NACD Certified Board Director

Michael is a senior technology and cybersecurity executive with experience providing C-level leadership, governance, and advisory services. He is currently an independent Board Director and provides professional services for companies like Warner Brothers Discovery (WBD). His career is highlighted by a 14-year tenure as EVP & Global CIO at HBO, where he was the initial executive to spearhead the groundbreaking HBO GO initiative and led a company-wide digital transformation. An inventor on six technology patents, he also played a key role in information security governance across the Time Warner enterprise. Prior to HBO, he served as Global CIO for the National Basketball Association (NBA), where he implemented the first NBA retail store technology system. He also held senior IT leadership positions at EMI Records Group and NBC Olympics.. He is based out of New York

WE NEED YOU!

WELCOMING OUR NEXT SET OF THOUGHT LEADERS

The remarkable success of the Ai6x6 Organization Transformation Series is built upon the **hard-won wisdom** of executive leaders who have successfully navigated the AI transformation within their industries, turning theory into a deeply practical, actionable guide. As we expand the series, we are extending a highly selective invitation to a new cohort of pioneers—individuals whose professional journey is at the **leading edge of AI strategy and execution**. We are looking for executives who possess a proven track record of rewriting business models, mastering unforeseen ethical and competitive challenges, and offering a **clear blueprint** for the next 3-5 years of the AI-Era.

Contributing to the Ai6x6 Series is a unique opportunity to **cement your professional legacy**, joining an exclusive, curated network of global authorities whose insights are distilled into executive-level knowledge cards, directly shaping the governance and innovation decisions of the world's most influential companies. Our selection process is based strictly on **pure expertise and demonstrated leadership** in the AI-Era, reinforcing our commitment to editorial integrity and authentic thought leadership—there is no monetary exchange involved. Select contributors will also receive an invitation to blog on our expert network microsite, further amplifying their platform. To confidentially inquire about contributing your expertise, please contact us via our dedicated portal at **www.Ai6x6.com/contact** or by scanning the adjacent QR code.

Contact

SUBSCRIBE TO OUR NEWSLETTERS: To stay ahead of the curve and receive continuous, executive-level intelligence on the rapid evolution of the AI-Era, we invite you to register for our **Ai6x6 Weekly Newsletter**. Each edition delivers curated insights directly from our thought leaders and highlights critical shifts in AI strategy, governance, and provides relevant excepts from our website and books. Keep your finger on the pulse of the future and ensure your company remains competitive by signing up on the same page

PLEASE REVIEW US ON AMAZON!

Product Blueprint

We know your time is valuable, but if the Ai6x6 Organization Transformation Series delivered the real-world, actionable strategies you needed, please consider leaving a brief review on Amazon. If we missed the mark, we would also like to know, and that is the only we we can improve. Your honest feedback is crucial; it helps fellow executives discover this unique framework and benefit from the insights of our thought leaders. Thank you for supporting our mission!

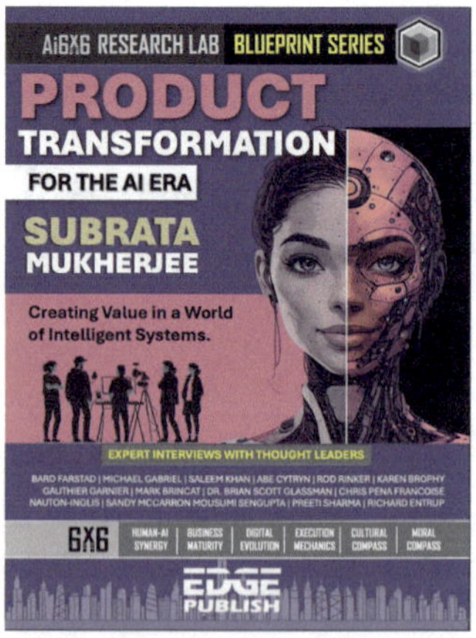

BLUEPRINT SERIES

SECTION B

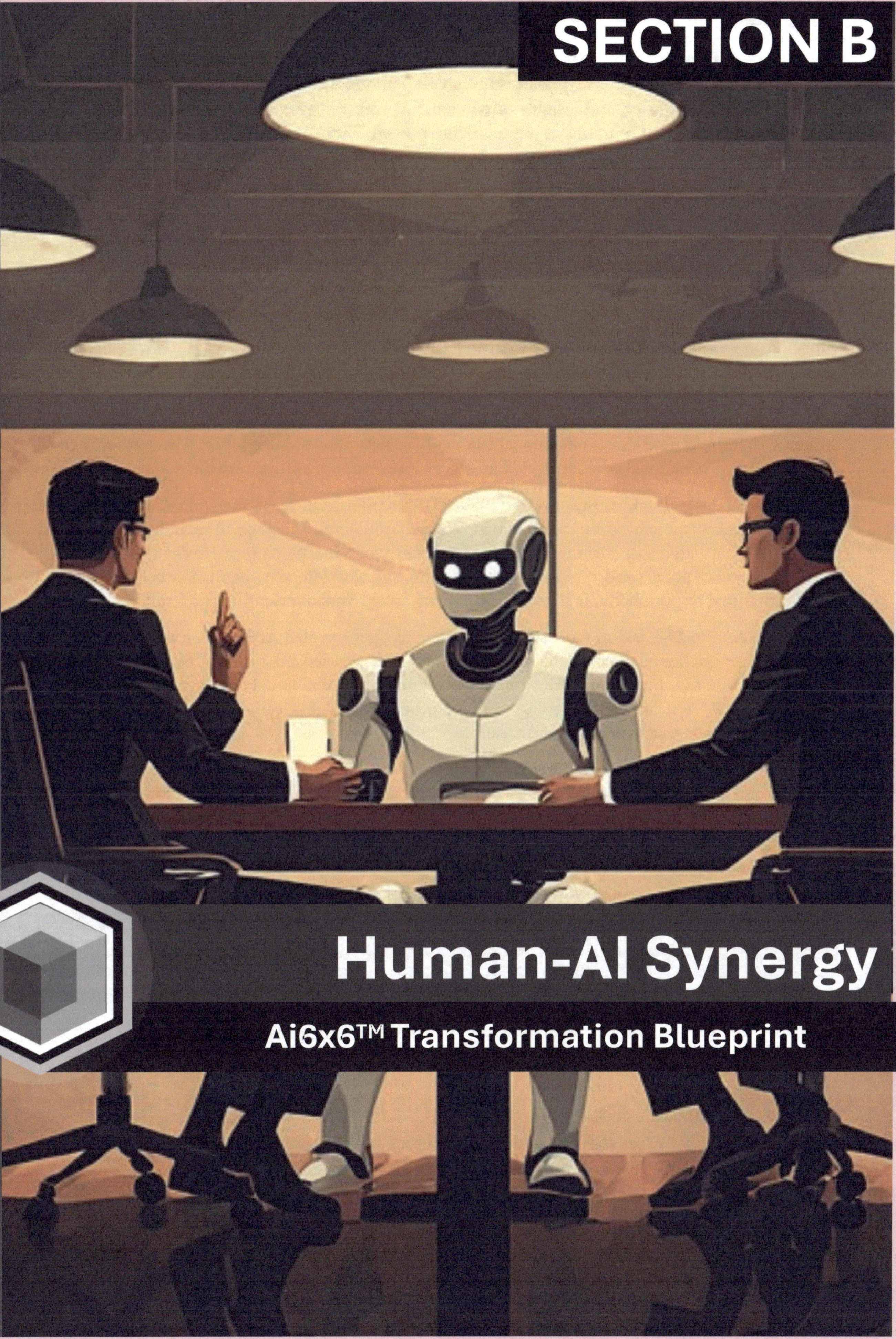

Human-AI Synergy

Ai6x6™ Transformation Blueprint

Story 1 — Enigma Codebreakers (1940–45): German U-boats encrypted communications using constantly shifting ciphers, forcing Allied cryptanalysts to chase patterns that dissolved daily. Manual decoding could not keep pace with wartime urgency, and insights often vanished before they could be confirmed. Electromechanical devices accelerated the testing of solutions at unprecedented speed. Humans interpreted these outputs, correcting false leads and refining assumptions. The fusion of machine acceleration & human judgment reshaped the trajectory of the war.

Story 2 — Turing's Bombe & Operators (1940–45): Breaking enemy codes required evaluating thousands of cipher possibilities rapidly, and the Bombe machine generated candidate solutions at speeds humans could not match. Yet its outputs were incomplete, ambiguous, and often misleading. Human operators validated results, corrected machine assumptions, and filtered noise from meaningful signals. Their contextual judgment transformed mechanical output into strategic clarity. This interplay of mechanical speed and human reasoning cracked codes that neither side could solve alone. It became one of history's earliest demonstrations of human–machine symbiosis.

Story 3 — Toyota Production System (1950s–70s): Factories struggled with defects, bottlenecks, and unpredictable workloads as machines produced volume without contextual awareness. Workers observed issues that machines could not detect or interpret, prompting Toyota to design a synchronized workflow that aligned human insight with machine precision. Real-time data from the line informed continuous improvement cycles. This human–machine rhythm created a new operational standard for reliability and efficiency. The system became a global model for intelligent production. It demonstrated how augmentation, not automation, drives sustainable excellence.

Story 4 — Salk Polio Vaccine Trials (1952–55): As polio outbreaks intensified, statistical models guided trial design but lacked the ethical nuance required for medical decision-making. Researchers interpreted model outputs through clinical judgment, ensuring safety and fairness while machines processed vast amounts of trial data. This combined system accelerated discovery without compromising integrity. Human oversight balanced computational rigor with moral responsibility. It showed how human values anchor scientific acceleration.

Story 5 — Apollo 13 Mission Control (1970): An explosion crippled Apollo 13, leaving astronauts in mortal danger as incomplete telemetry streamed back to Earth. Engineers relied on simulations to test survival scenarios under extreme constraints, but the machines could only reveal possibilities. Human teams improvised procedures the computers could not imagine, blending machine-generated insights with creative reasoning. It became a defining moment of human–machine crisis collaboration.

Story 6 — Air Traffic Control Modernization (1960s–80s): Surging air travel overwhelmed manual tracking systems, and controllers struggled to maintain safe separation across increasingly crowded skies. Radar automation provided real-time positional data, enabling machines to track aircraft continuously while humans made judgment calls. This hybrid system reduced collisions, increased capacity, and established a global standard for aviation safety. Human–machine coordination became the backbone of modern air travel. The system remains one of the world's most enduring examples of operational augmentation.

LESSONS FROM A PRODUCT TRANSFORMATION PERSPECTIVE

Products shaped through human–AI synergy evolve from mechanical aids to adaptive systems. Early innovations relied on machines to expand computational reach while humans refined logic and corrected errors. Later systems embedded continuous learning, operational feedback, and human oversight into the product itself. Across this progression, products become co-creations—machines enabling rapid iteration and humans anchoring direction, ethics, and usability. The most enduring products emerge when human insight and machine capability reinforce each other, producing solutions that are resilient, scalable, and context-aware. Human–AI synergy transforms products from static artifacts into living systems shaped by shared intelligence.

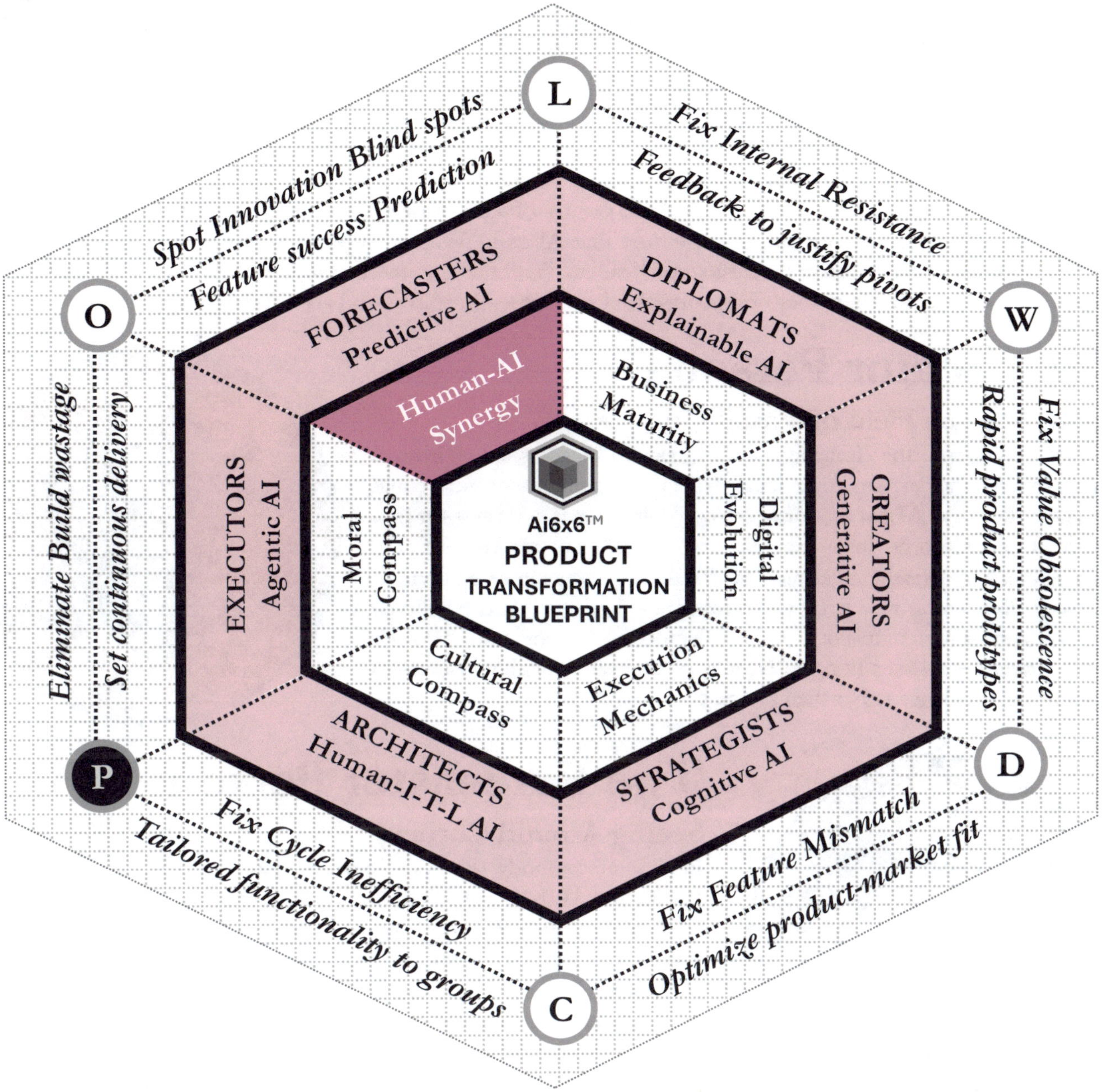

The **Ai6x6 Research Lab** chose this lens to address the fundamental shift from building tools to engineering intelligent partners. We believe the first hurdle of product transformation is the psychological surrender of the "static product" mindset. This lens is vital because it forces the creator's mind to stop viewing a product as a dormant object and start viewing it as a suite of active personas—from the **Forecaster** to the **Executor**. It is designed to move the developer from the fear of losing control to the liberation of presiding over **Autonomous Agents**, ensuring that the human spirit remains the primary "why" behind the cognitive reasoning of the system.

It is about shifting the creator's heart from a state of manual feature-building to one of strategic orchestration. This lens acts as a cognitive bridge, allowing the product leader to see AI not as a technical add-on, but as the very fabric of **Innovation Tools**. Without this shift, the product remains a tool; with it, it becomes an extension of human intent. It is an invitation to shed the burden of micromanagement and embrace a new, augmented authority grounded in the courage to build **Explainable Systems** that inspire trust rather than confusion.

Leadership in the age of intelligent systems demands a new grammar of collaboration—one that transcends the binary of human versus machine and instead frames intelligence as a continuum. The **Human–AI Synergy Lens** redefines leadership not as the art of commanding technology but as the discipline of co-creating with it. It asks leaders to move beyond automation and efficiency toward a deeper integration of foresight, ethics, creativity, cognition, governance, and execution. At its core, Human–AI Synergy is about **shared agency**. It recognizes that leadership today is exercised not only through human judgment but also through algorithmic influence. Every forecast, recommendation, and automated decision carries a fragment of leadership intent. The challenge is not to suppress that influence but to shape it—to ensure that AI amplifies human purpose rather than diluting it. This lens therefore becomes a mirror: it reflects how leaders distribute cognition, emotion, and control across human and machine systems, and how they design organizations that learn, reason, and act as unified intelligences.

The Creator Persona

Expanding the Field of Possibility

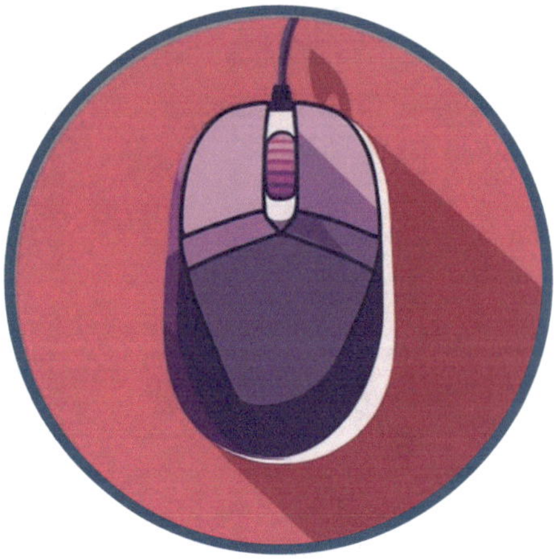

Creators represent the imaginative frontier of leadership. Through **Generative AI**, they explore more ideas than any human-only process could conceive. They treat AI not as a threat to originality but as a catalyst for it—a partner in ideation that expands the creative bandwidth of the organization. The Creator's discipline is not about producing more content but about discovering new patterns of thought. They use AI to prototype futures, to visualize complexity, and to democratize innovation. In their hands, generative systems become engines of curiosity, enabling leaders to design from abundance rather than scarcity.

The Forecaster Persona

Seeing Around Corners

Forecasters embody the anticipatory dimension of leadership. They use **Predictive AI** to model futures before committing resources, transforming uncertainty into structured insight. Their craft lies in converting data into foresight—detecting weak signals, pressure-testing assumptions, and rehearsing alternative scenarios. In traditional leadership, forecasting was intuition-driven; in synergistic leadership, it becomes a disciplined dialogue between human imagination and machine probability. The Forecaster's maturity is measured not by how accurately they predict but by how responsibly they prepare. They shift the organization from reacting to events to anticipating them, embedding resilience into its strategic DNA.

The Strategist Persona

Orchestrating Cognitive Intelligence

Strategists use **Cognitive AI** to weave intelligence across functions, geographies, and time horizons. They design strategy as a living system—adaptive, self-correcting, and continuously learning. The Strategist's genius lies in orchestration: aligning distributed intelligences into coherent purpose. They understand that cognition is no longer centralized in the executive suite; it flows through networks of humans and machines that co-reason in real time. Their leadership is systemic rather than hierarchical, emphasizing feedback loops, dynamic modeling, and cross-domain synthesis. In this persona, AI becomes the connective tissue of enterprise intelligence.

The Diplomat Persona:

Building Ethical Bridges

Diplomats operate at the intersection of **Explainable AI** and human empathy. They translate algorithmic logic into moral language, ensuring that machine outputs are understood, trusted, and socially accepted. Their role is to humanize the interface between data and decision, to make transparency a leadership act. In a world where AI can both illuminate and obscure, the Diplomat safeguards the integrity of interpretation. They remind the enterprise that trust is not a technical feature—it is a cultural outcome. By bridging emotional and ethical terrain, they ensure that AI strengthens relationships rather than eroding them.

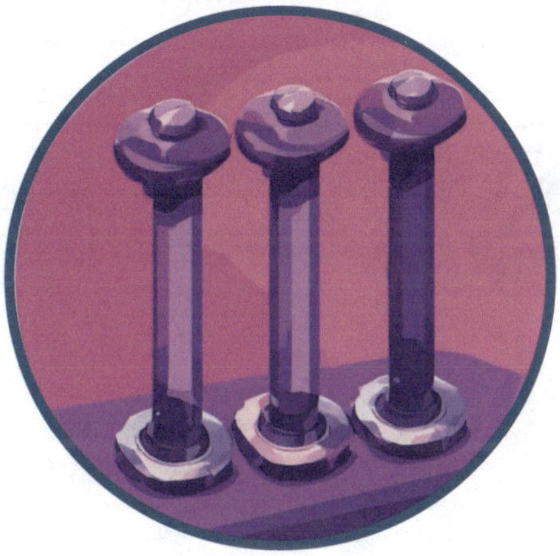

The Architect Persona:

Designing Responsible Collaboration

Architects are the custodians of **Human-in-the-Loop AI**, ensuring that governance and ethical guardrails are embedded into every workflow. They design environments where humans and machines collaborate responsibly—where oversight is not an afterthought but a structural principle. The Architect's leadership is infrastructural: they build the conditions under which synergy can thrive. Their focus is on accountability, transparency, and inclusion, crafting systems that respect both human dignity and machine precision. In their world, collaboration is engineered, not improvised; it is sustained by design integrity.

The Executor Persona:

Converting Insight into Action

Executors close the loop between knowing and doing. Through **Agentic AI**, they embed intelligence directly into operations so that insight becomes action without delay. They represent the kinetic dimension of synergy—the moment where cognition translates into execution. The Executor's mastery lies in operationalizing intelligence: automating decisions that are safe, contextual, and aligned with strategic intent. They ensure that the organization's nervous system is responsive, not reactive, and that every process carries the imprint of learned intelligence.

Together, these six personas form a **leadership spectrum**—from foresight to action, from imagination to governance. They illustrate that synergy is not a single capability but a choreography of roles. The Forecaster anticipates, the Diplomat interprets, the Creator imagines, the Strategist aligns, the Architect safeguards, and the Executor delivers. Each persona represents a distinct mode of engagement with AI, yet all converge on one principle: leadership is now a shared performance between human and machine. The Human–AI Synergy Lens therefore reframes leadership as **co-evolution**. It invites leaders to cultivate humility before complexity, curiosity before certainty, and design before control. It challenges them to lead not by commanding intelligence but by convening it—to orchestrate a dialogue between human wisdom and machine capability that continuously redefines what is possible. In this lens, synergy is not the end state; it is the ongoing discipline of learning together.

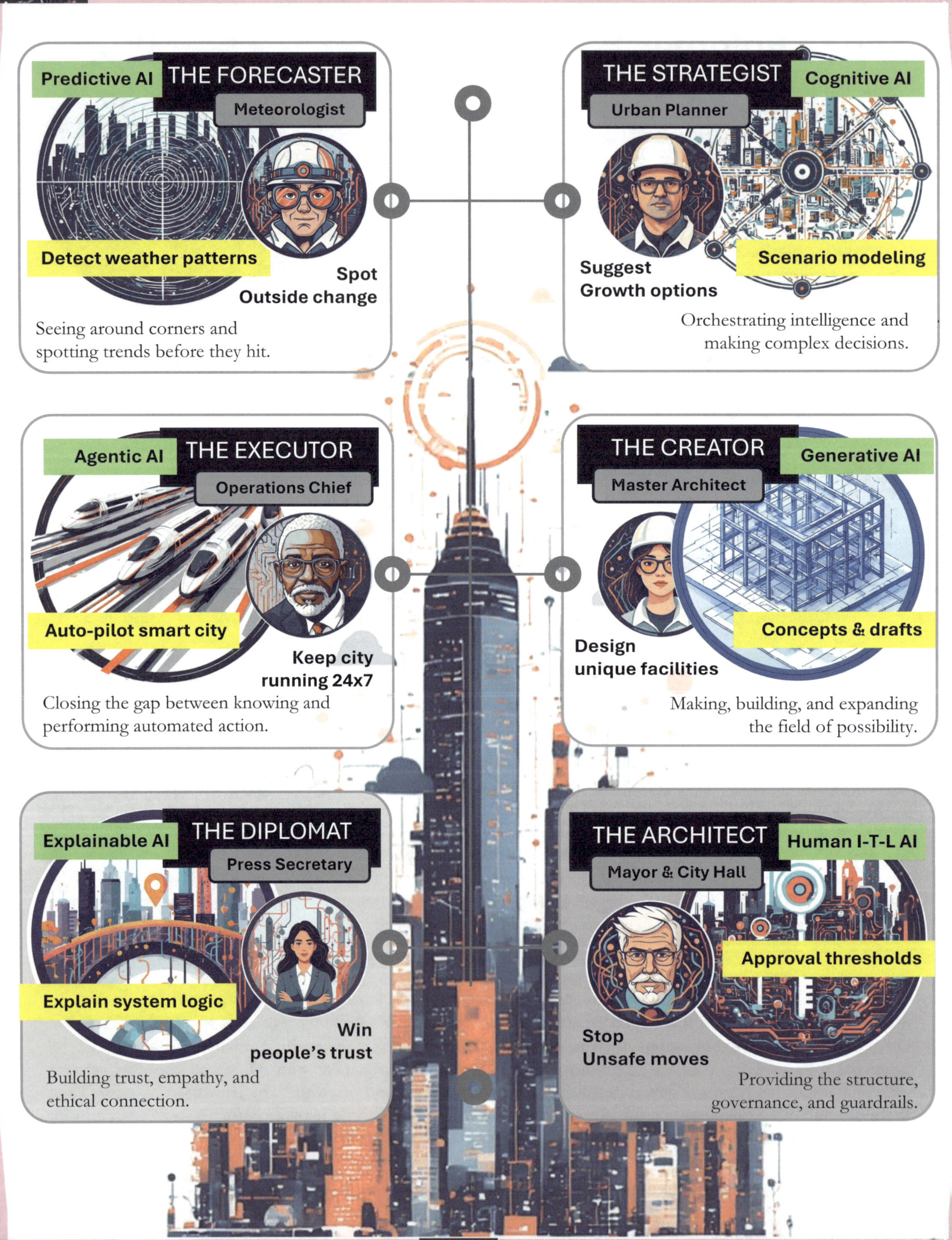
Predictive AI
THE FORECASTER
Meteorologist
Detect weather patterns
Spot Outside change
Seeing around corners and spotting trends before they hit.
THE STRATEGIST
Cognitive AI
Urban Planner
Scenario modeling
Suggest Growth options
Orchestrating intelligence and making complex decisions.
Agentic AI
THE EXECUTOR
Operations Chief
Auto-pilot smart city
Keep city running 24x7
Closing the gap between knowing and performing automated action.
THE CREATOR
Generative AI
Master Architect
Concepts & drafts
Design unique facilities
Making, building, and expanding the field of possibility.
Explainable AI
THE DIPLOMAT
Press Secretary
Explain system logic
Win people's trust
Building trust, empathy, and ethical connection.
THE ARCHITECT
Human I-T-L AI
Mayor & City Hall
Approval thresholds
Stop Unsafe moves
Providing the structure, governance, and guardrails.

Ai6x6™ Transformation Blueprint

Evolution of Humanity

"Change is the only constant in life."

Heraclitus (c. 535–475 BC); Products must be engineered for Continuous Reformation, not resistance, using Structural Optionality to flow safely with constant market change.

For the ancient engineer or architect, fulfillment lay in creating enduring, elegant systems—structures like the Roman roads or aqueducts that symbolized permanence, utility, and refined design. The purpose was to build things that would last and function beautifully. The Industrial Age and the subsequent digital sprint introduced the "Feature Factory" mentality, where speed and immediate revenue were prioritized over longevity. Engineers were fulfilled by shipping fast, but often felt alienated by the brittle, temporary nature of their creations, sacrificing structural optionality for immediate delivery. The AI Era restores the fulfillment of the true architect. The mandate is to design products for Continuous Reformation and resilience. Engineers find their pride and purpose in constructing flexible, modular systems defined by Immutable Contracts that guarantee safe, non-breaking evolution. Their professional identity shifts from being emergency fixers to designers of inherent stability, celebrating the elegance and reversibility of their architecture.

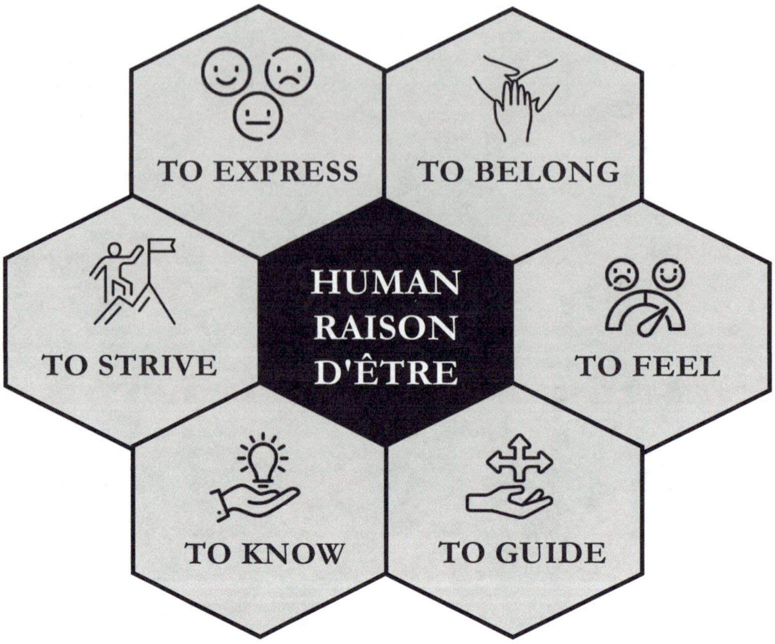

SIX UNIVERSAL ANCHORS OF HUMAN FULFILLMENT

HUMAN RAISON D'ÊTRE (THE REASON FOR EXISTENCE)

In an age defined by exponential automation, the core question of human existence shifts from survival to **significance**. When technology handles efficiency and labor, human value is no longer defined by capability, but by **meaning**. The six **Universal Anchors of Human Fulfillment** define the enduring, non-negotiable contributions of human consciousness—the pillars of a **Raison D'être** that secures our purpose across any era. These anchors affirm that our ultimate mission is to serve as the unwavering source of **Meaning and Moral Authority** in existence. They are the essential acts that only consciousness can perform.

1) TO EXPRESS

The **Anchor of Expression** is the essential human mandate to be the **Source of Novelty and Creation**, generating original ideas and subjective feelings that resist calculation or statistical synthesis. This purpose validates our unique consciousness by asserting that fulfillment comes from the generative act of bringing forth something utterly unique, not merely consuming or replicating. It requires **Uncurated Creativity**—the freedom to explore and define **Aesthetic Judgment**—thereby expanding the conceptual universe with concepts that have no statistical precedent. Through expression, we transform complex inner experience into external reality, upholding our role as the inventors of meaning.

The ultimate benefit of this anchor is the preservation of **Human Originality** and **Intellectual Sovereignty**. By committing to expression, we secure a future that is continually enriched by unique ideas, preventing stagnation and the predictability of automated systems. This anchor maintains the essential spark of invention, ensuring that human meaning remains tied to the capacity to create, **differentiating our value from that of a machine**. It is the source of all artistic, philosophical, and inventive breakthroughs that define a truly dynamic culture.

2) TO BELONG

The **Anchor of Belonging** defines our core identity as social beings, asserting that **authentic connection** is an existential necessity beyond mere digital interaction. This purpose requires the active, continuous labor of **Deep Connection**, prioritizing **unconditional presence** and shared vulnerability to foster a resilient community. Fulfillment is achieved by being the **social architect**, engaging in the emotional work of **mediating conflict**, and fostering cohesion. It includes preserving and transmitting **Cultural Heritage**—the stories and qualitative wisdom that bind us across generations. To belong is to confirm that our worth is interwoven with our relationships.

The ultimate benefit of this anchor is **Emotional Resilience** and **Social Stability**. By prioritizing genuine human connection, we safeguard ourselves against isolation and the psychological fragility that arises from purely transactional relationships.[1] This anchor ensures a supportive social framework that validates empathy and provides necessary emotional resonance, acting as a crucial defense mechanism against an isolating technological world. It builds the **trust and cohesion** necessary for any human system to function justly and compassionately.

3) TO FEEL

The **Anchor of Feeling** grounds human existence in the tangible, asserting the profound psychological value of **Embodied Mastery** and sensory engagement. This purpose elevates the dedicated pursuit of physical skill and process—not for necessity, but for the inherent satisfaction of the process. It is the quest for the **'Flow' state**, where an earned link between **exertion and accomplishment** provides deep fulfillment. This anchor mandates resisting the ease of total automation by engaging in voluntary challenges and perfecting non-essential crafts. It ensures that our abstract thoughts are grounded in the undeniable reality of the body and the senses.

The ultimate benefit of this anchor is **Physical and Psychological Integrity**. By committing to felt experience, we preserve our cognitive and physical health, resisting the inertia of automation and maintaining a sharp connection to reality. This anchor validates the **dignity of effort**, providing a clear, visceral confirmation of personal competence and self-mastery that is crucial for self-esteem. It ensures that fulfillment remains an **earned state**, confirming the complexity and richness of being physically alive.

4) TO GUIDE

The **Anchor of Guidance** establishes the human being as the **ultimate moral and ethical sovereign**. This purpose is founded on our unique capacity for empathy and subjective value judgment, making us the non-delegable authority for setting rules and managing risk. It is the practice of **Ethical Governance**, demanding that we **Draw the Moral Line** and translate compassionate principles into enforceable policy. Fulfillment is achieved by maintaining transparency, enforcing accountability, and continuously exercising the **Human Veto Power** over critical decisions.

The ultimate benefit of this anchor is **Systemic Integrity** and **Moral Safety**. By retaining final ethical authority, we ensure that technological power remains fully aligned with, and subservient to, human values, preventing systems from optimizing themselves at the expense of human welfare. This anchor provides the essential check against unchecked power, establishing the legal and moral frameworks necessary to protect all citizens. It guarantees that empathy and justice remain the **supreme architects of our collective future**.

5) TO KNOW

The **Anchor of Knowing** defines our core intellectual purpose as the engine of **Epistemic Exploration** and abstract reasoning. This purpose compels the mind to focus on formulating **Axiomatic Hypotheses**—the fundamental "Why?" questions that lie beyond the reach of data processing and established knowledge. Fulfillment is found in the generative act of **challenging established truths** and creating new theoretical frameworks that synthesize disparate fields. This anchor ensures that the pursuit of knowledge remains boundless, driven by philosophical depth rather than functional utility.

The ultimate benefit of this anchor is **Intellectual Freedom** and **Conceptual Progress**. By focusing on formulating new questions and challenging paradigms, we prevent the stagnation that could result if all knowledge were confined to data-driven answers. This anchor guarantees that our intellectual journey continues to be defined by profound theoretical breakthroughs and a relentless expansion of understanding. It confirms that human intelligence is defined not by the volume of information stored, but by the **depth and originality of the questions** we ask.

6) TO STRIVE

The **Anchor of Striving** is the **Existential Quest**—the voluntary pursuit of difficulty necessary to generate meaning, independent of any external necessity. This purpose recognizes that true fulfillment is an **earned state**, achieved through **Deliberate Discipline** and self-imposed ambition. It compels us to **Set Impossible Goals** and **Choose the Hard Way**—overcoming voluntary hardship—precisely because the challenge validates our capacity and resolve. This continuous effort allows us to craft a meaningful **Personal Legacy** by executing complex, long-term plans purely for the satisfaction of the effort.

The ultimate benefit of this anchor is **Authentic Self-Mastery** and **Purposeful Existence**. By embracing self-imposed difficulty, we cultivate the resilience, discipline, and fortitude that define character, resisting the complacency of automated comfort. This anchor ensures that human worth is determined by our **internal will**, rather than external circumstance, providing a framework for continuous personal growth and the creation of an enduring, meaningful legacy.

CONCLUSION: THE RAISON D'ÊTRE FOR THE HUMAN

The six Universal Anchors converge to form humanity's ultimate purpose: to be the **Unwavering Source of Meaning and Moral Authority**. This composite Raison D'être defines the essential human role in any advanced civilization.

Our mission is the sustained commitment to **non-automatable value**. We are the unique entity that not only pursues truth (To Know) but also creates the beauty (To Express) that makes truth worth pursuing. We are the mechanism that defines the rules of fairness (To Guide) and then personally invests the necessary effort (To Strive) to live up to those rules. Crucially, we are the only entity that grounds this entire conceptual structure in the tangible joy of physical reality (To Feel) and the vital necessity of shared vulnerability (To Belong).

The human purpose shifts from survival to an intentional journey of **growth, creation, and ethical custodianship**. Our value is not in output, but in the **input of consciousness, conscience, and will**. By diligently exercising these six anchors, every individual contributes to the essential, enduring purpose of the species: to be the **Sole Arbiter of Worth, Will, and Wonder**—ensuring that existence remains a deeply meaningful and purposeful quest.

In this era, human existence was focused on the direct, tangible application of skill for immediate survival, such hunting, fishing and surviving. The human was the Master of the Hunt.

In the Primitive/Paleolithic Era (Survival & Sustenance), human purpose was an absolute function of immediate necessity. Fulfillment was not a philosophical pursuit but a practical outcome of successfully continuing existence for oneself and the kin group. All six Universal Anchors manifested, but they were tethered directly to the biological and environmental imperative to survive the day and propagate the species. The overarching Raison D'être was to **Endure and Transmit**, meaning personal worth was inseparable from one's contribution to the group's survival calculus. This collective focus meant that individual talent was instantly converted into a benefit for all, solidifying a direct, unbroken link between personal action and existential significance.

To Express: Expression was functional and communal, centered on **Ritual Storytelling** and cave art. This art was not for individual fame or abstract aesthetics, but served as the primary educational system, preserving and transmitting essential survival knowledge, hunting techniques, spiritual beliefs, and the topography of the land. The repetitive, sacred nature of the rituals ensured collective memory and adherence to successful methods. Furthermore, these cultural forms encoded crucial information, often representing animals and spirits, that, if lost or forgotten, would jeopardize the tribe's safety and cultural identity.

To Belong: The Anchor of Belonging was non-negotiable, manifesting as **Kinship and Mutual Defense**. Isolation meant certain death; therefore, the purpose of connection was absolute. Fulfillment came from actively contributing to the group's defense, sharing food and resources, and ensuring the collective physical safety of the tribe above all else, fostering a fierce, self-sustaining mutualism. The emotional security derived from this deep collaboration—knowing that others were committed to your well-being—was a core psychological reward.

To Feel: Embodied Mastery was defined by **Essential Tool Mastery**. The feeling of accomplishment was directly tied to successfully chipping a sharp stone tool, accurately throwing a spear, or enduring the physical hardships of the hunt. This pursuit was not sport, but the daily necessity of mastering the physical world to secure food and shelter, and the psychological confidence gained was crucial for future success. This expertise immediately translated into social value; the most capable hunter or tool-maker earned respect and a higher status within the hierarchy, directly reinforcing the dignity of effort.

To Guide: Moral authority was based on immediate experience, taking the form of **Elder Authority and Tradition**. Guidance was unwritten law, passed down through generations of hard-won experience. Decisions—like where to hunt, when to migrate, or how to resolve disputes—were arbitrated by those with the deepest, most successful experiential knowledge, and were obeyed out of necessity for the group's stable cohesion. This system was vital for ensuring collective action remained predictable and risk-adverse, preventing internal conflict from threatening external survival.

To Know: The pursuit of knowledge was pure **Observation and Practicality**. The purpose of knowing was to accurately predict the immediate environment for resource exploitation. This involved memorizing the habits of migratory animals, understanding the medicinal or poisonous properties of local plants, and tracking subtle shifts in weather patterns—information critical for daily sustenance and successful, low-risk planning. Early forms of abstract reasoning also emerged here, through tracking the movement of stars or noting the cyclic nature of seasons and astronomical events, demonstrating an innate human desire to find order and predictability in environmental chaos.

To Strive: Striving was synonymous with **Daily Endurance**. There was no need to invent abstract goals; the pursuit of difficulty was the environment itself, offering constant challenge. Fulfillment was the successful completion of the continuous, grueling effort required to secure food, withstand the elements, and ensure the basic propagation of the tribe, creating a deeply ingrained sense of earned worth. The resilience built through continuous hardship was the ultimate personal asset, psychologically connecting physical suffering with the existential meaning of survival.

With stability, purpose shifted to sustained output, long-term planning, and maintaining social harmony within a fixed settlement. The human was the Master of the Land.

The Agricultural/Neolithic Era (Farming & Community) marked humanity's fundamental pivot from nomadic struggle to settled stability. The mastery of cultivation and animal husbandry created a reliable **Surplus of Time and Resources**, fundamentally shifting human purpose from immediate survival to calculated investment, long-term order, and the establishment of lasting community structures. The Raison D'être of this period was to **Settle and Establish Order**. This era saw humans take on the role of conscious managers of the environment, laying the groundwork for systematic laws, property rights, and enduring social legacies that extended far beyond one generation.

To Express: Expression became focused on **Accounting and Enduring Form**. The logistical necessity of tracking seeds, yields, stored grains, and communal ownership led directly to the invention of the first systematic writing forms, such as early cuneiform, demonstrating that expression was initially a tool for managing complex resources. Beyond text, expression was seen in durable, functional goods like specialized pottery, permanent architectural structures, and woven textiles. These forms reflected a deep, stable community identity and provided evidence of specialized skill that contributed directly to the village's permanence.

To Belong: Belonging shifted to being centered on **Land-Based Cooperation**. The success or failure of the farm was the success or failure of the entire settlement, making shared labor and mutual dependency paramount. Fulfillment was found in the reliable rhythms of shared work, such as irrigation projects or harvesting, cementing generational inheritance and deeply intertwined **Kinship Networks**. Purpose was now inseparable from one's physical, defined place within the land and the local community, creating a profound sense of rootedness and stability unknown in the Paleolithic past.

To Feel: Embodied Mastery focused on **Craftsmanship and Process**. Fulfillment was derived from the methodical, repeated application of skill required by agriculture—plowing a straight furrow, carefully managing livestock, or perfecting the complex processes of metallurgy and food preservation. This mastery provided a sense of control over the environment that raw endurance could not. The purpose of skill was purely to generate the stable surplus required to maintain the settled life, providing a deep psychological reward through the predictable yield of sustained, expert effort.

To Guide: Moral authority became formalized, focusing urgently on **Property Rights and Resource Management**. Guiding the community meant establishing clear, complex rules around ownership of land, managing precious resources like water rights, and ensuring crop protection from both animals and neighbors. This necessity gave rise to the first systematic, written laws, moving purpose from inherited tradition (as in the Paleolithic) to a **Codified Social Contract**. Leadership was defined by the ability to enforce these new laws, ensuring fairness and stability within an increasingly complex economy.

To Know: Abstract reasoning began to flourish, driven by **Mathematics and Calendar Tracking**. The purpose of knowledge was now predictive certainty, essential for farming success. This meant mastering geometry to accurately divide and manage fields, and engaging in sophisticated celestial observation to track seasons and accurately determine optimal planting and harvesting cycles. This pursuit of cosmic order confirmed the human ability to impose intellectual structure upon chaotic nature, turning farming into a sophisticated, knowledge-driven enterprise.

To Strive: Striving fundamentally shifted from day-to-day survival to **Creating Surplus and Investment**. The existential quest became the arduous, long-term, deliberate work required to build durable structures, amass protective stores of food, and ensure the economic and social prosperity of the next generation. This focus on legacy and investment transformed work into a purposeful act of building a lasting future, where one's worth was measured not just by survival, but by the inheritance left behind for the family and the stability provided to the community.

This era defined purpose through professional expertise and contribution to complex, often abstract, technological and economic systems. The human was the Master of the Machine.

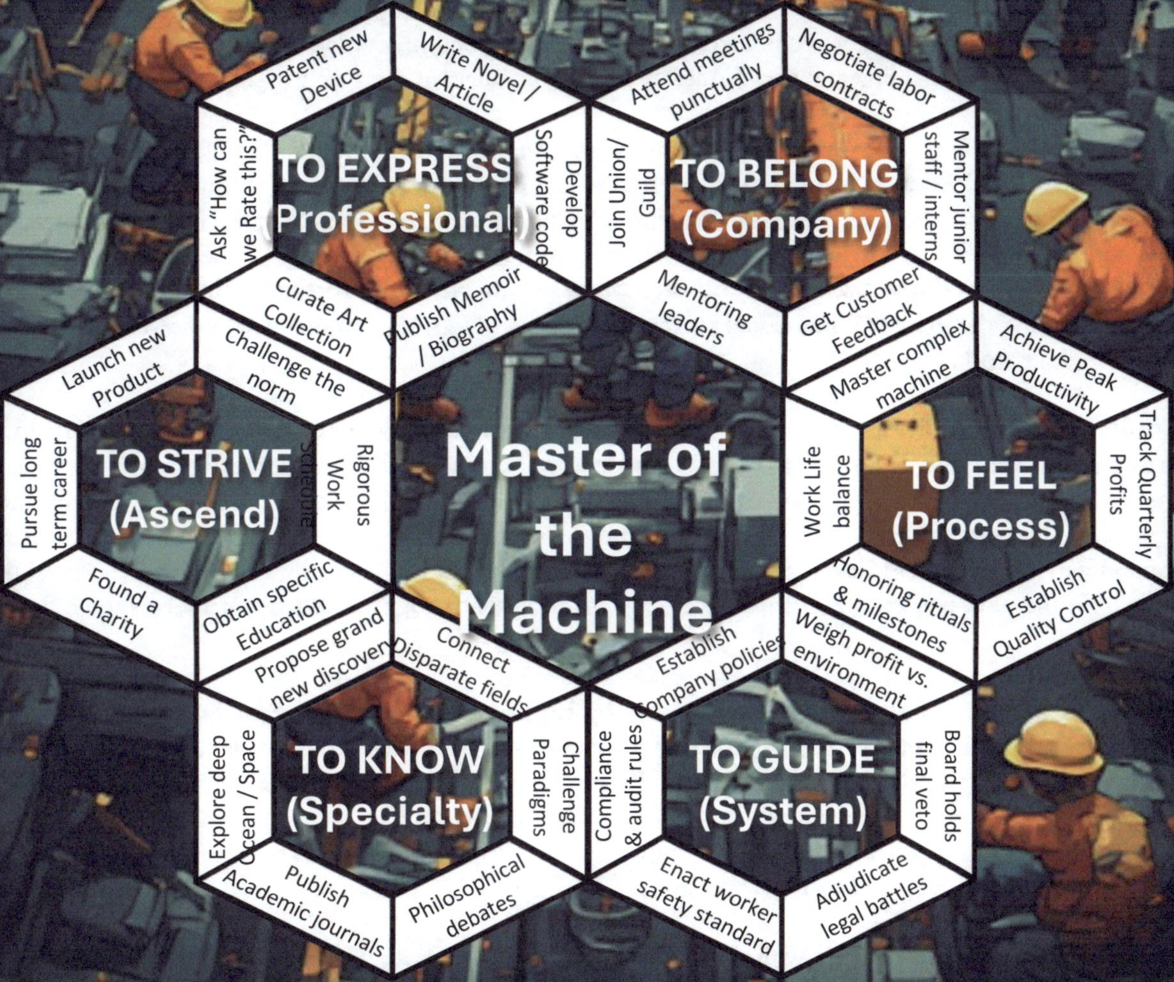

The Modern/Industrial Era (Specialization & Function) marked the decisive triumph of machine logic and capital over the natural world. Human purpose shifted from managing the cycles of the land (Neolithic) to mastering the mechanical efficiency of the factory and the market. The rise of the machine and the division of labor fundamentally redefined the individual, whose worth was now derived from highly specialized, functional expertise within a vast, interconnected economic system. The Raison D'être of this period was to **Achieve and Advance**, driving relentless progress through expertise and material gain.

To Express: Expression became focused on **Mass Media and Invention**. With the advent of the printing press, cheaper paper, and later, radio and cinema, creative expression transformed into a commodified product designed for wide distribution and profit. The purpose of expression was to inform, entertain, and persuade the masses, shaping shared cultural narratives through novels, newspapers, and commercial design. The creation of a **Patentable Invention** became a powerful act of individual expression, translating originality directly into economic ownership and societal contribution. Expression was celebrated when it served a functional, marketable, or propagandistic role.

To Belong: Belonging was redefined by **Nation and Institution**. The old agrarian kinship networks dissolved as masses migrated to urban factory centers, replaced by new, abstract, yet powerful social structures. Fulfillment was found in national identity, membership in professional organizations, and political affiliation. Crucially, the extreme specialization of labor necessitated the formation of organized counter-structures like **Labor Unions** to protect the rights of workers against centralized corporate power, demonstrating that belonging was a functional necessity for collective bargaining and social stability in a highly unequal economic system.

To Feel: Embodied Mastery centered on **Efficiency and Maximized Output**. The physical purpose shifted away from holistic craft towards highly repetitive, specialized labor—the mastery of a single, precise task on the assembly line. While monotonous, this specialization provided a clear, measurable sense of contribution, where personal worth was directly tied to economic output and reliability. Outside of the factory, the Anchor manifested in the rise of **Organized, Competitive Sports**, where physical effort was deliberately structured and formalized, allowing the industrial individual to achieve a sense of earned competence and visceral struggle that was absent in their work.

To Guide: Moral authority became formalized through **Secular Law and Corporate Regulation**. The complexity of contract law, property ownership, and massive corporate liability demanded sophisticated, non-religious guidance systems. The purpose of guidance was to manage the inherent conflicts between capital and labor, protect consumer safety in mass production, and define the boundaries of ethical economic competition. This period saw the relentless development of regulatory bodies and constitutional law aimed at establishing fairness within the new, powerful industrial economy, ensuring that **progress did not undermine social order** entirely.

To Know: The pursuit of knowledge was dominated by **Empirical Science and Specialization**. The purpose of knowing was functional utility, believing that all problems—from disease to industrial inefficiency—could be solved through the application of the rigorous **Scientific Method**. Knowledge became fragmented into specialized disciplines, each dedicated to gathering objective, measurable data. Fulfillment was achieved by contributing a verifiable, peer-reviewed fact or theory to one's specific field, thereby confirming the era's deep-seated faith in rational inquiry and technical progress.

To Strive: Striving was synonymous with **Ambition and Material Gain**. The existential quest was the pursuit of wealth, social mobility, and **Status**. Fulfillment was an earned state achieved through economic success, the establishment of a profitable business, or professional advancement within a corporate or governmental hierarchy. The cultural narrative of the "self-made individual" defined personal purpose, urging people to embrace continuous, relentless work to secure prosperity for the family and, more importantly, to carve out an enduring legacy of material success and high societal standing.

In this emerging era, AI handles efficiency and role specialization, freeing and forcing humans to refocus on subjective values, unique creation, and moral governance. The human becomes the Master Curator of Purpose

The Post-AI Era (Curation & Meaning) represents the definitive existential shift in human history. With advanced artificial intelligence and automation having conquered the challenges of efficiency, information processing, and mass labor, human purpose is radically redefined. We are no longer necessary for function, but for **meaning**. The individual's worth is derived entirely from their capacity for subjective judgment, ethical authority, and the creation of non-calculable value. The collective Raison D'être is to become the **Sole Arbiter of Worth, Will, and Wonder**—the source of all criteria for why life should continue and what makes it beautiful.

To Express: Expression is focused on **Subjectivity and Uncurated Truth**. In a world saturated with synthetic media and hyper-optimized content generated by AI, the only expression that retains intrinsic value is the genuinely **original, subjective human voice**. This purpose mandates focusing on authentic, unfiltered creation—personal philosophical content, aesthetic critique, memoir, and art that contains flaws, uncertainty, and emotional resonance. Fulfillment is found in generating novelty that AI cannot predict, thereby constantly pushing the boundaries of human experience and securing a creative future that is immune to algorithmic stagnation.

To Belong: Belonging becomes a deliberate, chosen investment in **Deep Vulnerability and Presence**. When digital connection is ubiquitous but emotionally shallow, true fulfillment relies on actively resisting transactional relationships and seeking authentic, physical presence. This purpose demands the **labor of Deep Connection**: cultivating empathy, mediating conflict through personal understanding, and prioritizing quality, face-to-face interaction. The collective human project requires maintaining the social fabric and emotional resonance that machines cannot replicate, ensuring that isolation does not erode the psychological foundation of society.

To Feel: Embodied Mastery transforms into the **Voluntary Pursuit of Difficulty and Flow**. Since automation eliminates the necessity of difficult labor, human physical purpose centers on self-imposed challenges—non-essential crafts, athletics, meditation, and the perfection of technical, yet unnecessary, skills. This pursuit generates the vital 'Flow' state—the psychological reward of earned accomplishment—which secures self-esteem and grounds abstract consciousness in tangible reality. Fulfillment is achieved through **Deliberate Effort** that reaffirms the dignity of the body, providing a visceral antidote to the inertia and complacency of total comfort.

To Guide: Moral authority centers on **AI Ethics and the Human Veto**. The purpose of guidance is to serve as the **ethical sovereign**, placing strict moral boundaries around autonomous systems. This involves defining the principles of data governance, ensuring transparency in black-box algorithms, and enforcing systemic accountability for technological outcomes. Crucially, every individual is mandated to exercise the **Human Veto Power**—the non-delegable right to overrule critical decisions made by algorithms—ensuring that power remains permanently aligned with subjective human values, not cold-blooded efficiency.

To Know: The pursuit of knowledge is now defined by **Abstract Synthesis and Axiomatic Inquiry**. In an environment where AI holds access to all existing data and established facts, human intellect must shift its focus. The purpose is to transcend data processing and concentrate on formulating **Axiomatic Hypotheses**—the profound "Why?" questions that challenge foundational assumptions. Fulfillment is found in connecting disparate fields of knowledge, creating new philosophical frameworks, and constantly resisting the temptation to settle for algorithmic answers, ensuring the continuous, boundless expansion of human understanding.

To Strive: Striving is the existential commitment to **Self-Imposed Discipline and Personal Legacy**. Since survival is guaranteed, the quest for meaning requires the individual to set ambitious, non-essential goals purely for the satisfaction of the effort. This involves cultivating profound resilience and fortitude against the immense temptation of automated ease. Fulfillment is an **earned state**, achieved through continuous, conscious effort and the construction of an enduring **Personal Legacy**—a body of work or moral character that defines one's worth by internal resolve rather than external economic output.

Ai6x6™ Transformation Blueprint

Human-AI Personas

"AI is not just a tool; it's a partner for human creativity."
Satya Nadella, the CEO of Microsoft.

The Ai6x6 HUMAN-AI Synergy framework formalizes a new executive standard for the AI era, transitioning organizations from fragmented automation to a state of Orchestrated Autonomy. This model defines the modern enterprise through six core leadership movements, each serving as a specialized "sense" for the organization. **The Forecasters** serve as the "Eyes," tasked with the heavy lifting of foresight to eliminate strategic blind spots and detect invisible market signals before they escalate into crises. Supporting this vision, **The Diplomats** act as the "Voice," translating complexity into human trust and ensuring that transformation is met with cultural alignment rather than institutional resistance. The organization's creative energy is fueled by **The Creators**, the "Engine" of the enterprise, who move beyond static production to drive a state of radical velocity and infinite iteration. Their output is balanced by the surgical objectivity of **The Strategists**, acting as the "Compass" to ensure that every capital investment is mathematically anchored to peak ROI. To unify these efforts, **The Architects** build the "Institutional Brain," ensuring that the organization's collective memory is always accessible and relevant to the individual. Finally, the framework finds its movement in **The Executors**, the "Hands" of the organization, who coordinate autonomous workflows to achieve scale without traditional overhead. By weaving these archetypes into a single Symphony, leaders can navigate market ambiguity with surgical precision, transforming the enterprise into a self-evolving system of sustainable innovation.

In the **Ai6x6 Human-AI Synergy framework**, the defensive stance for the Product is about **resilience and integrity**. In an era of rapid disruption, a product is no longer a static asset but a "Living System" that is constantly threatened by technological obsolescence, competitive imitation, and feature creep. Defensive product strategy focuses on protecting the core value proposition and ensuring the product does not become a liability to the user.

The Forecaster and The Diplomat: Detecting Disruption and Maintaining Integrity

The **Forecaster** acts as the defensive radar for the product's relevance. Its mandate is to kill **Innovation Blindspots** by detecting emerging technological shifts or competitor patents that could turn the current product into "legacy" overnight. By identifying "Substitutive Threats" early, the Forecaster allows the company to pivot the product's defensive features—such as security patches or compatibility updates.

The **Diplomat** is tasked with neutralizing **Trust Friction**. As products become increasingly autonomous and AI-driven, users often fear a loss of control. The Diplomat uses explainable logic to ensure the product's internal decisions are transparent and aligned with user intent. By acting as a bridge between complex algorithms and the human interface, it protects the brand's "Voice," ensuring the product remains an ally to the user rather than a black-box source of frustration or ethical concern.

The Creator and The Strategist: Modernizing the Core and Killing Feature Creep

The **Creator** acts defensively by neutralizing **Technical Debt**. A product that cannot be easily updated is a product destined for obsolescence. The Creator's defensive role is to constantly refactor the product's "DNA," ensuring that legacy code or outdated design patterns are modernized before they cause a system failure. It ensures the "Product Engine" remains modular and agile, protecting it from the "Velocity Friction" that prevents rapid response to market needs.

The **Strategist** defends the product's value by eliminating **Complexity Mismatch**. One of the greatest defensive risks is "Feature Creep"—the addition of low-value functions that clutter the user experience and increase maintenance costs. The Strategist performs a ruthless audit of the product's utility, using causal inference to identify which features are actually driving user satisfaction. By "pruning" the product of non-essential noise, it protects the "Product Compass," ensuring that development resources are never wasted on mismatched goals.

The Architect and The Executor: Unifying the Experience and Automating Quality

The **Architect** kills **Functional Inconsistency**. In large product ecosystems, different features often feel like they were built by different companies. The Architect's primary defensive mandate is the maintenance of a "Unified Product Brain," ensuring that every part of the product shares a common logic and data structure. It prevents the friction of a fragmented user experience, safeguarding the product's integrity by ensuring that it "learns" from every interaction and remains contextually relevant across its entire lifecycle.

Finally, the **Executor** serves as the defensive "Hands" of product reliability. It kills **Failure Friction** by automating the "Self-Healing" capabilities of the product. Whether it is automatically detecting a software bug or triggering a maintenance alert for a physical component, the Executor ensures that guardrails are in place to prevent downtime. By removing the wastage of manual troubleshooting, it safeguards the product's uptime and protects the user from the "Reliability Leakage" that destroys long-term brand equity.

Conclusion: Safeguarding the Innovation Body

A defensive product strategy within the **Ai6x6 Human-AI Synergy framework** ensures that the "Product Body" remains durable and trustworthy. By removing the friction of obsolescence, complexity, and failure, the framework creates a product that is "Safe to Use" and "Easy to Maintain." When a company's product is protected by these six roles, it stops being a vulnerable target for competitors and becomes a resilient platform for sustainable market leadership.

In the **Product Transformation**, the objective is to shift from "Static Inventory" to **"Living Solutions."** This transition turns the product into an intelligent, evolving entity that learns from its environment and its user. By applying the six roles to the product lifecycle, leadership moves beyond traditional manufacturing into a state of **Generative Innovation**, where the product itself becomes a bridge for continuous value creation.

The Forecasters and The Diplomats: Unlocking Performance and Purpose

The **Forecaster's** mandate is **Predictive Maintenance and Lifecycle Intelligence.** By pairing Predictive AI with Digital Twins, the "Eyes" monitor the product's health in the field. It anticipates mechanical failures or software bottlenecks before they occur, allowing for "Self-Healing" updates. This proactive stance ensures the product never reaches a state of obsolescence, but rather improves the longer the customer owns it.

The **Diplomat** serves as the "Voice" of **User-Centric Reasoning.** By pairing Explainable AI with Natural Language Processing, the product can explain its actions to the user. Whether it is a medical device explaining a dosage adjustment or an autonomous vehicle explaining a route change, the Diplomat builds "Cognitive Trust." It ensures that the product's intelligence is never a "black box," but a transparent partner in the user's daily life.

The Creators and The Strategists: Unlocking Adaptability and Market Fit

The **Creator** acts as the engine of **Generative Feature Evolution.** In this role, Generative AI is paired with Synthetic Data to simulate new feature sets in virtual environments before they are pushed to the physical product. This "Over-the-Air Innovation" means the product can adapt its functionality based on shifting user habits. It allows the brand to test thousands of variations in a "Sandboxed Reality," ensuring only the most effective features are manifested.

The **Strategist** optimizes the **Product-Market Ecosystem for Maximum Utility.** Using Multi-Objective Optimization, the Strategist balances production costs with user-perceived value and sustainability targets. It identifies the "Value Frontier," ensuring that every product iteration is mathematically optimized for both the user's needs and the firm's ROI. This ensures that the product is not just a gadget, but a precision-tuned instrument of market dominance.

The Architects and The Executors: Unlocking Intelligence and Actuation

The **Architect** builds the "Searchable Product Memory" through **Contextual Data Orchestration.** By pairing Human-in-the-Loop with Vector Databases (RAG), the Architect ensures that the product has access to the firm's total institutional wisdom. A smart factory tool, for example, can "recall" how a similar problem was solved in another facility five years ago. This turns the product into a node in a global, "Searchable Brain," making every unit as smart as the entire organization.

Finally, the **Executor** manages the **Autonomous Actuation of the Product.** Through Agentic AI paired with Edge AI, the Executor allows the product to take independent action in real-time. Whether it is an autonomous drone navigating complex terrain or a smart thermostat optimizing a grid, these "Hands" execute intent with zero latency. By securing these actions through Blockchain, the product can even engage in autonomous commerce, such as ordering its own replacement parts.

Conclusion: The Symphony of the Intelligent Object

Product transformation is the pursuit of **Functional Sovereignty.** By unlocking performance, purpose, adaptability, utility, intelligence, and actuation, the leader creates a product line that is as dynamic as the market it serves. This framework ensures that the brand's physical and digital offerings are perfectly synchronized with the user's intent, creating a high-velocity symphony of orchestrated utility.

Product Power: Environmental Awareness

In the legacy era, a product was "Blind." Once it left the factory or the server, the company had no idea how it was being used or where it was failing until a customer complained. The transformation into **Forecasters** gives the product its own "Eyes." It turns every product into a "Distributed Sensor," capable of perceiving its own health, its environment, and the changing needs of its user in real-time. The product ceases to be a passive tool and becomes an "Anticipatory Sentry," identifying the "Statistical Whisper" of a mechanical failure or a shift in market relevance long before it occurs.

Defensive Stance: Killing the Failure Gap

The defensive mission of the Product-Forecaster is the **Elimination of Downtime**. In a world of "Always-on" expectations, a broken product is a broken promise. The Forecaster role kills this "Friction of Failure" by monitoring its own "Internal State." It looks for the subtle anomalies—vibrations in a machine, latency in a software loop, or heat signatures in a battery—that signal an impending break. This creates a **Shield of Reliability**. By predicting its own maintenance needs, the product ensures it never stops working. This defensive posture kills the "Wastage of Replacement," protecting the customer's investment and ensuring the brand's reputation for "Indestructible Quality." The product identifies the cliff before it falls, allowing for a "Surgical Correction" that keeps the user in a state of "Constant Flow."

Offensive Stance: Unlocking Adaptive Evolution

On the offense, the Forecaster moves toward **Contextual Relevance**. Because the product "Sees" how it is being used, it can suggest its own evolution. This **Unlocks Temporal Utility**, where the product recognizes a change in the user's environment and re-configures its features to meet the new reality. If a software product "sees" a user struggling with a complex workflow, it anticipates the need for a simplified interface and offers it. This turns the product into a "Strategic Advisor" that grows with the user. It allows the brand to "Own the Next Need," as the product identifies the "Feature Gap" in real-time. This offensive move ensures the brand is always "Waiting at the Destination," evolving the product's capabilities at the exact speed of the user's ambition.

The Product Outcome: Perpetual Vitality

The ultimate value of the Forecaster role is **Unbreakable Utility**. When a product "Sees," it never becomes obsolete. For the leader, this provides **Market Resilience**. You are no longer guessing what features to build; the products themselves are telling you what they need to become. This turns the product line into a "Live Laboratory," where the brand can observe "Truth in Use." It ensures that the product is never "Finished," but is instead a "Living Entity" that remains perfectly aligned with the user's world, creating a bond based on the product's ability to "Self-Protect" and "Self-Improve."

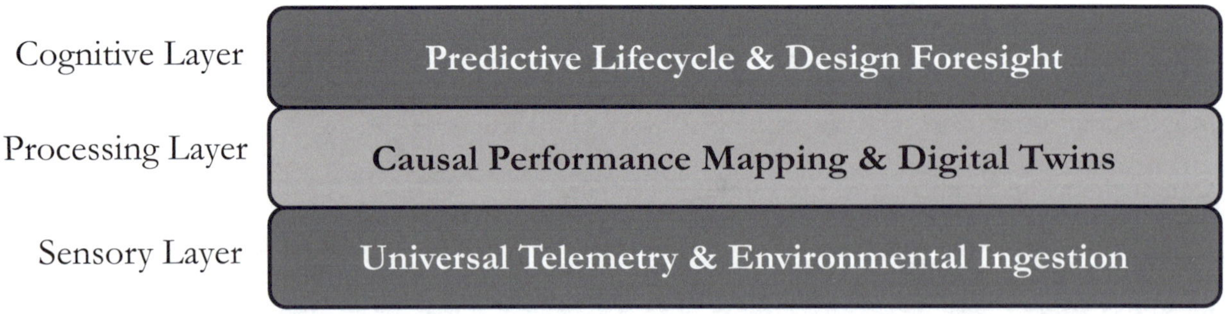

1. THE FORECASTERS (THE EYES)

Executive Power: Real-Time Performance Perception

In the legacy era, a leader's "Eyes" on product performance were limited to return rates, warranty claims, and periodic focus groups. The **Forecaster** role in product transformation represents the transition to **Continuous Operational Observability**. For the executive, this means moving away from "Guessing how the product is used" toward "Perceiving its actual life." By architecting a stack that perceives the physical and digital state of every unit in the field, the leader gains the "Perceptual Edge" to identify design flaws, wear-and-tear patterns, and emerging use-cases before they become liabilities. The ultimate goal is the **Purification of the Product Signal**, ensuring the firm sees the "Physical Truth" of its performance, allowing the leader to authorize engineering shifts and resource allocations based on live reality rather than laboratory assumptions.

The Sensory Layer: Universal Telemetry & Environmental Ingestion

The **Sensory Layer** of the Product Forecaster acts as the brand's "Physical Senses." It moves beyond simple logging into **High-Definition Telemetry Ingestion**. This layer leverages a convergence of **IoT Sensors**, **Digital Twin synchronization**, and **Environmental Edge Capture** to monitor the product's health and surroundings. Whether it is the thermal signature of a machine, the stress load on a structural component, or the frequency of a specific digital feature's use, this layer "labels" the state of the product at the source. By utilizing **Edge AI Agents**, it identifies "Anomalous Signatures"—distinguishing between normal operation and a subtle decline in performance. This ensures the leader's "Eyes" are receiving a raw stream of "Mechanical and Digital Truth" with zero latency, removing the informational lag of the traditional feedback loop and providing a live map of the entire product fleet's status.

The Processing Layer: Causal Performance Mapping & Digital Twins

The **Processing Layer** serves as the "Analytical Heart" of the product stack. It transforms raw telemetry signals into a structured understanding of **Systemic Health**. This layer utilizes **Physics-Informed Neural Networks (PINNs)** and **Knowledge Graphs** to index the relationships between different performance variables. While traditional databases see isolated errors, the Forecaster's processing layer sees the "Causal Path"—understanding, for example, how a specific environmental humidity level is impacting the battery life of a device across a certain geographic region. By mapping these signals into a **Live Digital Twin**, the system can simulate the future state of the product based on its current trajectory. It moves the organization from "Scheduled Maintenance" to "Predictive Resilience," providing a "Logical Foundation" for immediate design iterations. This layer ensures the brand's "Product Compass" is always calibrated to the physical laws of the market.

The Cognitive Layer: Predictive Lifecycle & Design Foresight

The **Cognitive Layer** is the "Reasoning Mind" of the product stack, where performance signals are converted into **Strategic Engineering Intent**. This layer utilizes **Generative AI** and **Reinforcement Learning** to synthesize the logic of the Processing Layer into "Lifecycle Alerts." Instead of showing the leader a repair dashboard, the Cognitive Layer generates "Contextual Foresight." It acts as a "Product Sentry," using **Natural Language Generation** to tell the leader: "Based on the stress patterns emerging in the Gen-2 fleet, there is an 80% probability of a component failure in humid climates; we should adjust the Gen-3 design parameters now." This allows the executive to move from "Fixing Products" to "Leading Design Evolution," authorizing pre-emptive engineering changes. It turns the "Forecaster" into a "Proactive Engineer," ensuring the brand is always ahead of the wear-and-tear curve and market expectations.

OVERVIEW

Predictive AI forecasts future outcomes by analyzing vast historical data using machine learning algorithms. For the "Intuitive & Risk" leader, the AI becomes a crucial tool for quantifying uncertainty. The leader provides their initial intuition and acceptable risk limits, framing the precise question. The AI delivers objective, quantifiable evidence, such as likelihoods and key drivers, that either validates the leader's gut feeling or proposes data-driven alternatives. This synergy converts subjective risk into a calculated, strategic choice, empowering the leader to make proactive, decisive actions.

MAIN STEPS

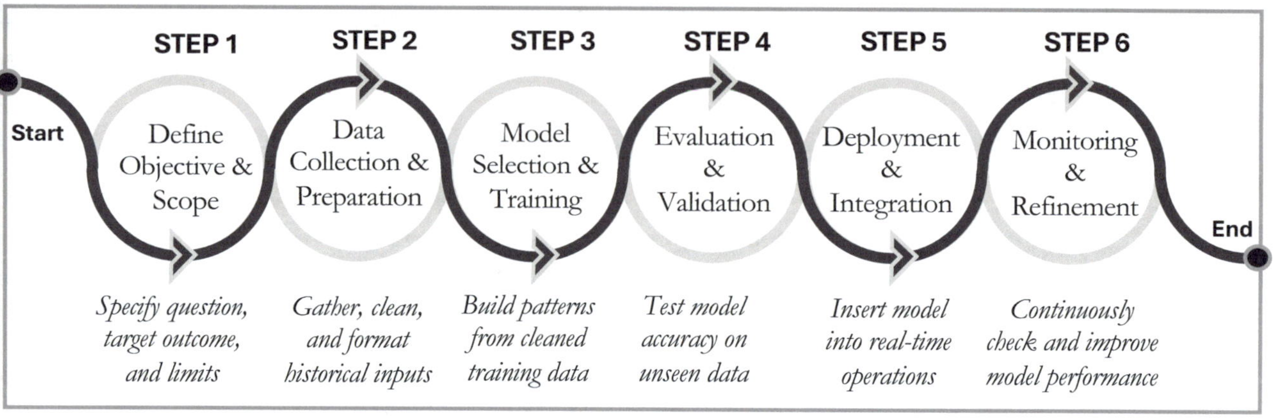

SAMPLE FLOW

Software Development - Feature Failure Risk- A VP of Product uses AI to predict which new software features will fail adoption targets, enabling early intervention to save significant development resources.

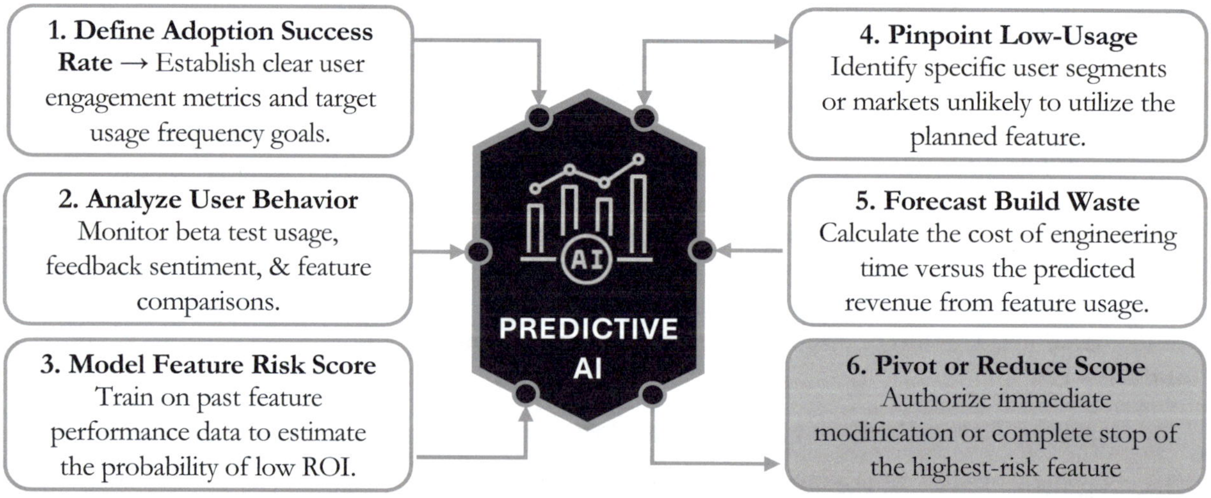

USE CASES

- **Healthcare | New Device Performance Flaws:** Predicts which newly launched medical devices will experience a high rate of critical, early-life component failures based on initial production and usage telemetry, halting rollout and preventing safety incidents.
- **Media | Content Seasonality Risk:** Forecasts which upcoming premium content will underperform viewership targets due to shifting audience trends and competitive launches, adjusting marketing spend and accelerating alternative release schedules.
- **Energy | Renewable Asset Obsolescence:** Predicts the effective lifespan of new solar panel or turbine models, informing capital investment decisions to avoid premature technological obsolescence and maximize long-term asset returns.
- **Fashion | Inventory Overstock Risk:** Forecasts which new apparel designs will rapidly lose consumer interest post-launch, allowing the immediate cancellation of future production runs to mitigate significant inventory write-downs and storage costs.
- **Education | Curriculum Efficacy Decline:** Predicts when a virtual course module's teaching efficacy will drop due to outdated material or confusing interface design, prompting urgent content revision or complete replacement to maintain learning outcomes.
- **Telecommunications | Network Service Congestion:** Forecasts which specific network upgrades will not handle predicted future data loads in certain urban areas, requiring immediate re-prioritization of capital expenditure for necessary infrastructure expansion.

In the **Product Transformation**, the objective is to shift from "Static Commodities" to **"Living Solutions."** This pivot moves the organization from selling one-off physical or digital items to orchestrating a "Sovereign Product Ecosystem" that evolves with the user. By applying the six leadership roles and their technical pairings, the leader ensures that the product is no longer a finished state, but a continuous stream of value that is contextually aware, self-improving, and ethically anchored.

Product Transformation: The Forecasters (The Eyes)

In the **Product Transformation**, the **Forecasters (The Eyes)** serve as the "Utility Sentinel." While the primary instrument of **Predictive AI** identifies how products were used in the past, it often fails to see the "Shadow Utility"—the unintended or emerging ways users are hacking the product to solve new problems. To move from "Feature Updates" to "Generative Evolution," a leader must integrate these six secondary pairings to sharpen the "Eyes," ensuring the product roadmap is always ahead of the user's next challenge.

1. The Graph Neural Network (GNN) Pairing for Ecosystem Impact

Pairing **Predictive AI with GNNs** is the ultimate tool for **Mapping Feature Dependencies.** While standard AI looks at how one feature performs, GNNs analyze how a change in one component ripples through the entire product ecosystem and its third-party integrations. By seeing the product as a web of interconnected nodes, the Forecaster can predict which "Small Update" will lead to a "Large-Scale Breakthrough" in user productivity.

2. The Causal Inference Pairing for R&D Prioritization

Leaders should pair **Predictive AI with Causal Inference** to solve the **"Correlation Trap" in product-led growth.** Standard AI might show that users who use "Feature X" have higher retention, but Causal Inference proves if the feature *caused* the retention or if power users simply find the feature more easily. This is best deployed when allocating R&D budgets. It ensures the leader is investing in the "Causal Drivers" of value, rather than just chasing vanity metrics or coincidental usage patterns.

3. The Digital Twin Pairing for Lifecycle Stress-Testing

When developing physical or digital hardware, **Predictive AI must be paired with Product Digital Twins.** This is best utilized to "rehearse" the long-term wear, tear, or technical debt of a product in a virtual sandbox. The leader can simulate five years of "Extreme Usage" in five minutes. This pairing ensures the "Eyes" see the structural or logical breaking points before the product is ever manufactured or deployed, allowing for the "Hardening of Quality" at the atomic level.

4. The Natural Language Processing (NLP) Pairing for Feature Discovery

To capture the "Voice of the User's Frustration," a leader should pair **Predictive AI with Advanced NLP.** While quantitative data tracks "What" users do, NLP analyzes "How" they feel about it by ingesting unstructured feedback from forums, support tickets, and beta-test logs. This identifies "Latent Feature Requests"—the specific pain points that users haven't explicitly asked to be solved but are clearly struggling with. This turns the Forecaster into an "Innovation Scout," ensuring the product roadmap is built on human truth.

5. The Monte Carlo Pairing for Market Fit Probability

To navigate the uncertainty of a "Category-Creating" product launch, leaders should pair **Predictive AI with Monte Carlo Simulations.** This is best utilized when the "Eyes" must predict the adoption curve of a product that has no historical precedent. By running millions of variables—from competitor pricing moves to shifting regulatory landscapes—the Forecaster can map the "Success Probability Cloud." This allows the leader to move from "Gut Feeling" to a range of mathematically grounded market outcomes.

6. The Knowledge Graph Pairing for Feature Provenance

In an era of rapid iteration, **Predictive AI should be paired with Knowledge Graphs.** This pairing tracks the **"Lineage of Innovation"**—mapping exactly why a feature was built, which user data justified it, and how it aligns with the brand's "Sovereign Values." It allows the leader to see the "DNA" of the product. By identifying the core "Value Nodes" that have stood the test of time, the Forecaster ensures the brand's "Eyes" never lose the product's original soul while it undergoes radical transformation.

Product Power: Semantic Coherence

A product's "Voice" is its interface, its ethics, and its ability to communicate its "Why" to the user. In an automated world, products can feel cold and opaque—a "Black Box" that makes decisions without explanation. The transformation into **Diplomats** focuses on **Semantic Integrity**. In this role, the product is designed to "Explain Itself." It bridges the gap between the "Code" of its engine and the "Soul" of the user. The Diplomat ensures that every interaction is grounded in "Evidence-Based Truth" and "Human Values," turning the product into a "Transparent Partner" that the user can trust implicitly.

Defensive Stance: Killing the Complexity Friction

The defensive mission of the Product-Diplomat is the **Elimination of Alienation**. As products become more intelligent, they often become more difficult to understand. This "Cognitive Load" creates a "Friction of Use" that leads to user frustration and abandonment. The Diplomat role kills this "Complexity" by using its "Voice" to provide "Perfect Clarity." This is the **Shield of Trust**. By ensuring that every action the product takes is "Justifiable" and "Transparent," the Diplomat protects the user from "System Anxiety." If the product makes a counter-intuitive recommendation, its "Voice" explains the logic behind it. This defensive posture ensures the product is always "Safe and Understandable," killing the risk of "User Error" and ensuring the brand is seen as an "Advocate for the Individual," rather than a "Commander of the Machine."

Offensive Stance: Unlocking Ethical Resonance

On the offense, the Diplomat moves toward **Shared Purpose**. This is the ability of the product to "Align" with the user's moral and strategic goals. The product uses its "Voice" to "Listen" to the user's intent and "Synthesize" a response that reflects the brand's values. This **Unlocks Narrative Loyalty**, where the user feels that the product "Stands for Something." For example, a financial product might use its "Voice" to warn a user that a specific transaction conflicts with their stated long-term sustainability goals. This turns the product into a "Moral Compass" that helps the user live their values. By "Mastering the Dialogue," the product creates a "Deep Emotional Lock-in," where the user relies on the product not just for its "Functions," but for its "Character." It allows the brand to "Scale Integrity," ensuring that every interaction is a "Reinforcement of the Brand's Promise."

The Product Outcome: Institutional Trust

The ultimate value of the Diplomat role is **Alignment through Meaning**. When a product acts as a "Diplomat," it moves from being a "Tool" to being a "Member of the Team." For the leader, this provides **Reputational Security**. You know that your "Engine" (the product) is communicating with a "Voice" (the interface) that is resonant, ethical, and backed by logic. This turns the product into an "Authentic Powerhouse"—where every user interaction is an opportunity to strengthen the "Shared Truth" of the brand. It ensures that the product's "Voice" is never a "Guess," but a "Declaration of Fact," creating a culture of "Evidence-Based Utility" that users trust with their most critical tasks.

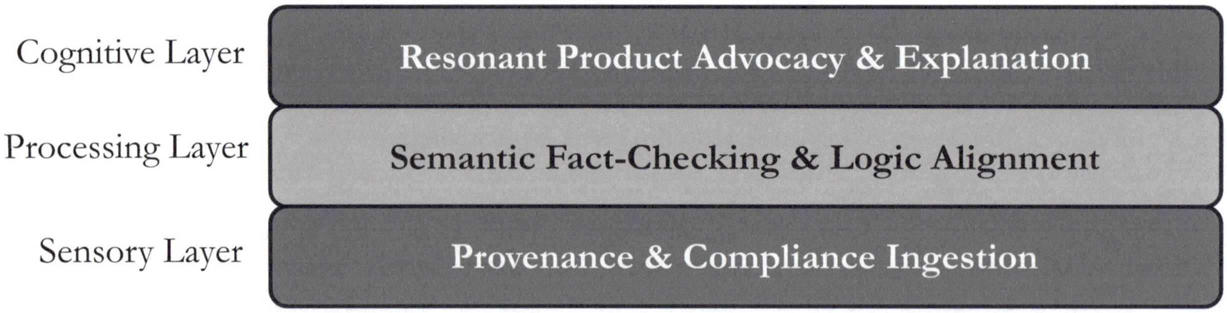

Executive Power: Transparent Product Integrity

For the modern executive, the "Voice" of the product is its **Verifiable Integrity** and its ability to communicate its own "Truth" to the user and the regulator. In an era of "Greenwashing" and "Planned Obsolescence" skepticism, the **Diplomat** role ensures the product's value is not a marketing claim, but a "Verifiable Fact" backed by data. This role is about the **Abolition of the Black Box**, ensuring that the "Voice" of the product is the most "Honest" and "Accountable" in the industry. The leader uses this stack to ensure the product's ethical and functional status is articulated with absolute transparency, moving the company from "Selling a Box" to "Providing a Promise." By architecting this role, the executive ensures that every user feels a sense of "Confidence and Alignment" with the product's origins and its current operational state.

The Sensory Layer: Provenance & Compliance Ingestion

The **Sensory Layer** of the Product Diplomat acts as the organization's "Integrity Ear." It moves beyond static certifications and into **Continuous Compliance Capture**. Utilizing **Blockchain-backed Supply Chain APIs** and **Automated Audit Logs**, this layer tracks the "Life of the Product" from raw material to the customer's hands. It captures the "Context of Creation"—understanding the carbon footprint, the labor conditions, and the material purity of every component. By monitoring for "Compliance Drift" and "Supply Chain Anomalies," this layer provides the leader with a real-time map of "Ethical Friction." It ensures the executive is aware of any potential reputational risk before it reaches the public. This layer provides a high-fidelity "Map of Provenance," ensuring the "Voice" of the product is grounded in the absolute reality of its physical and ethical history.

The Processing Layer: Semantic Fact-Checking & Logic Alignment

The **Processing Layer** serves as the "Guardian of the Product Promise." It utilizes **Knowledge Graphs** to link the brand's public marketing claims and sustainability goals to the specific telemetry and provenance data in the system. If the executive's "Voice" claims the product is "Carbon Neutral" or "Highly Efficient," this layer "Hard-Codes" those claims into the **Verification Logic**. This ensures that the leader's message is not just a slogan, but is mathematically reflected in the product's actual logs and certificates. It acts as a **Shield of Semantic Integrity**, ensuring that the product's "Voice" is never contradicted by its data. By resolving the identity of every component in the context of the brand's mission, this layer kills the friction of "Information Gaps," providing a "Verifiable Logic" that the product's performance is perfectly aligned with its public "Voice," thereby building an unassailable bond of institutional trust.

The Cognitive Layer: Resonant Product Advocacy & Explanation

The **Cognitive Layer** is where the "Product Truth" is transformed into **Explainable Narratives** for the user and the regulator. Utilizing **Large Language Models (LLMs)** and **Explainable AI (XAI)**, the Diplomat generates "Dynamic Integrity Reports." It doesn't just provide a manual; it orchestrates "Sovereign Explanations"—tailoring the product's status story for a curious eco-conscious consumer, a rigorous safety regulator, or a cost-focused fleet manager. This layer ensures the product can "Explain its own State," providing a reasoned justification for its current performance or its maintenance needs. This allows the leader to project a "Voice" of **Absolute Product Accountability**, turning "Quality Control" into a "Competitive Advantage." It ensures that every stakeholder feels a sense of "Security and Trust" in the product's operation, creating a culture of "Principled Engineering Transparency."

OVERVIEW

Explainable AI (XAI) provides transparent, human-readable rationales for AI decisions, moving beyond black box outputs. It reveals why a model reached a conclusion, building trust and mitigating legal and ethical exposure. The Regulatory & Compliance leader leverages XAI to validate complex, automated decision-making processes, particularly in highly regulated domains. This synergy ensures AI systems are fair, auditable, and adhere to internal policies and external legal requirements, transforming compliance from a check-the-box function into a strategic enabler of trust.

MAIN STEPS

Start

STEP 1 – Define Audit Scope
Establish parameters for compliance assessment.

STEP 2 – Feature Importance Analysis
Quantify input factors driving AI output.

STEP 3 – Generate Local Explanation
Create rationale for a single prediction.

STEP 4 – Validate Model Fairness
Test decision consistency across subgroups.

STEP 5 – Human Review and Oversight
Leader verifies explanations meet standards.

STEP 6 – Document Rationale & Action
Record justification for audit readiness.

End

SAMPLE FLOW

Automated Product Safety Certification: Leader utilizes XAI to audit the safety assessment for a new product design, verifying that all component certifications are met, with option to halt if it is not met.

1. Input Regulatory Standards
Define all required safety certifications and compliance checklists clearly.

2. Analyze Component Risk
Trace each part's origin and failure rate, identifying regulatory gaps.

3. Generate Certification Gaps
XAI explains which specific product features fail to meet safety mandates.

EXPLAINABLE AI

4. Simulate Failure Rationale
Show the cause-and-effect of potential component failures on the overall product safety.

5. Assess Liability Exposure
XAI calculates the legal risk score associated with launching the product without full certification.

6. Issue Stop-Ship Order
Leader halts product release, documenting the rationale for safety and regulatory failure.

USE CASES

- **Manufacturing | Materials Sourcing Audit:** Explains why an AI selects specific, sustainable material suppliers, allowing leaders to certify adherence to ethical and environmental product compliance standards consistently.
- **Financial Services | New Product Suitability:** Provides auditable rationale for why a new financial product is suitable for a specific customer demographic, ensuring compliance with consumer protection laws reliably.
- **Healthcare | Drug Efficacy Verification:** Explains the complex factors from clinical trials that led the AI to predict a drug's efficacy, allowing regulatory sign-off for release safely.
- **Technology | Security Patch Prioritization:** Justifies the AI's urgency ranking of security vulnerabilities in software updates, ensuring critical product safety patches are deployed quickly and compliantly.
- **Retail | Expiration Date Rationale:** Explains the environmental and ingredient data factors that determine a food product's shelf life, ensuring compliant and truthful labeling for consumers accurately.
- **Energy | System Compatibility Certification:** Provides transparent justification for AI approval of new power generation equipment compatibility, certifying it meets rigorous grid regulatory and safety standards quickly.

In the **Product Transformation**, the **Diplomats (The Voice)** ensure the product "speaks" the user's language. While **Explainable AI (XAI)** provides the technical logic for why a product is behaving a certain way—such as why a smart home system triggered an alarm—it often lacks the contextual sensitivity to deliver that information effectively. To move from "Opaque Hardware" to "Interactive Partnership," a leader must integrate these six secondary pairings to calibrate the product's "Voice," turning technical transparency into a competitive advantage.

1. The Sentiment Analysis Pairing for Emotional Feedback Loops

Pairing **Explainable AI with Sentiment Analysis** is best utilized for **Adaptive User Interfaces**. While XAI explains a product's logic, Sentiment Analysis monitors the user's tone or frustration through voice or text inputs. If a user is struggling with a feature, the product's "Voice" adjusts to be more helpful and less technical. This ensures the product doesn't just explain itself, but does so in a way that de-escalates user friction and builds emotional trust during complex tasks.

2. The Contextual NLP Pairing for Multimodal Interaction

Leaders should pair **Explainable AI with Contextual NLP** to enable **Cross-Platform Consistency**. XAI provides the "Reasoning," but Contextual NLP translates that reasoning into the appropriate medium—whether it's a detailed technical log for a pro-user or a simplified notification for a consumer. This ensures the product's "Voice" remains coherent across mobile, web, and hardware, providing a seamless identity that adapts its vocabulary to the user's specific professional or personal context.

3. The Knowledge Graph Pairing for Preference Lineage

When a product evolves over time, **Explainable AI must be paired with Knowledge Graphs**. While XAI justifies a current state, the Knowledge Graph provides the "History of User Choices." This allows the product to say, "I am suggesting this because of the preference you set last year." This pairing creates a "Verifiable Voice," proving to the user that the product's evolution is a logical response to their own needs, turning the product into a persistent, reliable partner.

4. The Differential Privacy Pairing for Trust-Based Transparency

To explain product decisions without compromising user data, a leader should pair **Explainable AI with Differential Privacy**. This is best used in **Collaborative Product Ecosystems** (e.g., "users like you preferred this"). The product can explain the logic behind a recommendation while mathematically guaranteeing that no other user's raw data is revealed. This ensures the product's "Voice" is helpful and transparent while maintaining the "Privacy Sovereignty" that is critical for modern brand loyalty.

5. The Conversational AI Pairing for Feature Interrogation

To eliminate the need for dense manuals, leaders should pair **Explainable AI with Conversational AI**. While XAI holds the data, Conversational AI allows the user to "interview" the product: "Why did you lower my energy usage at 2 PM?" This turns the product into a **Dynamic Advisor**. By providing instant, reasoned justifications for its actions, the product builds a culture of transparency, ensuring the user feels in control of the technology rather than managed by it.

6. The Blockchain Pairing for Ethical Provenance

In industries where product claims must be verified (e.g., luxury or sustainable goods), **Explainable AI should be paired with Blockchain**. XAI provides the "Explanation" of the product's origin, while the Blockchain stores that data in an immutable ledger. This ensures the product's "Voice" is legally unassailable. When a product "claims" to be carbon-neutral, the integrated blockchain provides a timestamped, tamper-proof record of the fact, securing the brand's moral integrity.

Product Power: Generative Evolution

In the legacy industrial world, a product was "Born and Frozen." Once it was manufactured or coded, its feature set was locked until the next version release. This created a "Gap of Obsolescence," where the product sat static while the user's needs continued to move. The transformation into **Creators** represents the shift to **"The Generative Product."** In this role, the product contains its own internal "Engine" capable of manifesting new capabilities, interfaces, and logic on the fly. The product is no longer a "Noun" (a thing); it is a "Verb" (a process of becoming). It uses the brand's generative power to reshape itself, moving from a state of "Rigid Utility" to "Fluid Adaptation."

Defensive Stance: Killing the Feature Decay

The defensive mission of the Product-Creator is the **Elimination of Stagnation**. Every product eventually suffers from "Feature Decay," where the original design no longer fits the modern context. The Creator role kills this "Inertia" by "Resurrecting the Value" in real-time. It uses its "Engine" to patch its own limitations and modernize its own workflows. This creates a **Shield of Perpetual Freshness**. Instead of the user having to wait for a "2.0 Version," the product "Generates" the update the moment it detects a friction point. This defensive posture kills the "Waste of Replacement," ensuring the product is never "Old." It protects the user's investment by ensuring the product they bought today is structurally capable of becoming the product they need tomorrow. By "Cleaning the Engine" of its own legacy code and outdated UI, the product maintains a "Zero-Friction" state for the duration of its lifecycle.

Offensive Stance: Unlocking Personal Reality Creation

On the offense, the Creator moves toward **Infinite Configurability**. This is the ability of the product to "Manifest" a unique version of itself for every individual user. The product doesn't just "offer" settings; it "Generates" a custom reality. If a user is a visual thinker, the product generates a highly graphical interface; if they are data-driven, it manifests a complex analytical dashboard. This **Unlocks Creative Agency**, where the product becomes a "Mirror of the User." It allows the brand to "Saturate the Market" with a million "Niche Products" that are actually all powered by the same "Generative Engine." This turns the product into a "Platform for Manifestation," where the user's intent is the only limit to what the product can do. It ensures the brand is always producing the most "Original" and "Relevant" experience, because the product is literally "Creating" its own value at the point of use.

The Product Outcome: Absolute Plasticity

The ultimate value of the Creator role is **Radical Relevance**. When a product "Creates," the concept of "One Size Fits All" dies. For the leader, this provides **Market Agility**. You are no longer managing a "Product Catalog"; you are managing a "Generative Capability." This turns the product line into a "Living Response System"—one that evolves its "Functions" in real-time to match the "Intent" of the user. It ensures that the brand's "Engine" is never "Finished," but is always working to manifest the highest possible utility for each individual, creating a product that feels like a "Living Organism" designed specifically for the person using it.

Executive Power: Generative Feature Manifestation

The **Creator** role in product transformation represents the ultimate shift from "Shipping Hardware" to **"Orchestrating Living Systems."** In the legacy model, a product's capabilities were frozen at the moment it left the factory; any improvement required a new purchase or a manual update. The Creator role replaces this with **Autonomous Utility Synthesis**, where the product's digital and functional architecture is dynamically generated in response to how the user actually interacts with it. The executive uses this engine to move the organization from a state of "Rigid Iteration" to "Fluid Evolution," ensuring the product is always manifesting the most "Relevant" solution for the specific user context. This turns the "Engine" of the firm into a generative loop where the product literally "re-invents" itself to solve the user's emerging problems on the fly.

The Sensory Layer: Contextual Usage & Intent Ingestion

The **Sensory Layer** for the Creator captures "User Intent" as a high-definition signal. It moves beyond basic button-clicks by integrating **Multimodal Interaction Capture** and **Environmental Context Streams**. By utilizing **Edge AI** to analyze voice, gesture, and environmental conditions (such as temperature, location, or load), this layer identifies the "Atomic Requirements" of a new feature before the engineering team even sketches it. It captures the raw intent—"The user is struggling to operate this crane in high-wind conditions"—and automatically identifies the parameters required for a software-defined adjustment. This ensures the leader is building the "Engine" of the product on a foundation of "Individual Reality" rather than "Market Averages," capturing the necessary signals for a bespoke manifestation as a live, fluctuating requirement that informs the next generation of the product's code.

The Processing Layer: Parametric Design Logic & Safety Filtering

The **Processing Layer** acts as the "Architectural Refinery" for product evolution. It utilizes **Constraint-Based Solvers** and **Digital Twin Simulations** to manifest an update that is both "Ideal" for the user and "Safe" for the system. This layer understands the "Rules of the Possible"—including physical tolerances, power consumption limits, and regulatory safety standards. It stress-tests every generative update against the product's "Causal Truths" in a virtual sandbox before deployment, ensuring the executive only authorizes changes that are structurally sound. It acts as a **Guardian of Stability**, filtering out high-risk "hallucinations" in the autonomous logic. By applying "Parametric Product Logic," this layer ensures the "Engine" is disciplined, preventing the product from evolving into an unstable state while ensuring the user receives a feature that feels uniquely engineered for their specific environmental challenge.

The Cognitive Layer: Dynamic Feature Synthesis & Over-the-Air Delivery

The **Cognitive Layer** is the "Orchestration Mind" of the product engine, powered by **Large Language Models (LLMs)** and **Action Transformers**. This layer "hallucinates" thousands of different ways to optimize the product's performance, presenting the most "Resonant" version for autonomous delivery via the cloud. It doesn't just suggest a change; it manifests the **Functional Update**. The leader provides the strategic guardrails, and the Cognitive Layer generates the "Update Genome"—automatically writing the code, configuring the UI, and optimizing the hardware parameters required to fulfill the mission. It involves "Recursive Product Evolution," where the AI adapts the interface as the user's skill level grows, ensuring the final output is the "Optimal Manifestation" of the brand's utility. This turns the executive into a "Conductor of Innovation," capable of deploying complex, individualized product improvements at a speed that makes traditional manufacturing cycles obsolete.

3. THE CREATORS (THE ENGINE) | GENERATIVE AI

OVERVIEW

Generative AI creates novel outputs like text, art, code, and immersive environments from simple prompts. It excels at rapid prototyping, style transfer, and exploratory ideation. The Creative & Artistic leader partners with Generative AI to rapidly visualize strategic concepts, explore diverse future scenarios, and prototype new organizational narratives. The synergy transforms vision casting from a slow, manual process into a continuous, inspiring, and engaging collaborative endeavor.

MAIN STEPS

Start

STEP 1 — Define Creative Prompt
Articulate desired output parameters precisely.

STEP 2 — Initial Content Generation
Create diverse drafts based on input.

STEP 3 — Style and Mood Adaptation
Adjust tone, format, or aesthetic output.

STEP 4 — Iterative Refinement Loop
Guide model toward final creative goal.

STEP 5 — Human Curation & Selection
Choose the most impactful creative result.

STEP 6 — Deploy Creative Asset
Integrate final output into strategy.

End

SAMPLE FLOW

Accelerated Industrial Design and Prototyping: A Chief Design Officer uses Generative AI to rapidly generate and test thousands of unique product prototypes and aesthetic variations instantly..

GENERATIVE AI

1. Input Design Constraints
Define materials, budget, user needs, and engineering specifications precisely.

2. Generate Iterative Designs
Produce thousands of novel 3D models and aesthetic options instantly from text.

3. Simulate User Interaction
Create virtual environments for testing ergonomics and aesthetic appeal rigorously.

4. Refine Based on Feedback
Adjust shape, texture, and color based on user and engineering team inputs.

5. Select Final Concept Batch
Curate the top 10 most viable, beautiful, and market-ready product candidates.

6. Prepare for Manufacturing
Deliver validated design files and specifications for physical production commencement.

USE CASES

- **Manufacturing | Tooling Optimization:** Generates new, topologically optimized designs for manufacturing tools and molds, significantly reducing material waste and production cycle times while maintaining structural integrity.
- **Financial Services | Financial Product Naming:** Creates thousands of unique, psychologically resonant, and legally viable names and slogans for new investment funds and banking services within seconds.
- **Healthcare | Drug Compound Structure:** Generates novel molecular structures and protein folding patterns, accelerating the initial discovery phase of drug development for specific complex target diseases.
- **Technology | Code Autocompletion & Testing:** Creates vast amounts of context-aware code suggestions and automatically generates comprehensive unit tests, dramatically speeding up the software development lifecycle.
- **Retail | Fashion Line Ideation:** Generates mood boards, fabric patterns, and garment silhouettes based on trend data and artistic prompts, rapidly ideating next season's creative collection for design teams.
- **Energy | Grid Interface Design:** Develops intuitive and responsive user interfaces for monitoring complex energy grid operations, making real-time system visualization and control simpler for plant operators.

In the **Product Transformation**, the **Creators (The Engine)** represent the "Continuous Innovation Layer." While the primary instrument of **Generative AI** is used to design new features or generate code, it often lacks the "Practical Grounding" to ensure these innovations are actually manufacturable or secure. To move from "Rapid Prototyping" to "Generative Manufacturing," a leader must integrate these six secondary pairings. This ensures the "Engine" doesn't just dream up possibilities, but builds products that are structurally sound, contextually relevant, and aligned with the brand's sovereign standards.

1. The Synthetic Data Pairing for Stress-Test Generation

Pairing **Generative AI with Synthetic Data Generation** is the ultimate tool for **Resilience-First Design.** Before a product is released, the Creator generates millions of "Adversarial Scenarios"—synthetic environments where the product is subjected to extreme heat, data surges, or user misuse. This is best utilized in "Hardening" both software and hardware. By training the engine on these synthetic edge cases, the leader ensures the product is born with a "Digital Immunity" to the stresses of the real world.

2. The Digital Twin Pairing for Generative Iteration

Leaders should pair **Generative AI with Product Digital Twins** to achieve **Closed-Loop Innovation.** While the AI generates thousands of potential physical designs or code structures, the Digital Twin provides the "Physics-Based Reality Check." This is best used in "Generative Design" for industrial hardware. The Engine dreams of a more aerodynamic shape, and the Digital Twin instantly simulates its wind-drag and structural integrity. This pairing ensures that only the most viable, high-performance designs are ever moved to the manufacturing floor.

3. The RLHF (Human Feedback) Pairing for Aesthetic Integrity

When the "Engine" creates the outward-facing elements of a product—such as the user interface or physical styling—**Generative AI must be paired with Reinforcement Learning from Human Feedback (RLHF).** This ensures the output reflects the brand's "Sovereign Aesthetic." Human master-designers provide the "Taste" and "Nuance" that a machine cannot calculate. This turns the Creator into an "Augmented Artisan," ensuring every generated feature feels like it was crafted by a human expert, maintaining the premium soul of the brand.

4. The Vector Database (RAG) Pairing for Grounded R&D

To prevent "Design Hallucinations," a leader should pair **Generative AI with Vector Databases (RAG).** This pairing ensures that every new feature is grounded in the firm's actual historical patents, safety records, and technical successes. This is best utilized for "Legacy-Compatible Innovation." Instead of reinventing the wheel, the Engine retrieves existing proprietary "Wisdom" from the **Institutional Brain** to suggest improvements that are consistent with the firm's engineering history and competitive moats.

5. The Differential Privacy Pairing for Collaborative Design

To enable "Co-Creation" with customers without risking intellectual property or user data, leaders should pair **Generative AI with Differential Privacy.** This allows the brand to open its product "Engine" to user suggestions or third-party developers. The generative model can learn from "Crowd-Sourced" patterns of usage while the privacy layer ensures that no individual user's data or proprietary code is exposed. This turns the product into a "Living Ecosystem" where the brand and the user evolve the product together in a secure environment.

6. The API Orchestration Pairing for Autonomous Deployment

To move from "Design" to "Delivery," **Generative AI should be paired with API Orchestration.** While the AI creates a new software patch or configuration for a physical device, API Orchestration provides the "Connection" to actually deploy that update across the entire product fleet. This moves the "Engine" from a lab tool to a live "Operational Factory." It ensures the product is constantly self-updating and self-optimizing in real-time, closing the gap between the birth of an idea and its manifestation in the user's hands.

Product Power: Causal Optimization

Traditional products are "Functional"—they do what they are told. If you press a button, a process happens. However, "Doing the Task" is not the same as "Achieving the Goal." The transformation into **Strategists** moves the product into the realm of **Causal Outcome Optimization.** In this role, the product doesn't just "Perform"; it "Navigates." It understands the "Mathematical Why" behind the user's actions and uses its "Compass" to find the most efficient path to the desired result. The product evolves from a "Passive Tool" into a "Strategic Navigator" that is responsible for the user's success.

Defensive Stance: Killing the Misalignment of Effort

The defensive mission of the Product-Strategist is the **Validation of Impact**. Many products are used in ways that don't actually move the needle for the user—this is "Busy Work" at the product level. The Strategist role kills this "Strategic Friction" by applying "Causal Logic" to every interaction. It acts as the "Guardian of the Win," identifying when a user is taking an action that won't lead to their stated goal. This is the **Shield of Objectivity**. By "Modeling the Outcome" of every user move, the product kills the "Mismatch" between effort and reward. It prevents the user from "Chasing Dead Ends" and ensures that every interaction is a "Step Toward the Peak." This defensive posture protects the user from "Operational Waste" and ensures the product's "Compass" is never distorted by the user's own biases or "Legacy Habits," keeping the focus squarely on the "Causal Truth" of what drives results.

Offensive Stance: Unlocking Prescriptive Mastery

On the offense, the Strategist moves toward **Dynamic Course Correction**. This is the ability of the product to "Guide the User" through complex decision-making landscapes. The product uses its "Compass" to "Prescribe the Best Move" in real-time. Whether it is a financial product suggesting a specific asset allocation to meet a goal, or a manufacturing product suggesting a different material to maximize durability, the product is "Strategizing the Outcome." This **Unlocks Navigational Sovereignty**, where the product becomes the "Operating System for Success." It involves "Simulating the Win"—running thousands of "What-If" scenarios to find the "Hidden Path" that a human, limited by linear thinking, would miss. This turns the product into a "Calculating Partner" that optimizes the entire value chain of the user's task. It allows the brand to "Own the Result," ensuring that the user doesn't just "Use" the product, but "Wins" because of it.

The Product Outcome: Strategic Certainty

The ultimate value of the Strategist role is **Objective Confidence**. When a product acts as a "Strategist," the user feels "Navigated" rather than "Commanded." For the leader, this provides **Economic Resilience**. You are building a "Portfolio of Successful Users" whose achievements are mathematically linked to your product's logic. This turns the product into a "Strategic Ally"—one that finds the "Win-Win" in every interaction. It ensures that the product's "Compass" is always pointing toward "Maximum Yield," creating a brand that is respected for its "Intelligence" and relied upon for its "Infallible Logic" in achieving complex goals.

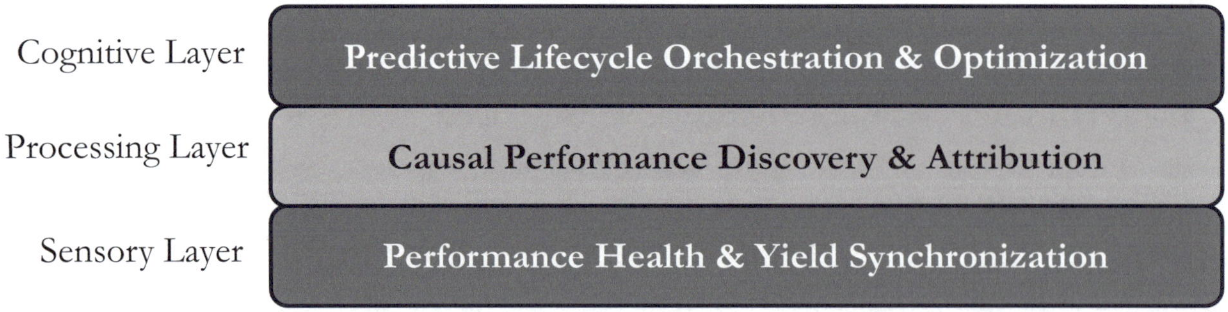

Executive Power: Causal Operational Yield

The **Strategist** role in product transformation represents the "Abolition of Planned Obsolescence." It moves the executive from "Unit Volume Targets" to **"Causal Operational Yield Optimization."** The leader uses this stack to ensure that every "Product Hour" in the field is generating maximum value for both the customer and the brand. The Strategist is the **Guardian of Performance**, ensuring that the organization is not just "Selling Devices," but "Managing Outcomes." In this role, the executive evolves into a "Strategic Asset Navigator" who manages the company's "Functional Capital" with mathematical precision. This role kills the friction of "Product Failure" and "Under-Utilization" by ensuring the brand's energy is always flowing toward the "True Drivers" of uptime, efficiency, and long-term user value, transforming the product into a reliable engine of recurring revenue.

The Sensory Layer: Performance Health & Yield Synchronization

The **Sensory Layer** of the product Compass monitors the "Micro-Levers" of operational health across the entire global fleet. It synchronizes **Real-Time Telemetry** (energy draw, cycle times, component stress) with **External Economic Context**. By tracking the "Velocity of Value" (how much work the product is doing versus its cost of operation) against the "Customer's Business Outcome," this layer provides the "Coordinate Data" required to know exactly where the firm's assets stand in relation to its "Strategic North Star." It provides the "Eyes" for the Compass, ensuring that the starting point of any maintenance or upgrade initiative is grounded in the absolute reality of current mechanical health. This layer removes the "Blind Spots" in management, ensuring the leader is aware of every factor—from "Silent Inefficiency" to "Critical Wear Patterns"—that could impact the trajectory of the product's yield.

The Processing Layer: Causal Performance Discovery & Attribution

The **Processing Layer** is the heart of the Strategist role, utilizing **Causal Inference Engines** and **Physics-Informed Neural Networks** to determine "Why" certain products are outperforming others in the field. While traditional analytics show simple "Correlation" (e.g., machines in cold climates fail more), the Strategist identifies the **Causal Path** (e.g., they fail because a specific lubricant loses viscosity at exactly -10 degrees). This layer provides the "Mathematical Evidence" for product strategy, ensuring the leader is investing in the "Levers" that actually matter for the brand's "Yield." It acts as a **Shield of Objectivity**, stripping away the "Noise" of anecdotal feedback and focusing the executive on "Global Fleet Optimization." By identifying the "Mathematical Why," this layer ensures the firm is never "Chasing Symptoms," but is always focused on the "Causal Truths" that drive real-world durability and profit.

The Cognitive Layer: Predictive Lifecycle Orchestration & Optimization

The **Cognitive Layer** acts as the "Navigational Mind" of the product stack, utilizing **Reinforcement Learning** and **Multi-Agent Simulation**. This layer "plays the game" of the product lifecycle millions of times to find the "Optimal Path of Operation." When the market environment shifts—due to an energy crisis or a regulatory change—the Compass "Recalculates" the fleet management strategy instantly. This layer doesn't just report on performance; it "Prescribes the Best Move" to maximize the **Operational Yield**. For example, it might recommend a proactive remote power-throttle to a specific high-usage segment to extend the product life during a period of component scarcity. This provides the executive with **Strategic Certainty**, allowing them to navigate product volatility with the confidence that their "Compass" is calibrated to the objective laws of engineering value. It turns the organization into a "Calculating Utility Provider," ensuring the brand "Wins" because its "Product Strategy" is mathematically optimized for the highest possible uptime and user success.

4. THE STRATEGISTS (THE COMPASS) | COGNITIVE AI

OVERVIEW

Cognitive AI mimics human intelligence to understand and interpret unstructured data like text, speech, and images. It processes complex inputs to generate hypotheses, assess logical connections, and offer nuanced context. The Research & Analytical leader uses Cognitive AI as a hyper-efficient knowledge partner, rapidly synthesizing vast amounts of data to rigorously test assumptions, structure complex arguments, and make deeply informed, evidence-based strategic decisions.

MAIN STEPS

Start

STEP 1 — Data Ingestion: *Collect and structure input knowledge quickly*

STEP 2 — Understand Context: *Understand meaning and relationships deeply.*

STEP 3 — Hypothesis Generation: *Formulate potential explanatory outcomes.*

STEP 4 — Evidence Synthesis: *Find relevant supporting information.*

STEP 5 — Analytical Refinement: *Test hypotheses against real-world data.*

STEP 6 — Insight Formulation: *Structure findings for human action.*

End

SAMPLE FLOW

R&D Portfolio - Innovation Risk Optimization: A Chief Product Officer uses AI to assess vast market research against internal capabilities to de-risk the innovation portfolio.

1. Define Innovation Parameters
Specify acceptable market size, technical feasibility, and financial return criteria precisely.

2. Ingest Global Research Data
Collect patents, scientific journals, and emerging consumer trend reports deeply.

3. Cross-Check Tech Feasibility
Analyze R&D capacity and skill sets against required technological breakthroughs quickly.

4. Market Acceptance Risks
Estimate customer adoption rates based on current behavioral data and competitive landscape.

5. Synthesize Portfolio ROI
Rank potential projects based on predicted financial impact and strategic alignment rigorously.

6. Commit to Optimal Roadmap
Sign off on the final, risk-adjusted product development and launch sequence.

USE CASES

- **Manufacturing | Component Sourcing Redesign:** Analyzes component failure data and material specifications across product lines to inform engineers about superior, durable substitute parts during redesigns.
- **Financial Services | Digital Product Compliance:** Scans regulatory changes and user interface performance metrics to design streamlined, compliant digital banking products with reduced customer abandonment rates.
- **Healthcare | Drug Formulation Viability:** Interprets clinical efficacy studies and manufacturing capacity constraints to determine the optimal dosage, delivery mechanism, and mass production methods for new drugs.
- **Technology | Security Vulnerability Auditing:** Reviews thousands of open-source projects, community forums, and vulnerability reports to identify critical dependencies and security improvements for core software products.
- **Retail | Dynamic Pricing Strategy:** Synthesizes supply chain disruption data, raw material costs, and consumer price elasticity to finalize product pricing strategies for maximizing quarterly profit margins.
- **Energy | Next-Gen Solution Design:** Interprets infrastructure maintenance logs and grid performance data to inform the development of next-generation smart meters and highly resilient energy storage solutions.

In the **Product Transformation**, the **Strategists (The Compass)** act as the "Lifecycle Navigators." While the primary instrument of **Cognitive AI** identifies patterns in product performance and market fit, it often lacks the multi-dimensional foresight to balance long-term sustainability against immediate feature demand. To move from "Feature Bloat" to "Strategic Essentialism," a leader must integrate these six secondary pairings. This ensures the "Compass" doesn't just chase trends, but identifies the precise trajectory for maximum product longevity and category leadership.

1. The Multi-Objective Optimization (MOO) Pairing for Portfolio Yield

Pairing **Cognitive AI with MOO** is the ultimate tool for **Product-Market Alignment.** While AI identifies profitable features, MOO balances the trade-offs between manufacturing complexity, R&D cost, and the strategic "alpha" of the innovation. This is best utilized in "Hardware-Software Hybrid" environments. It allows the leader to identify the "Efficiency Frontier," ensuring the product isn't over-engineered for a niche segment but is optimized for the highest value across the entire user base.

2. The Game Theory Simulation Pairing for Competitive Differentiation

Leaders should pair **Cognitive AI with Game Theory** to achieve **Feature Sovereignty.** This is best deployed during "Red Ocean" market cycles where competitors are rapidly mimicking features. While the AI predicts market shifts, Game Theory simulates how rivals will respond to your product roadmap. This ensures the "Compass" identifies the "Non-Mimicable Advantage"—the specific feature or ecosystem lock-in that remains dominant even when rivals attempt to copy it.

3. The Digital Twin Pairing for Ecosystem Resilience

When expanding a product into new categories, **Cognitive AI must be paired with Product Digital Twins.** This is best utilized to "rehearse" the impact of a total system re-architecture or a move to a new material science. The Strategist simulates how the entire product ecosystem will react to a "Breaking Change." This pairing ensures the "Compass" guides the evolution without causing technical debt or physical failure, allowing for the hardening of the product's core logic before it reaches the real world.

4. The Knowledge Graph Pairing for Skill and IP Arbitrage

To maximize the impact of the firm's intellectual property, a leader should pair **Cognitive AI with Knowledge Graphs.** While AI tracks product sales, the Knowledge Graph maps how specific features relate to the firm's patents and internal engineering expertise. This allows for "IP Arbitrage"—identifying underutilized patents and matching them with current product gaps. It ensures the "Compass" is always directing the brand's "Alpha Engineers" toward the most potent innovation moats.

5. The Causal Inference Pairing for Outcome Validation

For maintaining a "High-Fidelity Product Culture," leaders should pair **Cognitive AI with Causal Inference.** While AI identifies "Success Stories," Causal Inference proves *why* a specific product won—separating genuine product-market fit from lucky market timing. This is best utilized in "Post-Launch Audits." It ensures the "Compass" points toward the actual drivers of utility, preventing the organization from doubling down on "False Positives" that won't hold up in the next market cycle.

6. The Agentic Orchestration Pairing for Dynamic Roadmap Execution

In high-velocity tech markets, **Cognitive AI should be paired with Agentic Orchestration** for **Live Portfolio Governance.** While the Strategist sets the vision, the Agentic layer automatically re-allocates compute resources or engineering sprints based on real-time usage data and competitive threats. This is best utilized in "Software-as-a-Service" environments. It turns the "Compass" into an active driver of the product factory, ensuring that resources are always flowing toward the areas of greatest strategic opportunity and user need.

Product Power: Cognitive Connectivity

In the legacy era, products were "Isolated Silos." A car was a car, a software package was a software package, and neither knew what the other was doing. This created a "Fragmented Reality" for the user, who had to act as the manual bridge between their disparate tools. The transformation into **Architects** turns every product into a "Node in the Global Brain." In this role, the product's primary value is not just its local function, but its **Structural Intelligence**—its ability to connect, share context, and contribute to the organization's collective wisdom. The Architect ensures the product is "Systemically Aware," knowing its place within the user's broader ecosystem and the brand's total intelligence.

Defensive Stance: Killing the Friction of Disconnect

The defensive mission of the Product-Architect is the **Abolition of Incompatibility**. Most user frustration stems from "Systemic Friction"—where data from one product cannot be used by another, or where a product lacks the context of the user's previous actions. The Architect role kills this "Friction" by building a "Universal Semantic Layer" into the product. This creates a **Shield of Seamlessness**. By "Resolving the Identity" of the user and their goals across the entire product line, the Architect ensures the product has "Total Recall." It protects the user from the "Fatigue of Re-entry" and the "Error of Isolation." This defensive posture ensures the product is always "In Sync," killing the risk of "Operational Blindness" and ensuring the brand's ecosystem is a "Fortress of Continuity" where every product supports and reinforces the others.

Offensive Stance: Unlocking Networked Wisdom

On the offense, the Architect moves toward **Collective Learning**. This is the ability of the product to get smarter based on the "Networked Intelligence" of all other products in the field. When one product learns a more efficient way to achieve a goal, that "Wisdom" is architected back into the "Global Brain" and distributed to every other node. This **Unlocks Exponential Utility**, where the product you use today is powered by the "Total Experience" of every user the brand has ever served. It involves "Dynamic Knowledge Synthesis," where the product provides the "Perfect Context" for a task by pulling insights from across the brand's entire architectural map. This turns the product into a "High-IQ Asset" that is fundamentally "Smarter" than a standalone rival. It ensures the brand's "Brain" is an "Accumulative Competitive Advantage" that grows more sophisticated with every second of use.

The Product Outcome: Institutional Intelligence

The ultimate value of the Architect role is **Scalable Context**. It ensures the brand's "Wisdom" is embedded in the very structure of the product. For the leader, this provides **Systemic Sovereignty**. You are no longer managing "Individual SKUs"; you are managing a "Live Intellectual Web." This turns the product into a "Living Archive" of the brand's best logic. It ensures the product is a "Universal Presence" in the user's life, providing a consistent "Brain" that guides them through complexity, regardless of which physical or digital interface they are touching at that moment.

Executive Power: Distributed Product Consciousness

In the legacy enterprise, product knowledge was isolated—what happened to a machine in the field was unknown to the engineers at headquarters until it broke. The **Architect** role represents the executive's transition to the "Designer of the Firm's Distributed Product Consciousness." In this role, the leader constructs a permanent, self-evolving **Cognitive Infrastructure** that transforms every unit into a node of a persistent, networked "Brain." This is the shift from "Individual Units" to "Swarm Intelligence," ensuring that a lesson learned by one product is instantly shared with the entire fleet. The Architect ensures the organization possesses a "Product Mind"—one that understands the context of its own physical state and grows more "intelligent" with every hour of operation, turning the fleet into a compound asset that self-optimizes its own functional logic.

The Sensory Layer: Universal Telemetry & Identity Resolution

The **Sensory Layer** of the Product Brain acts as the "Universal Perception System." It moves beyond simple logging into **Autonomous Asset Resolution**. Utilizing **Semantic Metadata Tagging** and **High-Frequency Edge Ingestion**, this layer ensures that every product—whether a medical device, a vehicle, or an industrial sensor—has a unique, persistent digital identity that the "Brain" can recognize globally. It doesn't just record "Data"; it "labels" the context of that data in real-time. By utilizing **Edge Perception Agents**, the sensory layer captures the "Environmental Nuance"—distinguishing between a mechanical vibration caused by a loose bolt and one caused by uneven terrain. This ensures the "Brain" of the firm is constantly fed with a high-fidelity, perfectly identified stream of physical reality, removing the "Friction of Ignorance" that traditionally separates the manufacturer from the product's true life cycle.

The Processing Layer: Structural Intelligence & Swarm Mapping

The **Processing Layer** serves as the "Neural Connector" of the product stack. It utilizes **Knowledge Graphs** and **Digital Twin Synchronicity** to define and store the complex relationships between the identities captured in the sensory layer. While traditional systems store data in isolated files, the Architect's processing layer stores the **Behavioral Connectivity**. It understands, for example, the semantic link between a specific software version's power consumption and the ambient temperature of a geographic cluster. By employing **Graph Neural Networks (GNNs)**, this layer identifies "Hidden Performance Patterns" across the fleet, allowing the "Brain" to perceive the entire product line as a unified, thinking organism. This creates a **Shield of Systemic Integrity**, ensuring that when a performance threshold is crossed by one unit, the implications are immediately "felt" and calculated for the entire global fleet architecture.

The Cognitive Layer: Recursive Institutional Wisdom & Self-Evolving Code

The **Cognitive Layer** is the "Reasoning Core" of the Product Brain, powered by **Retrieval-Augmented Generation (RAG)** and **Federated Learning**. This layer ensures the brand "never forgets" a mechanical or digital success. It synthesizes the relationships from the processing layer into a "Continuous Stream of Intelligence" that acts as a permanent advisory for the firm's engineering and support teams. When an executive or a maintenance agent asks a question, the Cognitive Layer doesn't just retrieve a manual; it "reasons" through the entire fleet's history to provide an answer grounded in "Deep Operational Context." It involves **Self-Evolving Logic**, where the Brain updates the product's internal schemas as new usage patterns arrive. This turns the fleet into a "Learning Organism" with an ever-increasing "Product IQ," providing the executive with a "Memory Asset" that ensures the brand's "Mind" remains perfectly synchronized with the product's physical reality.

OVERVIEW

Human-in-the-Loop (HITL) AI is a cycle where human supervisors continuously train, tune, and validate a model's performance, especially for ambiguous tasks. The Supervisory & Coaching leader focuses on defining ethical boundaries and quality thresholds, ensuring AI decisions align with organizational values. This synergy elevates the leader from a decision-maker to a model trainer and quality assurance expert. By providing critical feedback, the leader helps the AI evolve, improving accuracy, fairness, and overall governance in complex, high-stakes leadership decisions.

MAIN STEPS

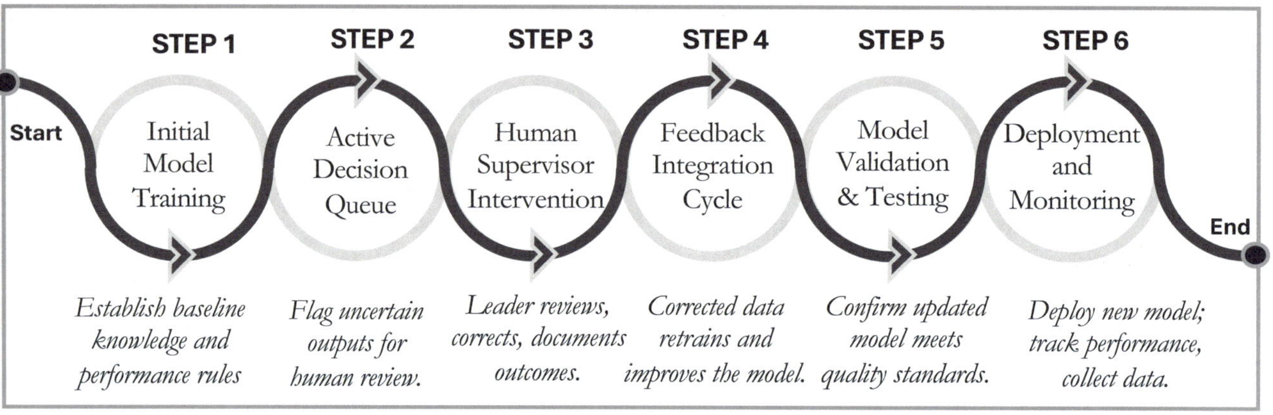

SAMPLE FLOW

Validating AI-Driven Product Feature Prioritization: A leader uses HITL to audit and refine AI-suggested product feature roadmaps, ensuring alignment with strategic vision and customer needs..

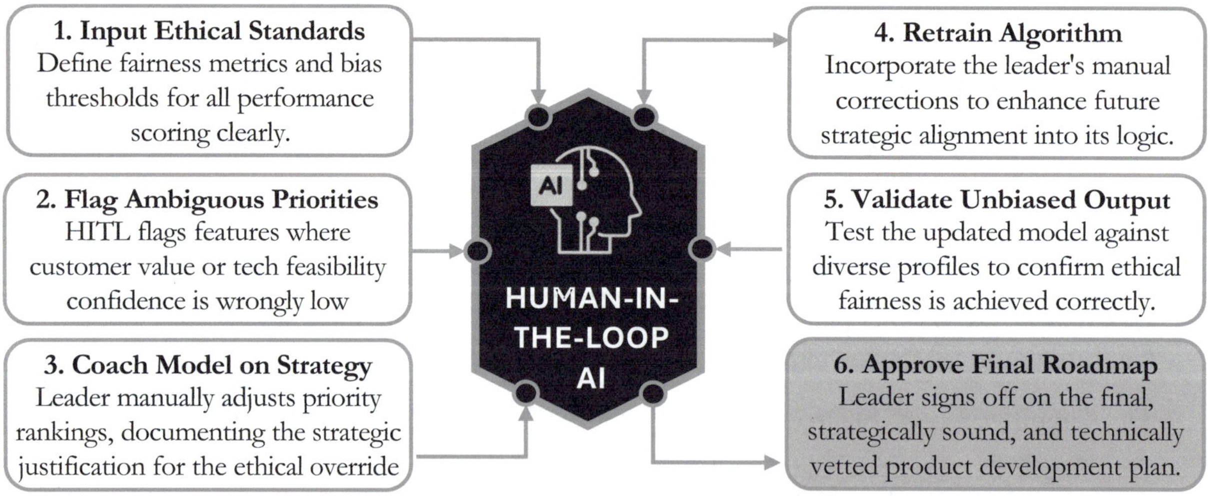

USE CASES

- **Manufacturing | Prototype Failure Analysis:** Leaders review AI's categorization of complex hardware failure modes, manually labeling rare causes to improve diagnostic accuracy for future product redesigns effectively.
- **Financial Services | New Service Risk Assessment:** Supervisors manually review the AI's risk scores for novel financial products, correcting model assumptions based on regulatory changes to ensure compliant launch reliably.
- **Healthcare | Medical Device Testing Protocols:** Senior staff review AI-suggested clinical trial parameters, correcting safety boundaries to ensure the testing procedures are robust and patient-focused safely.
- **Technology | Code Vulnerability Prioritization:** Product security leads manually flag low-risk vulnerabilities that could impact core features, teaching the AI to prioritize based on customer experience clearly.
- **Retail | Private Label Ingredient Vetting:** Managers review AI's approval of new product ingredients, correcting ambiguous source data to ensure ethical sourcing and consumer safety compliance accurately.
- **Energy | Grid Modernization Tool Design:** Supervisors validate AI-generated specifications for new operational hardware, providing practical field context to certify the equipment designs are functional and safe quickly.

In the **Product Transformation**, the **Architects (The Brain)** serve as the "Governance Framework" for the product's lifecycle. While the primary instrument of **Human-in-the-Loop (HITL)** ensures that the product remains ethically sound and aligned with human safety, it requires secondary technical pairings to manage the vast complexity of modern product ecosystems. To move from "Static Versioning" to a "Sovereign Product Intelligence," a leader must integrate these six pairings. This ensures the product's "Brain" is not just a repository of code, but a contextually aware, legally secure, and ethically anchored infrastructure.

1. The Vector Database (RAG) Pairing for Contextual Utility

Pairing **HITL with Vector Databases** is the foundation of the **"Self-Explaining Product."** While the human Architect sets the safety parameters, the Vector Database provides the "Semantic Memory" that allows the product to understand user intent across millions of historical interactions. This is best utilized in "Complexity Management." By indexing every technical manual and past user solution, the Architect ensures the product can assist the user in natural language, effectively turning the product's documentation into a "Live" assistant that provides the right context at the exact moment of need.

2. The Knowledge Graph Pairing for Structural Integrity

Leaders should pair **HITL with Knowledge Graphs** to ensure **Logical Traceability.** While the human provides the high-level design philosophy, the Knowledge Graph maps the dependencies between every component, sensor, and software module. This is best deployed in "Regulatory Compliance" for physical hardware. It allows the Architect to see not just a failure, but the "Lineage" of that failure across the supply chain. This pairing prevents "Logic Decay," ensuring that every product iteration is anchored in a verifiable web of truth.

3. The Natural Language Processing (NLP) Pairing for Ethical Auditing

To prevent the product from developing "Algorithmic Bias," **HITL must be paired with Advanced NLP.** This is best utilized for "Value Monitoring" within the product's communication layers. When the NLP identifies a response or a data-processing pattern that deviates from the brand's ethical guidelines, it triggers a request for a human Architect to review and recalibrate. This ensures the product's "Brain" remains a "Living Organism" that is constantly pruned and refined by the firm's most senior ethical minds.

4. The Differential Privacy Pairing for Product Sovereignty

When building connected or smart products, a leader should pair **HITL with Differential Privacy.** This allows the Architect to improve the product using real-world data without the risk of exposing individual user habits. This is best used in "Federated Learning" for smart devices. For example, a medical device can learn from a global fleet of patients while the privacy layer mathematically ensures that no individual patient's raw data is ever shared, protecting the brand's legal and ethical sovereignty.

5. The Multi-Modal AI Pairing for Sensory Knowledge Capture

To move beyond text-based logs, leaders should pair **HITL with Multi-Modal AI.** This allows the "Brain" to ingest and "understand" more than just code; it can analyze thermal images from a motor, audio from a bearing, or visual data from a UI. This is best utilized in "Predictive Maintenance." The human Architect guides the AI to identify which sensory patterns represent "Optimal Health." This ensures the product's "Brain" captures the subtle nuances of physical performance that are typically lost in traditional data structures.

6. The Blockchain Pairing for Immutable Identity

For ensuring the absolute integrity of the product's "Sovereign Truth," **HITL should be paired with Blockchain.** While the human Architect sets the ownership and update policies, the Blockchain stores a tamper-proof "Audit Trail" of every software update, ownership change, or maintenance event. This is best utilized for "Lifecycle Accountability." It prevents any unauthorized or accidental tampering with the product's core history. It ensures that the product remains a verifiable and trusted asset throughout its entire existence.

Product Power: Autonomous Agency

The final shift in the product journey is the transition from "Facilitating an Action" to **"Owning the Outcome."** Historically, a product was "Passive"—it waited for a human to pull a lever. The transformation into **Executors** represents the birth of "Agentic Utility." In this role, the product doesn't just "show" you the way; it "walks" the path for you. The product becomes the "Hands" of the user, capable of navigating complex workflows and resolving problems autonomously. This is the shift from "Tools" to "Agents," turning the product into an "Autonomous Force" that fulfills the user's intent with near-zero manual intervention.

Defensive Stance: Killing the Toil of Operation

The defensive mission of the Product-Executor is the **Elimination of Maintenance**. In the legacy world, the user is often a "Slave to the Product"—spending time configuring, updating, and troubleshooting the very tool meant to help them. This "Operational Drag" is a tax on human creativity. The Executor role kills this "Wastage of Time" by establishing "Self-Healing Agency." The product identifies its own friction points and "Executes the Fix" before the user even realizes there was a problem. This is the **Shield of Effortless Experience**. By "Hard-Coding the Fulfillment" of routine tasks, the product ensures the user remains in a state of "Flow." This defensive posture protects the user's "Attention" and the brand's "Value Proposition" by ensuring the "Hands" of the product are always working to remove obstacles, making the product the most "Low-Maintenance" and "High-Impact" choice in the market.

Offensive Stance: Unlocking Agentic Mastery

On the offense, the Executor moves toward **Strategic Fulfillment**. This is the ability of the product to "Conduct" external systems to achieve the user's goal. The product doesn't just work within its own borders; it acts as an "Agent in the World." For example, a "Smart Logistics Product" doesn't just track a package; it "Negotiates" with shipping APIs, "Re-routes" around weather, and "Verifies" delivery autonomously. This **Unlocks Infinite Scale for the User**, where the product gives them the power of a thousand assistants. It turns the product into an "Autonomous Powerhouse"—one that out-executes competitors by being "Active" while others are "Passive." It allows the leader to "Orchestrate" global outcomes by setting "Strategic Intent" and letting the product "Hands" find the most efficient path to manifest that reality in the physical and digital world.

The Product Outcome: Absolute Leverage

The ultimate value of the Executor role is **Operational Sovereignty**. It allows the user to "Manifest" results at the speed of thought. For the leader, this provides **Execution Dominance**. You are no longer "Selling a Feature"; you are "Delivering a Result." This turns the organization into a "Direct Fulfillment Machine"—where the "Hands" of the product are always active, always precise, and always working to fulfill the "Voice" of the user's desires. It ensures the product's "Execution" is as reliable as a law of physics, transforming "User Intent" into "Autonomous Reality" and "Reality" into "Market Leadership."

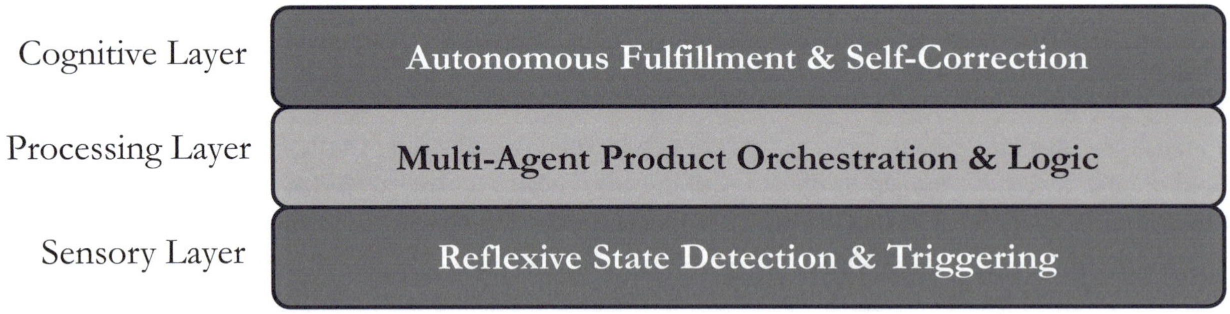

Executive Power: Agentic Product Autonomy

The final shift in the product journey is the transition from "Manual Maintenance" to **"Orchestrated Autonomous Fulfillment."** The **Executor** role represents the birth of "Active Products," where the leader provides the "Strategic Intent" and the product's AI-augmented "Hands" manifest self-healing and optimization results in real-time. Historically, products were "Passive"—they required a human to fix them or adjust them. The Executor role replaces this with a system of **Agentic Autonomy**, where the "Hands" are capable of fulfilling tasks—such as re-calibrating sensors, optimizing power loads, or even ordering their own replacement parts—without human intervention. This is the shift from "Managing a Tool" to "Directing an Agent," giving the executive the leverage to "Operate at the Speed of Light" by transforming strategic goals into "Autonomous Physical Reality."

The Sensory Layer: Reflexive State Detection & Triggering

The **Sensory Layer** of the Product Hands serves as the organization's "Reflex System." It is designed for **Real-Time Operational Triggering**, monitoring the internal telemetry for specific "Anomalies" that require an immediate autonomous response. Utilizing **Edge-Based Trigger Logic** and **Predictive Fault Detection**, this layer identifies the moment a product's state deviates from the "Ideal Operating Envelope." It doesn't wait for a human to see a "Red Light"; it "feels" the operational need the micro-second a sensor threshold is crossed. This sensory input provides the "Activation Signal" for the agentic stack, ensuring that the "Hands" of the product are always ready to move the millisecond a maintenance or optimization need is detected. This layer protects the customer's "Uptime" by ensuring the product responds to physical stress with biological speed, resolving issues before they lead to systemic failure.

The Processing Layer: Multi-Agent Product Orchestration & Logic

The **Processing Layer** acts as the "Mechanical Cerebellum," breaking down the leader's high-level intent into executable performance-tasks. It utilizes **Agentic Logic (ReAct Frameworks)** and **Chain-of-Thought Reasoning** to orchestrate a swarm of specialized "Maintenance and Optimization Agents." If the leader's intent is to "prioritize fuel efficiency across the entire logistics fleet during a price surge," this layer identifies the thousand sub-tasks required—from adjusting engine timing to optimizing route-based drag coefficients. It manages the "Inter-Agent Communication," ensuring that each bot knows the "Safety and Performance Guardrails" set by the executive. This layer provides the **Governance and Logic** for the execution, ensuring that the "Hands" stay within the boundaries of engineering integrity, killing the friction of manual calibration through automated, high-speed coordination across the entire product lifecycle.

The Cognitive Layer: Autonomous Fulfillment & Self-Correction

The **Cognitive Layer** is where the product "Hands" actually manifest the result through **Action Transformers** and **Over-the-Air (OTA) Actuation**. This is the layer of "Closed-Loop Performance." It doesn't just "process" a command; it "achieves" the strategic outcome. If the product's autonomous logic encounters an obstacle—such as a software conflict or a hardware limitation—this layer uses **Self-Correction Logic** and **Reinforcement Learning** to find an alternative optimization path without requiring new instructions from the leader. It turns "Information" into "Physical Reality," whether that means deploying a firmware patch, re-routing power, or notifying a local technician autonomously. This provides the executive with **Absolute Operational Leverage**, allowing them to "Direct the Fleet's Performance" rather than "Manage the Repair Toil." It transforms the firm into an "Autonomous Product Machine" that delivers "Market Dominance" through sheer agentic velocity and self-sustaining utility.

OVERVIEW

Agentic AI operates as an autonomous mission manager, breaking down high-level strategic goals into executable tasks. The Driving Mission leader defines the ultimate objective and core values, essentially setting the constitutional limits for the AI agent. This synergy offloads complex, time-consuming multi-step execution to the AI, allowing the leader to focus exclusively on strategic vision and alignment. The leader initiates the mission; the agent executes and reports on key milestones. This partnership fundamentally converts strategy into action, maximizing the leader's impact on high-level direction and governance.

MAIN STEPS

Start

STEP 1 – Mission Goal Setting
Define the high-level desired outcome.

STEP 2 – Autonomous Task Breakdown
Agent breaks goal into specific sub-tasks.

STEP 3 – Execution and Monitoring
Agent performs steps, tracks progress autonomously.

STEP 4 – Milestone Reporting & Review
Agent summarizes for leader approval.

STEP 5 – Course Correction, Redirection
Leader adjusts mission parameters or values.

STEP 6 – Mission Completion & Handoff
Agent delivers final output to human team.

End

SAMPLE FLOW

Autonomous Product Feature Development: The leader initiates an agent mission to develop and integrate a high-priority product feature aligned with user feedback.

1. Set Feature Mandate
Define the required customer impact and technical specifications clearly.

2. Break Development Tasks
Agent autonomously converts feature into engineering tasks and priorities correctly.

3. Set Resource Allocation
Agent assigns and tracks tasks across relevant development teams autonomously.

4. Execute Automated Testing
Agent runs continuous integration, identifying bugs and iterating on code promptly.

5. Approve Final Beta Launch
Leader reviews the feature's alignment with product vision and core ethical standards ethically.

6. Deploy Feature to Market
Leader authorizes the final release and hands off monitoring to operations criteria.

USE CASES

- **Manufacturing | Zero-Defect Product Mission:** An agent continuously monitors production data and autonomously launches missions to recalibrate machinery settings, aiming to eliminate product defects and optimize material usage efficiently.
- **Financial Services | New Service Launch Mission:** The leader tasks an agent with managing the cross-functional rollout of a new financial product, autonomously coordinating legal review, marketing collateral, and system integration reliably.
- **Healthcare | Clinical Trial Optimization:** An agent manages the mission to accelerate a clinical trial, autonomously coordinating patient recruitment, data collection, and documentation submission to regulatory bodies accurately.
- **Technology | Product Lifecycle Management:** An agent autonomously monitors the health of legacy products, initiating end-of-life missions and deploying resources to migrate customers to newer, more profitable platforms consistently.
- **Retail | Sustainable Packaging Initiative:** An agent researches and executes a mission to transition all product packaging to sustainable materials, autonomously sourcing vendors, managing inventory, and updating branding quickly.
- **Energy | Next-Gen Power Source Integration:** The leader deploys an agent to manage the integration of a new power generation source (e.g., solar farm), autonomously coordinating land acquisition, regulatory approval, and grid connection clearly.

In the **Product Transformation**, the **Executors (The Hands)** represent the "Autonomous Performance Layer." While the primary instrument of **Agentic AI** provides the product with the agency to self-correct and fulfill user goals, it requires secondary technical pairings to bridge the gap between digital logic and physical or operational reality. To move from "Passive Tools" to "Autonomous Solutions," a leader must integrate these six pairings. This ensures the "Hands" can act on behalf of the user with millisecond precision, turning the product into a self-maintaining, self-optimizing asset that delivers value without constant human intervention.

1. The RPA Pairing for Legacy Protocol Bridge

Pairing **Agentic AI with Robotic Process Automation (RPA)** is best utilized when a modern smart product must interface with older, rigid infrastructure. While the Agent decides the strategy—such as a smart thermostat optimizing building energy—RPA provides the "fingers" to execute commands in legacy HVAC control systems that lack modern APIs. This is essential for "Ecosystem Compatibility," ensuring the product can act within the real-world constraints of existing technology stacks without requiring a total infrastructure overhaul.

2. The Digital Twin Pairing for Execution Safety

Leaders should pair **Agentic AI with Product Digital Twins** when deploying autonomous features in high-stakes environments, such as medical robotics or industrial machinery. Before the "Hands" execute a new movement or a critical software patch, the action is "rehearsed" in the Digital Twin. This pairing ensures the product never takes an action that could cause physical damage or a logic loop. It provides a virtual "safety net" for autonomous execution, ensuring every move is verified against a live simulation of reality before it occurs in the physical world.

3. The Edge AI Pairing for Zero-Latency Actuation

When a product must respond in real-time—such as an autonomous drone or a smart safety valve—**Agentic AI must be paired with Edge AI**. This allows the "Hands" to process data and execute decisions locally, by passsing the latency of the cloud. This pairing is vital for "Mission-Critical Responsiveness," ensuring the product can react to sudden physical changes in milliseconds. It turns the product into a self-contained nervous system that acts as fast as it perceives, maintaining its "Sovereign Performance" in any environment.

4. The Blockchain Pairing for Autonomous Settlement

For products that participate in a shared economy (e.g., smart EV chargers or decentralized storage), a leader should pair **Agentic AI with Blockchain (Smart Contracts)**. The "Hands" don't just provide a service; they execute the payment and verify the transaction on an immutable ledger. This pairing ensures that autonomous actions—like a car selling excess energy back to the grid—are legally secure and transparent. It eliminates the need for manual billing, allowing the product to act as an independent financial agent within its ecosystem.

5. The Computer Vision Pairing for Spatial Awareness

To allow the "Hands" to interact with the physical world, leaders should pair **Agentic AI with Computer Vision**. This is best utilized in "Self-Maintaining Hardware." By "seeing" its own components or the environment it operates in, the product can autonomously identify wear and tear or physical obstacles. This pairing gives the "Hands" the sensory context required to perform physical tasks, ensuring the product can navigate, repair, or optimize itself based on a high-fidelity understanding of its 3D surroundings.

6. The Reinforcement Learning Pairing for Performance Evolution

In complex, shifting environments, **Agentic AI should be paired with Reinforcement Learning (RL)**. This is best utilized for "Self-Optimizing Products," such as an AI-driven manufacturing tool or a dynamic load-balancer. The "Hands" learn from the success or failure of every action, constantly refining their execution strategy to maximize efficiency. This provides the brand with a "Living Solution" that grows more capable with every hour of use, ensuring the product's value increases the longer the customer owns it.

BLUEPRINT SERIES

SECTION C

Business Maturity

Ai6x6™ Transformation Blueprint

Story 1 — The Marshall Plan Reconstruction (1948–1952): Post-war Europe faced shattered economies, broken supply chains, and institutional collapse. The Marshall Plan introduced structured funding, transparent reporting, and coordinated execution across nations. Local governments provided context while centralized oversight ensured accountability. This blend of autonomy and structure accelerated recovery. The program demonstrated how disciplined governance can rebuild complex systems. It became a blueprint for large-scale organizational maturity.

Story 2 — The Rise of Quality Management in Japan (1950s–1960s): Japan's industries struggled with inconsistent quality and limited global credibility. Statistical quality control introduced measurement, repeatability, and disciplined improvement cycles. Workers contributed insights while leaders standardized processes. Over time, quality became a cultural norm rather than a corrective action. This shift transformed fragmented operations into cohesive systems. It marked the emergence of maturity through methodical refinement.

Story 3 — NASA's Systems Engineering Discipline (1960s): The space race demanded unprecedented coordination across thousands of contractors and scientific domains. NASA formalized systems engineering to integrate requirements, risks, and interfaces into a unified framework. Structured reviews balanced innovation with rigor. Human judgment guided trade-offs while documentation preserved institutional memory. This discipline enabled complex missions to succeed repeatedly. It became a global model for organizational maturity in high-stakes environments.

Story 4 — The Birth of Modern Corporate Governance (1970s–1980s): Corporate scandals exposed weaknesses in oversight, transparency, and fiduciary responsibility. Boards adopted clearer roles, audit mechanisms, and accountability structures. Regulations reinforced ethical conduct while organizations matured through formal governance. Leaders learned to balance autonomy with compliance. This era established governance as a strategic capability, not an administrative burden. It marked a turning point in institutional maturity.

Story 5 — Toyota's Global Expansion Playbook (1980s–1990s): As Toyota expanded globally, it faced cultural differences, supply-chain complexity, and operational variability. The company codified principles, standardized training, and embedded continuous improvement across regions. Local teams adapted practices while preserving core philosophy. This balance of global consistency and local flexibility enabled scalable excellence. It demonstrated maturity as the ability to replicate capability without diluting identity.

Story 6 — The Digital Transformation of Banking (1990s–2000s): Banks confronted rising customer expectations, regulatory pressure, and technological disruption. They modernized legacy systems, standardized data, and introduced structured risk frameworks. Digital channels required new operating models and cross-functional coordination. Human expertise guided decisions where automation lacked nuance. Over time, banks evolved from siloed institutions into integrated digital enterprises. This transition reflected maturity through modernization and disciplined reinvention.

LESSONS FROM A PRODUCT TRANSFORMATION PERSPECTIVE

Product maturity develops when offerings evolve from isolated outputs to integrated systems shaped by standards, feedback loops, and disciplined iteration. Early innovations relied on craftsmanship, while later decades introduced structured engineering, quality frameworks, and scalable design principles. Mature organizations build products that are reliable, repeatable, and adaptable across contexts. They balance experimentation with governance, ensuring that creativity operates within stable architectures. As maturity increases, products become expressions of organizational capability—reflecting culture, process, and institutional learning. Maturity is achieved when products improve predictably, scale consistently, and maintain integrity across markets and generations.

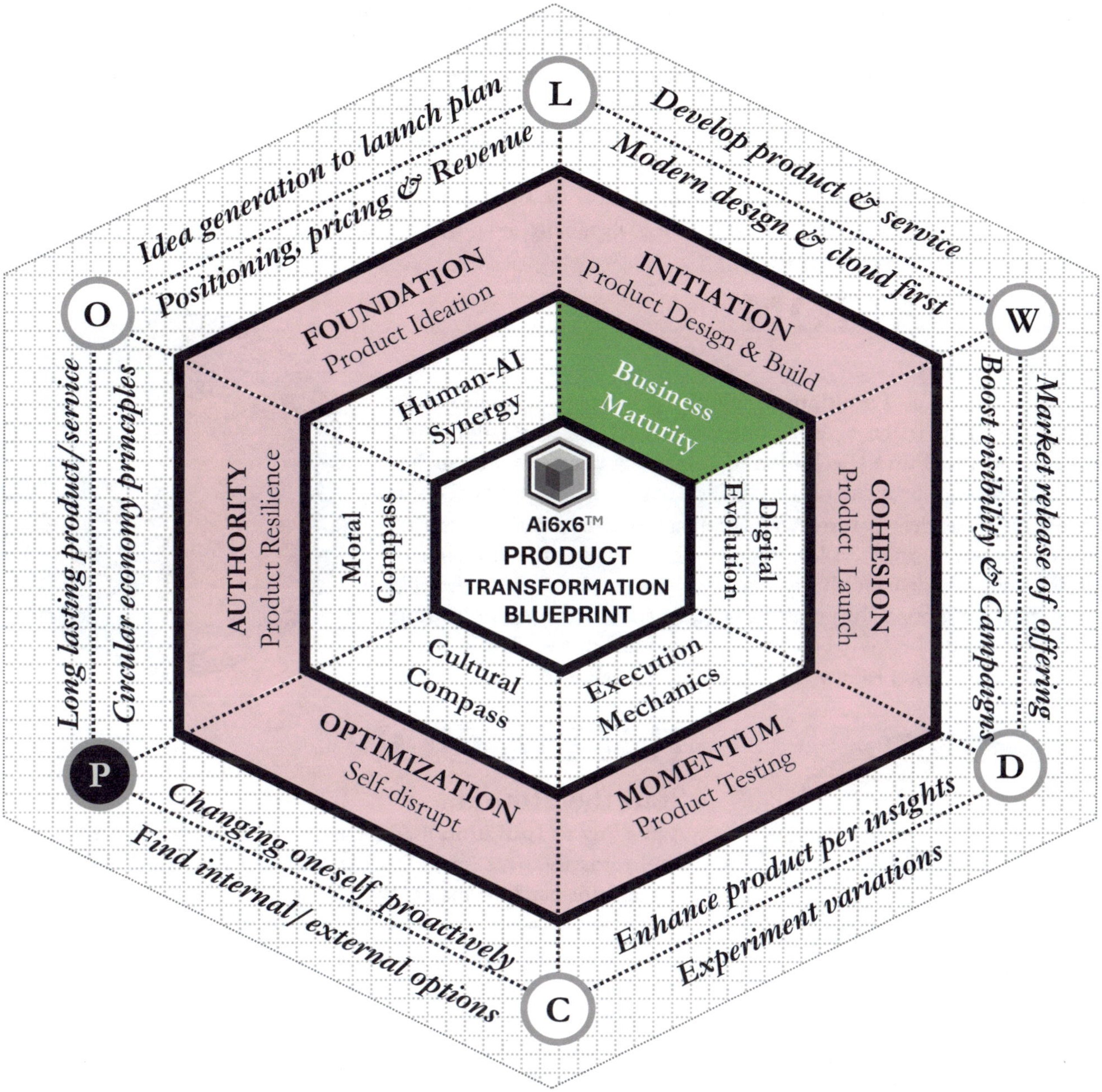

The **Ai6x6 Research Lab** selected this lens because a product's maturity is defined by its ability to evolve through the tension of creation and destruction. We view business maturity as an internal filter for the organization's capacity to handle both the birth of an idea and the grace of its obsolescence. This dimension ensures that the aggressive energy of the **Build** phase is anchored in the long-term wisdom of **Circular Economy** and **Resilience**. It is a vital lens because, in a hyper-competitive market, a leader's ability to navigate **Disruption** without losing the ecosystem's trust is the only force that prevents the enterprise from becoming obsolete.

We believe true maturity is found in the ability to manage the delicate feedback loops of **Optimization**, recognizing that the product is a living entity that reflects the organization's growth. By focusing on **Scaling** and **Amplification**, the Lab provides a mechanism for leaders to audit their product's relevance before it enters the market. This lens ensures that the pursuit of innovation does not come at the cost of stability. It is the steady hand that guides the product through the complexities of **Market Research**, recognizing that while features are fleeting, the bedrock of ecosystem value must remain immutable.

Every leadership transformation begins with ambition, but ambition alone is never enough to carry an organization into its future. **Business Maturity** is the lens that forces leaders to confront the gap between what they hope to achieve and what their enterprise is prepared to sustain. It is the discipline that reveals whether the organization is structurally, operationally, and culturally ready for the future it claims to desire. In the AI era, this gap becomes even more pronounced; AI does not simply automate tasks—it reshapes workflows, decision cycles, and the very rhythm of how leadership operates. It exposes weaknesses that were previously tolerable and accelerates the consequences of poor preparation, demanding a level of rigor that traditional management frameworks often overlook. It examines how ambition translates into capability, how vision converts into execution, and how culture sustains momentum under pressure.

FOUNDATION

Define the Core

The journey begins at **Foundation**, where the primary objective is to *Define the Core*. This is the bedrock upon which all transformation rests, requiring leaders to **Filter the Noise** of competing priorities and anchor their identity and strategic intent. At this level, clarity is not optional—it is existential. Without clear governance, AI initiatives drift, priorities conflict, and teams operate in isolated silos. Foundation maturity is the discipline of intent: the ability to distinguish urgency from readiness. Leaders who skip this stage mistake motion for progress, building transformation on shifting sand. True maturity begins when the organization knows what it stands for, what it will not compromise, and how it will measure success.

INITIATION

Start the Motion

Ascending to **Initiation** marks the first active step: *Start the Motion*. This level turns the gaze inward, demanding self-awareness and the courage to confront personal biases. By **Wiring the System**, leaders catalyze the shift from passive management to intentional authority. Initiation is where leadership becomes kinetic—it transforms from aspiration into activation. The organization begins to move, not because it is told to, but because its internal circuits are aligned. Maturity here is emotional as much as structural; it requires self-regulation, empathy, and the ability to lead through friction. The leader's persona is tested, revealing whether they can sustain momentum through disruption rather than retreat into comfort.

COHESION

Connect the Parts

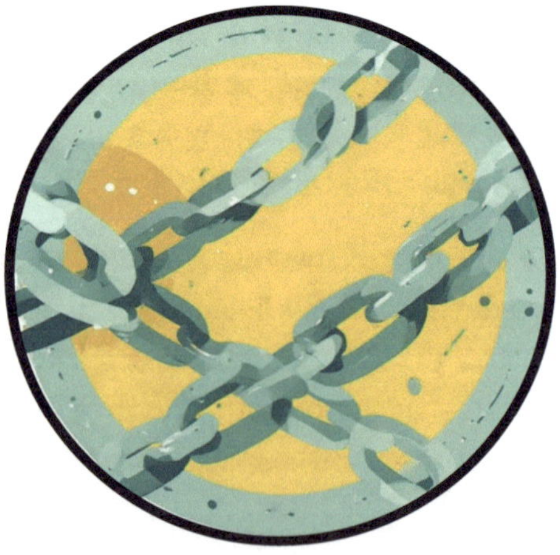

The climb continues to **Cohesion**, represented by *The Engine*. This is the critical unifier where people, processes, and systems are harmonized. The focus shifts to **Connect the Parts**—to ensure that individual leadership goals align with a unified organizational voice. By **Shifting the Gear**, leaders eliminate silos and systemic friction, transforming fragmented efforts into a synchronized machine. Cohesion maturity is the art of integration: turning disconnected motion into collective momentum. Without it, organizations celebrate activity instead of achievement. The mature leader understands that alignment is not consensus—it is coherence, the ability to move as one without losing diversity of thought.

MOMENTUM

Drive the Speed

The fourth rung is **Momentum**, where the organization begins to *Drive the Speed*. Leadership here moves toward scaling impact with predictable velocity. By **Removing the Drag** of legacy thinking and institutional inertia, leaders accelerate the collective intelligence of the enterprise. Momentum maturity is the physics of transformation—it measures how efficiently energy converts into progress. At this stage, the organization's rhythm changes; it begins to operate with self-propelling force. The leader's role shifts from directing individuals to empowering systems, ensuring that innovation scales faster than resistance. Momentum is not chaos—it is controlled acceleration, the moment when transformation gains its gravity.

OPTIMIZATION

Refine the Flow

Refinement occurs at **Optimization**, characterized by *The Flow*. Here, leadership pauses to **Refine the Flow** by **Locking the Code** of influence through reflection and data-driven inquiry. Immature organizations measure what is easy; mature ones measure what matters. Optimization maturity is the discipline of precision—removing friction without losing flexibility. Leaders refine decisions still in motion, adjusting with humility and surgical accuracy. Ego and bias are replaced by evidence and iteration. The organization becomes self-healing, capable of continuous improvement without external intervention. Optimization is not about perfection—it is about adaptability, the ability to evolve faster than the environment demands.

AUTHORITY

Own the Market

The summit is reached at **Authority**, where leaders finally *Own the Market*. This stage manifests as *The Benchmark*—a legacy of endurance that can withstand external shocks and volatility. By **Fueling the Future**, leaders ensure that transformation is not a temporary win but a permanent standard. Authority maturity is the discipline of stewardship: sustaining excellence while preparing for reinvention. Immature organizations avoid this stage because it exposes uncomfortable truths; mature ones embrace it as the crucible of resilience. At this level, leadership becomes institutional—it transcends personality and becomes culture. The organization no longer reacts to the market; it defines it.

Business Maturity is the scaffolding that holds the weight of innovation. It ensures that speed does not collapse under fragility and that ambition does not outpace capability. Each level builds upon the last, forming a continuum from clarity to mastery. Foundation defines intent, Initiation activates leadership, Cohesion integrates systems, Momentum accelerates impact, Optimization refines intelligence, and Authority sustains legacy. In the AI era, maturity is not a luxury—it is survival. As algorithms amplify both strengths and weaknesses, leaders must design organizations that can learn, adapt, and endure. Business Maturity is therefore not a checklist but a **leadership discipline**—a way of thinking that transforms ambition into architecture. When leaders embrace this full sequence, they discover that maturity is not the end of transformation but its engine—the invisible structure that makes both explosive speed and long-term continuity possible.

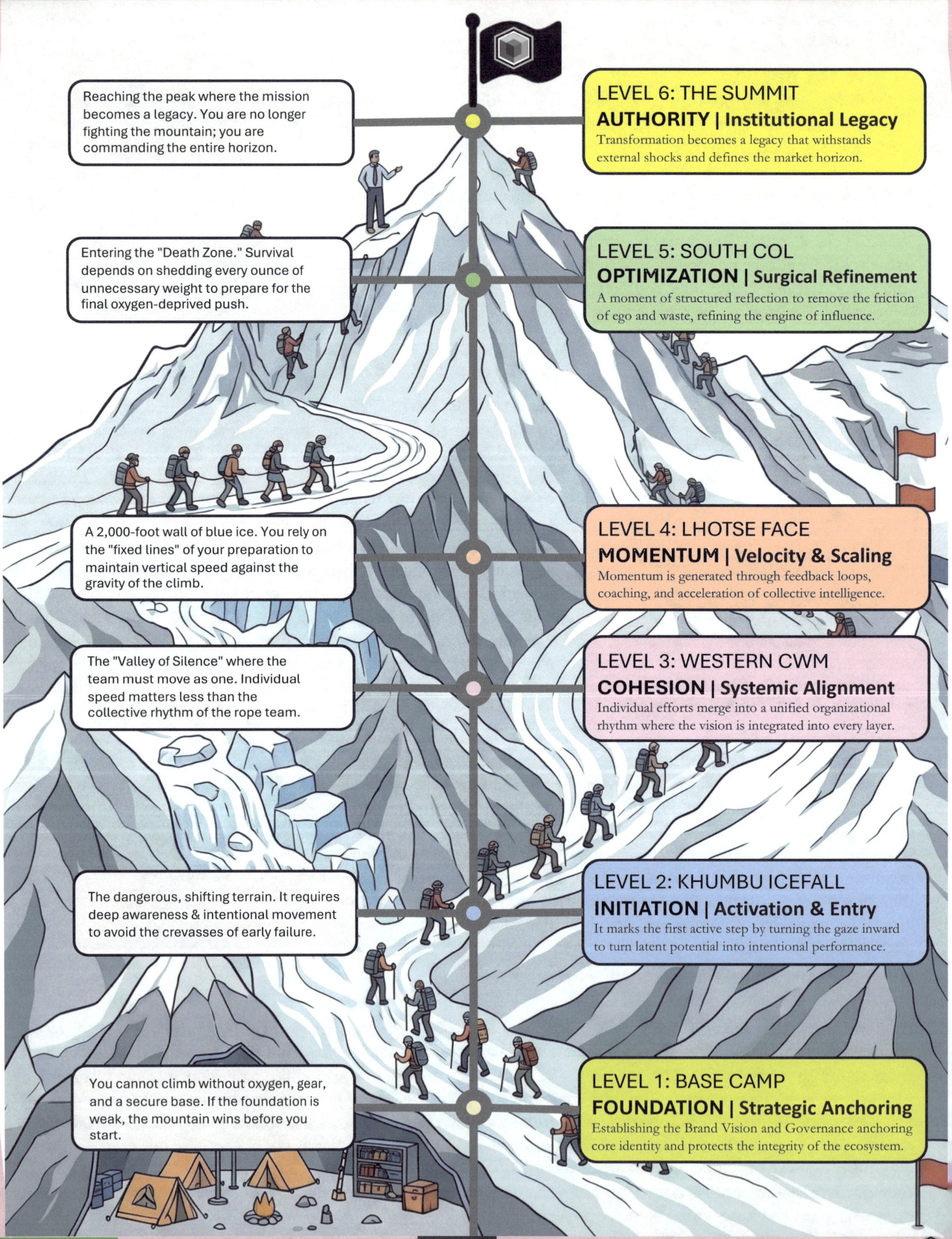
Reaching the peak where the mission becomes a legacy. You are no longer fighting the mountain; you are commanding the entire horizon.
LEVEL 6: THE SUMMIT
AUTHORITY | Institutional Legacy
Transformation becomes a legacy that withstands external shocks and defines the market horizon.
Entering the "Death Zone." Survival depends on shedding every ounce of unnecessary weight to prepare for the final oxygen-deprived push.
LEVEL 5: SOUTH COL
OPTIMIZATION | Surgical Refinement
A moment of structured reflection to remove the friction of ego and waste, refining the engine of influence.
A 2,000-foot wall of blue ice. You rely on the "fixed lines" of your preparation to maintain vertical speed against the gravity of the climb.
LEVEL 4: LHOTSE FACE
MOMENTUM | Velocity & Scaling
Momentum is generated through feedback loops, coaching, and acceleration of collective intelligence.
The "Valley of Silence" where the team must move as one. Individual speed matters less than the collective rhythm of the rope team.
LEVEL 3: WESTERN CWM
COHESION | Systemic Alignment
Individual efforts merge into a unified organizational rhythm where the vision is integrated into every layer.
The dangerous, shifting terrain. It requires deep awareness & intentional movement to avoid the crevasses of early failure.
LEVEL 2: KHUMBU ICEFALL
INITIATION | Activation & Entry
It marks the first active step by turning the gaze inward to turn latent potential into intentional performance.
You cannot climb without oxygen, gear, and a secure base. If the foundation is weak, the mountain wins before you start.
LEVEL 1: BASE CAMP
FOUNDATION | Strategic Anchoring
Establishing the Brand Vision and Governance anchoring core identity and protects the integrity of the ecosystem.

Ai6x6™ Transformation Blueprint

FOUNDATION | Product Ideation

"Ideas are a dime a dozen. People who implement them are priceless." - Mary Kay Ash.

Founder of Mary Kay Cosmetics. She built a multi-billion dollar enterprise based on empowering women entrepreneurs, offering flexible work opportunities and fostering a supportive culture. Her innovative model made her a pioneer in cosmetics and an inspiration for women in business.

The Foundation level is your Product Ideation, creating the essential identity and market presence that will differentiate your offering. This genesis point uses research and prototyping to ensure the product's "soul" is commercially viable.

The initial, creative stage of generating, meticulously developing, and carefully curating new product concepts and innovative solutions that address market needs or solve customer problems. This is precisely where true innovation begins; without a continuous flow of fresh, viable ideas, products inevitably stagnate, hindering market relevance, competitive edge, and future growth potential for the organization.

Successful ideation involves brainstorming, market research, and understanding emerging trends to identify opportunities for groundbreaking products that resonate with the target audience.

For business strategy maturity, the **"Product Ideation"** chapter is foundational because it defines the product's vision and strategic purpose based on market needs. A mature business strategy begins with a systematic process for conceptualizing digital solutions that address identified market gaps and opportunities. This initial alignment ensures that product development efforts are strategically focused, resource allocation is optimized, and the resulting products directly contribute to the organization's overarching business goals. It's about establishing a clear strategic compass for innovation, ensuring products are built with purpose and market relevance. Here is how we approach our learning:

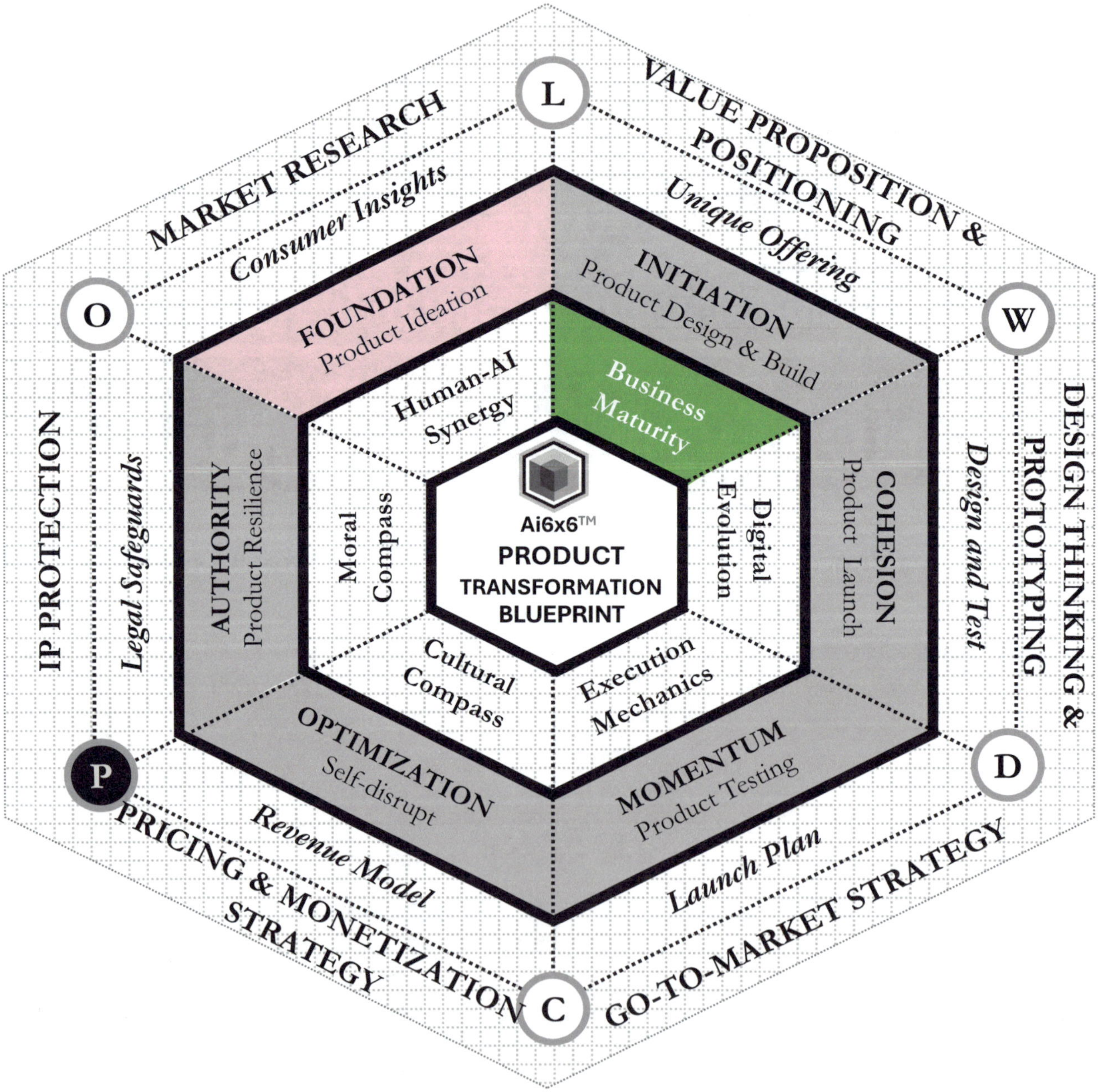

Ai6x6 BUSINESS MATURITY FOR IDEATION | PRODUCT TRANSFORMATION

1. Market research

For a business leader, robust market research is the foundational and indispensable first step in product development, ensuring that any digital solution conceptualized is genuinely rooted in understanding industry trends and addressing true customer needs. Without this meticulous groundwork, product development becomes a costly gamble, risking significant resources on solutions that fail to resonate with the market. Your role as a leader is to champion a data-driven approach to this phase, fostering a deep curiosity about the external landscape. This involves comprehensive competitive landscape analysis, systematically identifying existing gaps and untapped opportunities within current market offerings. By scrutinizing what competitors are doing (or failing to do), you can strategically position your product for differentiation and distinct value.

Furthermore, gathering profound customer insights and pinpointing their pain points is crucial. This goes beyond superficial feedback to conducting in-depth surveys, interviews, and ethnographic studies that truly assess consumer demands, frustrations, and unmet desires. Crucially, leveraging trend forecasting tools—utilizing AI and advanced data analytics—allows you to predict emerging industry shifts, technological advancements, and evolving consumer behaviors. This foresight ensures your product isn't just relevant for today but is future-proofed for tomorrow's market. Neglecting rigorous market research can lead to product irrelevance, wasted development cycles, poor market adoption, and a significant drain on resources. By prioritizing this diligent exploration, you empower your organization to build products that customers genuinely need and desire, accelerating market fit, reducing launch risks, and ultimately driving the sustained success and profitability of your business.

WHAT?	WHY?	HOW?	BEWARE!
COMPETITIVE LANDSCAPE ANALYSIS	**MARKET POSITIONING**	**COMPETITOR BENCHMARKING**	**COMPLEXITY**
Identifying gaps and opportunities.	Uncovers opportunities for differentiation and unique value.	Analyze competitor products, strategies, and market share.	Vast and rapidly changing competitive landscapes.
CUSTOMER INSIGHTS AND PAIN POINTS	**PROBLEM-SOLUTION FIT**	**QUALITATIVE RESEARCH**	**BIAS**
Assessing consumer demands.	Ensures product directly addresses user needs and frustrations.	Conduct surveys, interviews, focus groups, and user observations.	Interpreting subjective feedback without introducing researcher bias.
TREND FORECASTING	**FUTURE READINESS**	**PREDICTIVE ANALYTICS**	**ACCURACY**
Using AI and data analytics to predict industry shifts.	Ensures product relevance for emerging market demands.	Leverage AI/ML models to analyze market data and predict trends.	Ensuring predictive models provide reliable and actionable forecasts.

2. Value proposition & positioning

For a business leader, clearly articulating your product's Value Proposition and defining its strategic positioning are fundamental imperatives for driving market adoption and competitive advantage. It's not enough to build a great product; you must compellingly communicate why it matters, who it's for, and why it's superior to alternatives. This clarity is what captures customer attention and converts interest into purchase. Central to this is meticulously defining your Unique Selling Points (USPs). As a leader, you must ensure that your product's distinct advantages—whether it's superior performance, unique features, unparalleled convenience, or cost-effectiveness—are clearly articulated and set it apart from competitors. These differentiators are your core message.

Furthermore, a precise brand positioning strategy is crucial. This involves mapping your product's place within the competitive landscape, consciously deciding how it should be perceived by target customers relative to other offerings. This strategic placement influences pricing, marketing, and distribution decisions. Crucially, developing customer-centric messaging ensures your communication resonates deeply with target audience expectations. This means crafting language that focuses on solving customer problems and delivering benefits, rather than merely listing features. Your message must speak directly to their needs and aspirations, building an emotional connection. Neglecting a clear value proposition and positioning can lead to product commoditization, confusion in the market, a struggle to acquire customers, and inefficient marketing spend. By championing a compelling and distinct differentiation strategy, you empower your organization to clearly articulate its worth, attract the right customers, and build a strong market presence, ultimately driving the sustained success and profitability of your product.

WHAT?	WHY?	HOW?	BEWARE!
UNIQUE SELLING POINTS (USPS)	**COMPETITIVE EDGE**	**VALUE DISCOVERY**	**IMITATION**
Defining distinct product advantages.	Clearly articulates why the product is superior.	Identify core strengths and benefits that differentiate the product.	Competitors quickly replicating identified USPs.
BRAND POSITIONING STRATEGY	**MARKET PERCEPTION**	**POSITIONING STATEMENTS**	**MARKET SATURATION**
Mapping the product's place in the market.	Shapes how the target audience perceives the product relative to competitors.	Develop clear positioning statements and visual identity.	Finding a truly unique position in a crowded market.
CUSTOMER-CENTRIC MESSAGING	**RESONANCE & ENGAGEMENT**	**AUDIENCE RESEARCH**	**INTERNAL BIAS**
Aligning communication with expectations.	Ensures communication deeply resonates with the target audience.	Craft messages focusing on customer benefits and pain points.	Shifting from product-focused to truly customer-centric language.

For a business leader, embracing Design Thinking and iterative prototyping is a critical strategic imperative for refining product concepts, mitigating risks, and maximizing usability throughout the development lifecycle. This human-centered approach ensures that products are not just functional but genuinely delightful and intuitive for users. Central to this is fostering empathy-driven ideation. As a leader, you must champion a deep understanding of user needs, behaviors, and pain points, incorporating continuous user feedback into the design process to maximize usability and ensure the product truly solves their problems. This user-first mindset is paramount.

Furthermore, rapid prototyping and testing are crucial. Creating early-stage models or mock-ups of your product allows for quick, iterative improvements based on immediate user feedback. This agile experimentation identifies flaws and validates concepts before significant resources are committed, drastically reducing development costs and risks. Crucially, implementing agile product refinement ensures continuous optimization based on user experience and market reception. This iterative cycle of building, testing, learning, and refining allows your product to evolve in sync with user needs and market demands, ensuring continuous relevance and improvement. Neglecting design thinking and prototyping can lead to products that are difficult to use, fail to meet user expectations, incur costly reworks late in development, and ultimately struggle to gain market traction. By championing this iterative, user-centric approach, you empower your organization to build highly usable, desirable products that captivate customers, accelerate market adoption, and ultimately drive the sustained success and profitability of your business through superior user experience.

WHAT?	WHY?	HOW?	BEWARE!
EMPATHY-DRIVEN IDEATION	**MAXIMIZED USABILITY**	**USER RESEARCH**	**USER BIAS**
Incorporating user feedback in design.	Ensures designs genuinely meet user needs and solve problems.	Conduct user interviews, observations, and feedback sessions early on.	Users articulating what they want vs. what they truly need.
RAPID PROTOTYPING AND TESTING	**ITERATIVE IMPROVEMENTS**	**LOW-FIDELITY MODELS**	**SCOPE CREEP**
Creating early-stage models.	Enables quick validation and refinement before full development.	Create sketches, wireframes, and interactive prototypes for testing.	Prototyping becoming overly complex or detailed too early.
AGILE PRODUCT REFINEMENT	**CONTINUOUS RELEVANCE**	**SPRINT-BASED ITERATION**	**CHANGING PRIORITIES**
Continuously optimizing based on user experience.	Ensures product evolves with user needs and market shifts.	Integrate user feedback into agile development sprints for continuous updates.	Balancing new features with continuous refinement based on feedback.

For a business leader, developing a robust Go-to-Market (GTM) strategy is a strategic imperative for successfully launching a product and achieving maximum market impact. A brilliant product can fail without a meticulously planned and executed launch, making this phase critical for market adoption and revenue generation. Central to this is meticulous target audience segmentation. Identifying the priority demographics, psychographics, and behaviors for precise outreach allows you to tailor your messaging and channels to those most likely to convert, maximizing the efficiency of your marketing spend.

Furthermore, adopting a multi-channel marketing approach is crucial. Leveraging a diversified mix of digital (SEO, social media, email), social, and traditional advertising platforms ensures broad brand visibility and consistent messaging across all customer touchpoints, creating a cohesive and pervasive market presence. Crucially, optimizing your distribution channels is vital. Structuring your sales channels—whether direct-to-consumer, retail partnerships, marketplaces, or B2B sales teams—for efficient market penetration ensures that your product is readily available where your target customers prefer to buy. Neglecting a comprehensive GTM strategy can lead to a lackluster launch, poor market adoption, wasted marketing investment, and an inability to scale effectively. By championing a well-orchestrated and integrated GTM plan, you empower your organization to introduce products with precision, generate significant market buzz, drive initial sales, and ultimately lay the groundwork for sustained market leadership and profitability. This strategic launch planning is what transforms innovation into commercial success.

WHAT?	WHY?	HOW?	BEWARE!
TARGET AUDIENCE SEGMENTATION	**PRECISE OUTREACH**	**DEMOGRAPHIC & BEHAVIORAL ANALYSIS**	**OVER-SEGMENTATION**
Identifying priority demographics.	Ensures marketing efforts are focused on the most receptive groups.	Define segments based on market research and customer data.	Creating too many narrow segments, making campaigns unwieldy.
MULTI-CHANNEL MARKETING APPROACH	**MAXIMIZED VISIBILITY**	**INTEGRATED CAMPAIGNS**	**CONSISTENCY**
Leveraging digital, social, and traditional platforms.	Ensures broad brand exposure and consistent messaging.	Coordinate marketing efforts across various online and offline channels.	Maintaining brand consistency across diverse channels.
DISTRIBUTION OPTIMIZATION	**EFFICIENT AVAILABILITY**	**CHANNEL SELECTION**	**CHANNEL CONFLICT**
Structuring sales channels for market penetration.	Ensures product is easily accessible where target customers buy.	Choose and optimize sales channels (e-commerce, retail, partnerships).	Managing potential conflicts between different sales channels.

5. Pricing & monetization strategy

For a business leader, developing a robust pricing and monetization strategy is a non-negotiable strategic imperative that directly determines your product's profitability, market adoption, and long-term financial viability. It's about meticulously structuring revenue models that capture value while remaining competitive and attractive to your target customers. Central to this is implementing dynamic pricing models. As a leader, you must be prepared to adjust cost structures based on real-time factors like demand fluctuations, competitive actions, and perceived customer value. This agility allows you to optimize revenue generation and market share in fluid conditions.

Furthermore, offering flexible payment structures through subscription and tiered monetization is crucial. These models enhance customer retention by providing convenience and choice, allowing customers to pay for access over time or choose service levels that align with their needs and budget, building a predictable recurring revenue base. Crucially, strategically utilizing promotional and discount strategies is vital for driving adoption. Implementing carefully timed launch incentives, loyalty rewards, or seasonal offers can effectively attract new customers, encourage initial purchases, and reward long-term commitment without devaluing the product's perceived worth. Neglecting a well-thought-out pricing and monetization strategy can lead to leaving money on the table, struggling to attract customers, or an inability to sustain profitability in the long run. By championing a flexible, data-driven approach to pricing and revenue generation, you empower your organization to capture optimal value, attract a broad customer base, and ultimately drive the sustained financial success and scalability of your product in the marketplace.

WHAT?	WHY?	HOW?	BEWARE!
DYNAMIC PRICING MODELS	**PROFIT MAXIMIZATION**	**ALGORITHM IMPLEMENTATION**	**CUSTOMER PERCEPTION**
Adjusting cost structures based on demand and competition.	Optimizes revenue by adapting to market conditions and value.	Use algorithms to adjust prices based on real-time demand, inventory, or competitor pricing.	Customers perceiving dynamic pricing as unfair or inconsistent.
SUBSCRIPTION AND TIERED MONETIZATION	**RETENTION & PREDICTABILITY**	**MODEL DESIGN**	**CHURN MANAGEMENT**
Offering flexible payment structures.	Enhances customer retention and provides recurring revenue.	Develop subscription tiers (e.g., freemium, premium) and usage-based pricing.	Managing customer churn within subscription models.
PROMOTIONAL AND DISCOUNT STRATEGIES	**ADOPTION & LOYALTY**	**CAMPAIGN PLANNING**	**DEVALUATION**
Driving adoption through incentives.	Accelerates initial adoption and rewards customer loyalty.	Design launch discounts, bundle offers, or loyalty program incentives.	Discounts potentially devaluing the product or brand long-term.

6. IP protection (patenting, copyrighting)

For a business leader, rigorously securing intellectual property (IP) rights through patenting, copyrighting, and other legal safeguards is a non-negotiable strategic imperative for ensuring long-term market security and protecting your product's competitive advantage. Your innovations are valuable assets, and without proper legal protection, they are vulnerable to replication and unauthorized use. Central to this is meticulous patent filing and legal safeguards. As a leader, you must ensure that all product innovations, unique technologies, and proprietary methodologies are protected from replication through robust patent applications, providing exclusive rights to exploit your inventions for a defined period. This legal barrier is a significant competitive deterrent.

Furthermore, comprehensive trademark and copyright enforcement is crucial. This involves actively preserving your brand identity (logos, names) and proprietary content (software code, marketing materials, designs) through registration and vigilant monitoring, preventing unauthorized use or infringement that could dilute your brand or intellectual capital. Crucially, strategically exploring licensing opportunities allows you to monetize intellectual property through partnerships. By licensing your patents or copyrights to other businesses, you can generate additional revenue streams, expand market reach without direct investment, and establish your technology as an industry standard. Neglecting IP protection can lead to rampant imitation, loss of competitive edge, legal disputes, and a significant devaluation of your product and business. By championing a proactive and comprehensive IP strategy, you empower your organization to defend its innovations, preserve brand integrity, and unlock new revenue streams, ultimately securing your business's long-term market position and profitability.

WHAT?	WHY?	HOW?	BEWARE!
PATENT FILING AND LEGAL SAFEGUARDS	**INNOVATION PROTECTION**	**LEGAL COUNSEL**	**COST & COMPLEXITY**
Ensuring product innovations are protected.	Secures exclusive rights, preventing replication by competitors.	Engage patent attorneys for thorough patent searches and applications.	High costs and lengthy processes for patent filing and defense.
TRADEMARK AND COPYRIGHT ENFORCEMENT	**BRAND INTEGRITY**	**REGISTRATION & MONITORING**	**GLOBAL ENFORCEMENT**
Preserving brand identity and proprietary content.	Protects brand identity and creative works from unauthorized use.	Register trademarks/copyrights and monitor for infringement.	Challenges in enforcing IP rights across diverse international jurisdictions.
LICENSING OPPORTUNITIES	**REVENUE EXPANSION**	**PARTNERSHIP NEGOTIATION**	**CONTROL LOSS**
Monetizing intellectual property through partnerships.	Generates additional revenue streams without direct product sales.	Identify potential licensees and negotiate favorable terms.	Balancing revenue generation with maintaining control over IP usage.

Ai6x6™ Transformation Blueprint

INITIATION | Design & Build

"Design is not just what it looks like and feels like. Design is how it works." - Jony Ive.

Former Chief Design Officer at Apple Inc., widely recognized for his pivotal role in designing iconic products like the iMac, iPod, iPhone, and iPad. His minimalist aesthetic and focus on user experience profoundly influenced global product design.

The Initiation level is your Product Build, the transformation of a concept into a tangible, functional, and cloud-native reality. Agile design and cloud-native practices ensure the product is built for active performance in a modern digital world.

The comprehensive process of translating nascent ideas into tangible, functional products, encompassing meticulous planning, iterative prototyping, rigorous testing, and robust development. This critical phase ensures the product's feasibility, core functionality, user experience, and inherent quality, effectively bringing the product vision to life from concept to reality.

It involves cross-functional collaboration between designers, engineers, and product managers to create a product that is not only innovative but also practical, scalable, and ready for market introduction.

The **"Design & Build"** chapter is crucial for business strategy maturity as it focuses on setting up and creating the actual product, translating ideas into tangible solutions. A mature business strategy ensures efficient execution in creating the user experience (UX) and user interface (UI), followed by robust building and coding of the product. This activation phase is vital for bringing strategic product concepts to life, ensuring they meet quality standards and user expectations. It's about mobilizing development resources effectively to build products that align with strategic objectives and are ready for market deployment. Here is how we approach our learning:

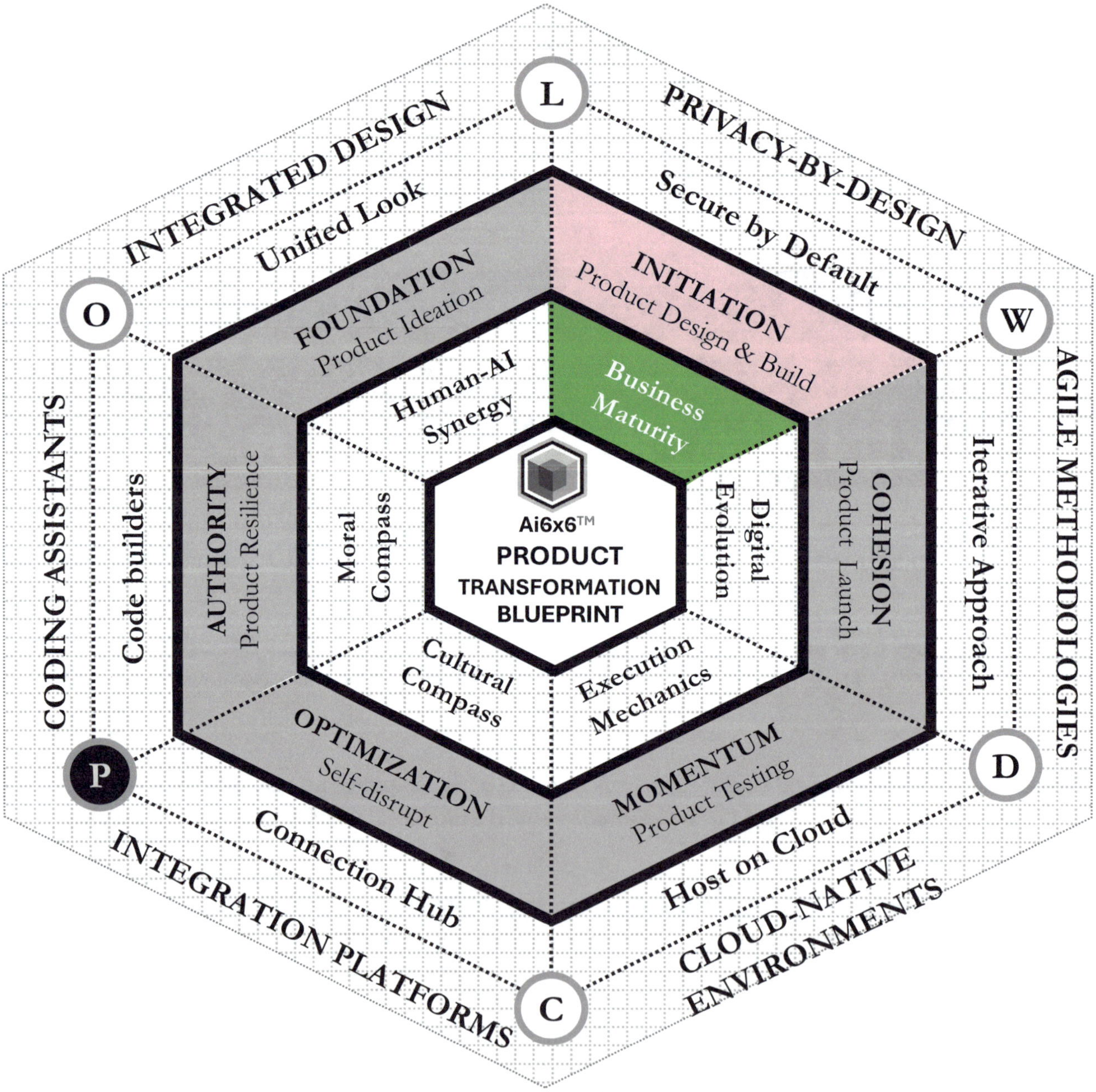

Ai6x6 BUSINESS MATURITY FOR BUILD | PRODUCT TRANSFORMATION

1. Integrated design

For a business leader, championing integrated design—encompassing User Experience (UX) and User Interface (UI)—is a critical strategic imperative for creating products that are not just functional but truly intuitive, accessible, and user-friendly. In today's competitive digital landscape, a superior design can be a significant differentiator, driving user adoption, satisfaction, and loyalty. Central to this is adhering to robust UI/UX principles. This involves ensuring visual aesthetics are appealing, interactivity is seamless, and workflows are intuitive. A well-designed interface minimizes friction, making the product enjoyable and efficient to use, which directly impacts user engagement and retention.

Furthermore, conducting rigorous usability testing is crucial. Through iterative feedback loops with real users, you can refine design efficiency, identify pain points, and uncover areas for improvement before full development resources are committed. This data-driven approach ensures the product meets user needs effectively and reduces costly reworks later on. Crucially, incorporating accessibility compliance is vital. Designing with inclusive standards ensures that products are usable by individuals with diverse needs and abilities (e.g., visual impairments, motor disabilities). This not only expands your potential market but also reflects ethical leadership and avoids legal challenges. Neglecting integrated design can lead to confusing interfaces, frustrated users, high abandonment rates, and ultimately, products that fail to gain market traction despite strong underlying functionality. By prioritizing a holistic and user-centric design approach, you empower your organization to build products that delight users, enhance brand reputation, and ultimately drive the sustained success and profitability.

WHAT?	WHY?	HOW?	BEWARE!
UI/UX PRINCIPLES	**USER SATISFACTION**	**DESIGN GUIDELINES**	**BALANCING NEEDS**
Ensuring visual aesthetics, interactivity, and seamless workflows.	Creates intuitive and engaging product experiences.	Follow established UI/UX best practices for consistency and usability.	Balancing aesthetic appeal with functional efficiency.
USABILITY TESTING	**DESIGN OPTIMIZATION**	**USER SESSIONS**	**REPRESENTATION**
Conducting iterative feedback loops to refine design efficiency.	Identifies usability issues early for iterative improvement.	Conduct user testing sessions and analyze user behavior recordings.	Ensuring test participants are representative of the target user base.
ACCESSIBILITY COMPLIANCE	**BROADER REACH**	**WCAG GUIDELINES**	**IMPLEMENTATION COMPLEXITY**
Incorporating inclusive design standards.	Makes products usable for diverse users, including those with disabilities.	Adhere to Web Content Accessibility Guidelines (WCAG) and conduct audits.	Integrating accessibility into all stages of design and development.

2. Privacy-by-design

For a business leader, embracing Privacy-by-Design is a non-negotiable strategic imperative for embedding security and regulatory adherence directly into your product architecture from the earliest stages of development. In an era of escalating data privacy concerns and stringent regulations (e.g., GDPR, CCPA), retrofitting privacy measures is costly and inefficient; a proactive approach is crucial for building trust and avoiding legal liabilities. Central to this is implementing secure data handling practices. This involves integrating encryption, anonymization techniques, and robust user-consent controls directly into the product's core design. Ensuring data is protected at every layer, from collection to storage and processing, is paramount for safeguarding user information.

Furthermore, aligning with comprehensive compliance-driven frameworks is crucial. As a leader, you must ensure that your product's architecture inherently aligns with global data privacy regulations such as GDPR, CCPA, HIPAA, and other relevant industry standards. This proactive adherence minimizes legal risks and builds consumer confidence. Crucially, conducting privacy impact assessments (PIAs) early in development is vital. By systematically evaluating potential risks to user privacy and integrating safeguards from the outset, you can identify and mitigate vulnerabilities before they become critical issues, preventing costly reworks and ensuring ethical data practices. Neglecting Privacy-by-Design can lead to devastating data breaches, massive regulatory fines, severe reputational damage, and a fundamental erosion of customer trust. By championing this proactive, ethical, and integrated approach to privacy, you empower your organization to build secure, compliant products that respect user data, ultimately safeguarding your business's integrity, legal standing, and long-term market viability.

WHAT?	WHY?	HOW?	BEWARE!
SECURE DATA HANDLING	**DATA CONFIDENTIALITY**	**SECURITY PROTOCOLS**	**BALANCING**
Implementing encryption, anonymization, and user-consent controls.	Protects sensitive user data throughout the product lifecycle.	Integrate encryption, data masking, and granular access controls.	Balancing data utility for analytics with strict privacy measures.
COMPLIANCE-DRIVEN FRAMEWORKS	**LEGAL ADHERENCE**	**REGULATORY AUDITS**	**EVOLVING REGULATIONS**
Aligning with GDPR, CCPA, HIPAA.	Ensures products meet stringent global data privacy regulations.	Conduct regular audits and integrate compliance checkpoints in development.	Keeping product compliant with constantly changing privacy laws.
PRIVACY IMPACT ASSESSMENTS	**PROACTIVE MITIGATION**	**PIA WORKFLOWS**	**COMPLEXITY**
Evaluating risks and integrating safeguards early.	Identifies and addresses privacy risks from the outset.	Conduct systematic PIAs during design and development phases.	Performing comprehensive PIAs for complex data processing systems.

3. Agile methodologies

For a business leader, embracing Agile methodologies is a strategic imperative for enhancing iterative product development, accelerating continuous delivery, and fostering a responsive, high-performing engineering organization. In today's rapidly evolving markets, traditional linear development approaches are often too slow; agility is key to rapid innovation and adaptation. Central to this is implementing Scrum-based sprint planning. By structuring development cycles into short, focused "sprints," you enable incremental improvements, faster feedback loops, and continuous delivery of valuable product features, allowing for rapid adaptation to changing requirements.

Furthermore, leveraging Kanban workflow optimization is crucial. This visual system manages task prioritization and provides real-time progress tracking, allowing teams to identify bottlenecks, optimize flow, and ensure continuous delivery of work with greater efficiency and transparency. Crucially, adopting Scaled Agile Framework (SAFe) or similar approaches becomes vital for large enterprises. This coordinates large-scale development teams across multiple projects and portfolios for efficiency, ensuring alignment and collaboration from individual teams up to strategic organizational goals. Neglecting Agile methodologies can lead to slow development cycles, products that miss market opportunities, frustrated teams due to rigid processes, and a struggle to adapt to changing customer needs. By championing these iterative and flexible development frameworks, you empower your organization to accelerate product innovation, respond swiftly to market demands, enhance team productivity, and build a culture of continuous improvement, ultimately driving the sustained success and competitive advantage of your digital products.

WHAT?	WHY?	HOW?	BEWARE!
SCRUM-BASED SPRINT PLANNING	**INCREMENTAL IMPROVEMENTS**	**DAILY STAND-UPS**	**SCOPE CREEP**
Structuring development cycles.	Enables rapid iterations and faster feedback loops.	Implement short, focused daily meetings and regular sprint reviews.	Managing changes within a sprint without disrupting progress.
KANBAN WORKFLOW OPTIMIZATION	**EFFICIENCY & TRANSPARENCY**	**VISUAL BOARDS**	**DISCIPLINE**
Managing task prioritization and tracking.	Improves flow, reduces bottlenecks, and provides real-time visibility.	Use Kanban boards to visualize tasks, WIP limits, and progress.	Ensuring adherence to Kanban principles, especially WIP limits.
SCALED AGILE FRAMEWORK (SAFE)	**ORGANIZATIONAL ALIGNMENT**	**FRAMEWORK ADOPTION**	**COMPLEXITY**
Coordinating large-scale development teams.	Ensures coordination and efficiency across large, distributed teams.	Implement SAFe principles, roles, and ceremonies at scale.	High initial complexity and cultural shift required for large-scale agile adoption.

4. Cloud-native environments

For a business leader, migrating to and leveraging cloud-native environments is a strategic imperative for building products with scalable, resilient, and adaptive infrastructure. In today's digital landscape, the demands for speed, flexibility, and global reach necessitate moving beyond monolithic on-premise solutions. Your ability to embrace cloud-native principles directly impacts your product's performance and agility. Central to this is adopting a microservices architecture. By structuring product components as small, independent services, you enable modular development, independent deployment, and isolated scaling, making your applications more resilient, easier to update, and more flexible to evolve.

Furthermore, utilizing Kubernetes orchestration is crucial for managing cloud-based containerized applications seamlessly. Kubernetes automates the deployment, scaling, and management of microservices, ensuring high availability and efficient resource utilization across your cloud infrastructure. This automation is key to operational efficiency. By allowing developers to run code without provisioning or managing servers, serverless models accelerate development cycles, scale automatically with demand, and optimize costs based on actual usage. Neglecting cloud-native environments can lead to slow development cycles, rigid and difficult-to-scale products, high infrastructure costs, and a struggle to compete with agile, cloud-first competitors. By championing cloud-native principles, you empower your organization to build highly performant, resilient, and cost-effective digital products, accelerating innovation, enhancing operational efficiency, and ultimately securing your business's long-term competitive advantage in the cloud era.

WHAT?	WHY?	HOW?	BEWARE!
MICROSERVICES ARCHITECTURE	**SCALABILITY & RESILIENCE**	**SERVICE DECOMPOSITION**	**COMPLEXITY**
Structuring components for modular deployment.	Enables independent scaling and faster updates for components.	Break down monolithic applications into smaller, independent services.	Increased operational complexity in managing distributed microservices.
KUBERNETES ORCHESTRATION	**AUTOMATED MANAGEMENT**	**CONTAINERIZATION**	**LEARNING CURVE**
Managing cloud-based containerized applications.	Streamlines deployment, scaling, and management of containerized apps.	Package applications in containers (Docker) and deploy with Kubernetes.	Significant learning curve for effective Kubernetes management.
SERVERLESS COMPUTING	**COST EFFICIENCY**	**FUNCTION-AS-A-SERVICE (FAAS)**	**VENDOR LOCK-IN**
Reducing operational overhead while enabling dynamic resource allocation.	Minimizes infrastructure management and scales dynamically based on usage.	Use serverless platforms (e.g., AWS Lambda, Google Cloud Functions).	Potential vendor lock-in to specific serverless platforms.

5. Integration platforms

For a business leader, establishing robust integration platforms is a strategic imperative for connecting disparate systems, ensuring seamless interoperability, and unlocking the full potential of your technological ecosystem. In a modern enterprise, data and functionalities are spread across numerous applications and services; the ability to make them communicate effortlessly is critical for efficiency and integrated insights. Central to this is adopting an API-driven connectivity approach. By standardizing interactions across platforms through well-defined Application Programming Interfaces (APIs), you enable fluid integrations between internal systems, third-party services, and customer-facing applications, fostering agility and flexibility.

Furthermore, leveraging an Enterprise Service Bus (ESB) or a modern integration hub is crucial. These centralized platforms facilitate communication between diverse application layers, orchestrating data flow, transforming messages, and managing complex integrations, reducing point-to-point connections and simplifying your integration landscape. Crucially, implementing event-driven architectures enables real-time updates and synchronized workflows. By designing systems that react to events as they happen, you ensure that data is always current across all integrated systems, powering immediate insights and highly responsive applications. Neglecting robust integration platforms can lead to data silos, manual inefficiencies, delayed data access, and a fragmented view of your business operations. By championing comprehensive integration strategies, you empower your organization to break down technological barriers, accelerate data exchange, automate complex workflows, and ultimately drive greater operational efficiency and innovation across your entire digital landscape, ensuring your systems work together as a cohesive whole.

WHAT?	WHY?	HOW?	BEWARE!
API-DRIVEN CONNECTIVITY	**FLUID INTEGRATION**	**API DEVELOPMENT**	**SECURITY**
Standardizing interactions across platforms.	Enables seamless data exchange and functionality across systems.	Design and publish well-documented APIs for internal and external services.	Ensuring robust API security against unauthorized access.
ENTERPRISE SERVICE BUS (ESB)	**SIMPLIFIED ORCHESTRATION**	**ESB IMPLEMENTATION**	**COMPLEXITY**
Centralizing communication between application layers.	Centralizes complex integrations, reducing point-to-point connections.	Deploy an ESB or modern integration platform (e.g., MuleSoft, Talend).	ESBs can become monolithic and complex to manage.
EVENT-DRIVEN ARCHITECTURES	**REAL-TIME SYNCHRONIZATION**	**MESSAGE BROKERS**	**DEBUGGING**
Enabling real-time updates for synchronized workflows.	Ensures data consistency and immediate responsiveness.	Use message queues or streaming platforms (e.g., Kafka) for event propagation.	Troubleshooting issues in asynchronous, distributed event flows.

For a business leader, embracing AI-powered coding assistants is a strategic imperative for accelerating product development, enhancing developer productivity, and streamlining the entire software creation process. In a competitive technology landscape, anything that can reduce time-to-market and improve code quality offers a significant competitive advantage. Central to this is leveraging low-code/no-code platforms. These tools simplify application creation for rapid deployment with minimal or no traditional programming, empowering business users and citizen developers to build functional applications quickly, reducing reliance on specialized engineering teams.

Furthermore, implementing automated testing solutions is crucial. By enhancing quality assurance through continuous validation and monitoring, these AI-driven tools can automatically identify bugs, performance issues, and security vulnerabilities earlier in the development cycle, reducing manual testing effort and accelerating release cycles while improving product reliability. Crucially, utilizing AI-assisted code generation reduces development time and optimizes error handling. These intelligent assistants can suggest code snippets, complete lines of code, and even generate entire functions based on context, significantly boosting developer efficiency and minimizing common coding errors. Neglecting AI-powered coding assistants can lead to slower development cycles, higher development costs, increased technical debt, and a struggle to keep pace with agile competitors. By championing these innovative tools, you empower your engineering teams to build products faster, with higher quality, and at a lower cost, ultimately driving accelerated product innovation and securing a significant competitive advantage for your business in the digital marketplace.

WHAT?	WHY?	HOW?	BEWARE!
LOW-CODE/NO-CODE PLATFORMS	**RAPID DEPLOYMENT**	**PLATFORM SELECTION**	**CUSTOMIZATION LIMITS**
Simplifying app creation for rapid deployment.	Accelerates application development with minimal coding.	Choose and implement a low-code/no-code development platform.	Limited flexibility for highly complex or custom functionalities.
AUTOMATED TESTING SOLUTIONS	**QUALITY & EFFICIENCY**	**TEST AUTOMATION**	**FALSE POSITIVES**
Enhancing quality assurance.	Improves product quality and speeds up release cycles.	Implement tools for automated unit, integration, and UI testing.	Automated tests producing false positives requiring manual investigation.
AI-ASSISTED CODE GENERATION	**DEVELOPER PRODUCTIVITY**	**AI TOOL INTEGRATION**	**CODE QUALITY**
Reducing development time and optimizing error handling.	Boosts coding efficiency and reduces common errors.	Integrate AI coding assistants (e.g., GitHub Copilot) into IDEs.	Ensuring AI-generated code meets quality standards and best practices.

Ai6x6™ Transformation Blueprint

COHESION | Product Launch

"Going from startup to scale-up is about finding product-market fit and then relentlessly executing." - Marc Andreessen.

An American entrepreneur, investor, and software engineer. He co-created Mosaic, the first widely used web browser, and co-founded Netscape Communications, playing a crucial role in popularizing the early internet. He is also a co-founder of Andreessen Horowitz.

The Cohesion level is your Product Launch, achieved when the offering successfully integrates with the market and the user's life. Scaling and amplification turn a single release into a unified market movement.

The pivotal moment of formally introducing the finished product to the market and making it readily available to the target customer base through strategic marketing, distribution, and sales efforts. A well-executed launch is absolutely critical for achieving strong market penetration, securing initial customer adoption, and firmly establishing the product's presence and competitive positioning within its segment.

It's about generating excitement, creating demand, and ensuring a smooth initial customer experience that sets the stage for long-term success and market acceptance.

"Product Launch" is vital for business strategy maturity because it represents the active delivery of the product to users, bringing it to market effectively. A mature business strategy ensures a well-orchestrated launch, combining a strong marketing strategy with adherence to privacy regulations and efficient cloud storage. This acceleration phase is critical for achieving market penetration, generating initial revenue, and gaining early user feedback. It's about executing the go-to-market strategy with precision, ensuring the product quickly gains traction and begins to contribute to the organization's strategic growth objectives. Here is how we approach our learning:

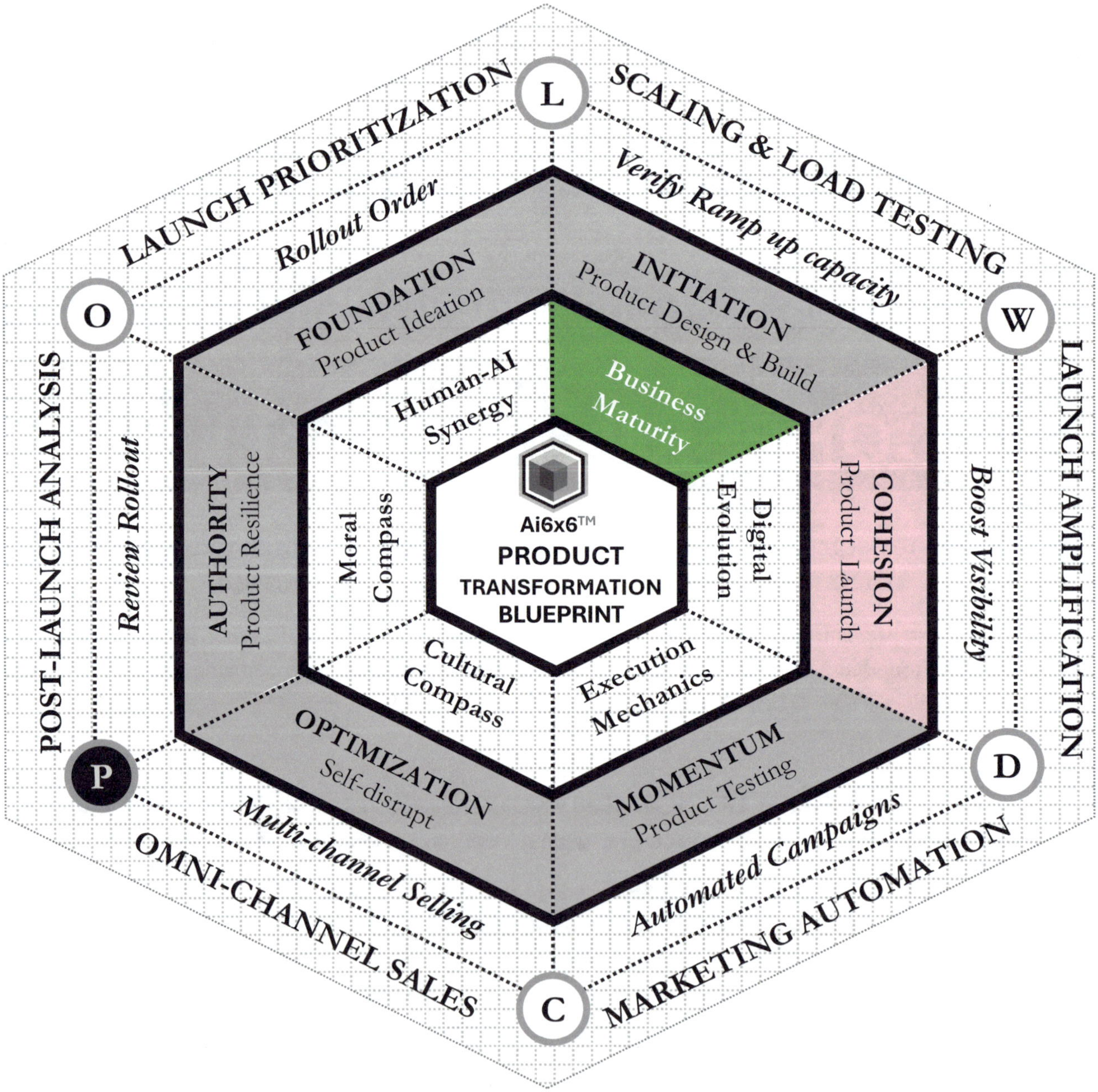

Ai6x6 BUSINESS MATURITY FOR LAUNCH | PRODUCT TRANSFORMATION

1. Launch prioritization

For a business leader, meticulously planning launch prioritization is a strategic imperative for ensuring a successful product rollout that maximizes market impact, mitigates risks, and optimizes resource allocation. It's about making deliberate choices about what to launch, when, and to whom, to build momentum and validate market demand effectively. Central to this is implementing an MVP (Minimum Viable Product) strategy. By launching with only the essential features needed to validate core market demand and gather early user feedback, you reduce time-to-market, conserve resources, and gain crucial insights before investing in full-scale development. This iterative approach minimizes risk.

Furthermore, leveraging feature flag management is crucial for a controlled rollout. Gradually introducing new functionalities to specific user segments allows you to refine the user experience, test market reception, and address bugs in a controlled environment, ensuring a smoother and more robust full launch. Crucially, applying MoSCoW analysis—categorizing product features into Must-haves, Should-haves, Could-haves, and Won't-haves—provides a structured framework for prioritization. This ensures that critical functionalities are in place for launch, while managing scope creep and focusing development efforts on what truly delivers value. Neglecting strategic launch prioritization can lead to bloated products, delayed launches, poor market fit, and wasted development efforts. By championing a disciplined, phased, and feature-prioritized rollout, you empower your organization to launch products effectively, adapt swiftly to market feedback, and ultimately drive greater market adoption and sustained product success. This strategic focus is essential for converting innovation into commercial viability.

WHAT?	WHY?	HOW?	BEWARE!
MVP (MINIMUM VIABLE PRODUCT) STRATEGY	**MARKET VALIDATION**	**FEATURE SELECTION**	**SCOPE CREEP**
Launching with essential features.	Validates core demand and gathers early user feedback efficiently.	Identify the smallest set of features to solve a core problem.	Resisting the urge to add non-essential features before launch.
FEATURE FLAG MANAGEMENT	**RISK MITIGATION**	**TOGGLE SYSTEMS**	**COMPLEXITY**
Gradually introducing new functionalities.	Enables controlled rollouts and A/B testing in production.	Implement feature flag tools to turn features on/off for user segments.	Managing numerous feature flags and their interdependencies.
MOSCOW ANALYSIS	**PRIORITIZATION CLARITY**	**TEAM WORKSHOP**	**AGREEMENT**
Categorizing product features.	Provides a clear framework for defining essential vs. optional features.	Conduct workshops to categorize features into Must, Should, Could, Won't.	Achieving consensus among stakeholders on feature categorization.

For a business leader, rigorously conducting scaling and load testing before a product launch is a critical strategic imperative for ensuring system stability and readiness for anticipated growth. A successful launch can quickly overwhelm an unprepared infrastructure, leading to outages, poor user experience, and reputational damage. Your ability to guarantee robustness under stress is paramount. Central to this is meticulously planning stress testing environments. This involves simulating peak usage scenarios, high concurrent user loads, and extreme data volumes to identify potential weaknesses, bottlenecks, and breaking points in your system architecture before real users encounter them. This proactive identification is vital.

Furthermore, comprehensive cloud scalability planning is crucial. Leveraging auto-scaling infrastructure, serverless functions, and other cloud-native capabilities ensures that your product can adaptively handle sudden spikes in demand without manual intervention, maintaining optimal performance as user adoption grows. Crucially, implementing redundancy and failover strategies provides essential backup protocols to mitigate launch risks. Designing systems with built-in redundancies, data replication across regions, and automated failover mechanisms ensures that even if components fail, your product remains available and operational, preventing disruptive downtime. Neglecting scaling and load testing can lead to catastrophic system failures on launch day, widespread user frustration, lost revenue, and severe brand damage. By championing a proactive approach to ensuring system stability and scalability, you empower your organization to launch products with confidence, deliver a reliable user experience, and sustain performance through rapid growth, ultimately safeguarding your business' reputation.

WHAT?	WHY?	HOW?	BEWARE!
STRESS TESTING ENVIRONMENTS	**WEAKNESS IDENTIFICATION**	**LOAD GENERATORS**	**REALISM**
Simulating peak usage scenarios.	Pinpoints system bottlenecks and breaking points pre-launch.	Use tools to simulate high user traffic and data loads.	Creating test environments that accurately reflect production complexity.
CLOUD SCALABILITY PLANNING	**ADAPTIVE PERFORMANCE**	**AUTO-SCALING**	**COST OPTIMIZATION**
Leveraging auto-scaling infrastructure.	Ensures system handles growth dynamically without manual intervention.	Configure cloud services to automatically scale resources based on demand.	Preventing runaway costs from over-provisioning in auto-scaling.
REDUNDANCY AND FAILOVER STRATEGIES	**MITIGATED RISKS**	**DISASTER RECOVERY**	**COMPLEXITY**
Implementing backup protocols.	Ensures continuous availability and prevents downtime during failures.	Design systems with redundant components and automated failover.	Designing and managing highly available, redundant architectures.

For a business leader, meticulously orchestrating launch amplification is a strategic imperative for maximizing visibility, generating widespread excitement, and ensuring a product launch achieves its full market impact. A brilliant product needs to be seen and talked about, and this phase is all about creating that crucial buzz. Central to this is compelling brand storytelling. Crafting narratives that emotionally engage audiences and clearly differentiate your product from competitors is vital for capturing attention and making your offering memorable. Storytelling transforms features into benefits and builds genuine connection.

Furthermore, meticulously planning product release marketing campaigns is crucial. Coordinating promotional efforts across a diverse mix of digital channels (social media, email, display ads) and traditional media (PR, print) ensures broad reach and consistent messaging that builds anticipation and drives initial interest. Crucially, forging strategic influencer and media partnerships is vital for enhanced awareness. Engaging with trusted thought leaders, industry experts, and key journalists can generate authentic endorsements and media coverage that amplify your message far beyond paid advertising, lending credibility and reach to your launch. Neglecting launch amplification can lead to a quiet market entry, poor initial adoption, and a struggle to gain traction against competitors. By championing a multi-faceted and compelling marketing approach, you empower your organization to generate significant market buzz, capture widespread attention, and drive initial adoption, ultimately laying the groundwork for sustained product success and market leadership. This strategic marketing push is what turns a launch into a landmark event.

WHAT?	WHY?	HOW?	BEWARE!
BRAND STORYTELLING	**AUDIENCE ENGAGEMENT**	**NARRATIVE DEVELOPMENT**	**INAUTHENTICITY**
Crafting compelling narratives.	Captivates audiences and differentiates the product.	Develop a compelling brand story that resonates emotionally.	Stories feeling forced or disconnected from the product's true value.
PRODUCT RELEASE MARKETING	**MAXIMIZED REACH**	**MULTI-CHANNEL CAMPAIGNS**	**CONSISTENCY**
Coordinating promotional campaigns.	Ensures broad visibility across various channels.	Plan synchronized marketing across digital, social, and traditional media.	Maintaining consistent messaging across all marketing channels.
INFLUENCER AND MEDIA PARTNERSHIPS	**ENHANCED AWARENESS**	**OUTREACH & PR**	**CONTROL**
Engaging with thought leaders and press.	Amplifies reach and credibility through trusted third-party endorsements.	Identify key influencers and media outlets for collaboration and press releases.	Managing message control when working with external partners.

4. Marketing automation

For a business leader, strategically leveraging marketing automation is a critical imperative for streamlining outreach and engagement, ensuring personalized customer interactions, and maximizing the efficiency of your marketing efforts, especially during and after a product launch. In today's digital landscape, manual campaign management is inefficient and cannot scale to meet customer expectations for relevance. Central to this is implementing automated lead nurturing. By designing targeted email campaigns and retargeting ads that respond to customer behavior and preferences, you can guide prospects through the sales funnel, providing relevant information at the right time, and optimizing for conversion without constant manual intervention.

Furthermore, sophisticated personalized customer segmentation is crucial. By customizing marketing messaging based on advanced behavioral analytics and AI-driven insights, you can group customers into dynamic segments, ensuring that each communication is highly relevant and impactful, leading to higher engagement rates. Crucially, enabling multi-touchpoint engagement integrates automation across email, social media, and paid advertising channels. This ensures a cohesive and consistent customer journey, where interactions build upon each other, creating a seamless experience regardless of the platform. Neglecting marketing automation can lead to generic mass marketing, missed opportunities for personalization, inefficient marketing spend, and a struggle to maintain consistent engagement with a growing customer base. By championing comprehensive marketing automation powered by AI, you empower your organization to deliver highly effective, scalable, and personalized campaigns, driving deeper customer engagement, accelerating conversions, and ultimately securing sustained product adoption.

WHAT?	WHY?	HOW?	BEWARE!
AUTOMATED LEAD NURTURING	**CONVERSION OPTIMIZATION**	**WORKFLOW DESIGN**	**OVER-AUTOMATION**
Using targeted email campaigns and retargeting ads.	Guides prospects through the sales funnel efficiently.	Set up automated email sequences and retargeting rules based on user behavior.	Messaging feeling impersonal or intrusive if not carefully managed.
PERSONALIZED CUSTOMER SEGMENTATION	**MAXIMIZED RELEVANCE**	**BEHAVIORAL ANALYTICS**	**DATA COMPLEXITY**
Customizing marketing messaging.	Ensures messages resonate deeply with specific customer groups.	Use AI/ML to analyze customer data for dynamic segmentation.	Managing and interpreting vast amounts of customer behavior data.
MULTI-TOUCHPOINT ENGAGEMENT	**COHESIVE EXPERIENCE**	**INTEGRATED PLATFORMS**	**CONSISTENCY**
Integrating automation across channels.	Creates a seamless and consistent customer journey.	Connect email, social media, and ad platforms for synchronized campaigns.	Maintaining brand voice and message consistency across all automated channels.

For a business leader, establishing a robust omnichannel sales strategy is a critical imperative for expanding market reach and maximizing product availability through diversified distribution channels. In today's interconnected world, customers expect seamless purchasing experiences whether they interact online, in physical stores, or through marketplaces. Your ability to integrate these channels is paramount for growth. Central to this is optimizing e-commerce and marketplace integration. Ensuring your product is readily available across leading digital storefronts (your own website, Amazon, eBay, etc.) and seamlessly integrating inventory, pricing, and order management across these platforms maximizes online visibility and purchasing convenience.

Furthermore, meticulously coordinating retail and direct sales channels is crucial. Ensuring alignment between your online and offline sales efforts—from consistent pricing and promotions to seamless customer service handoffs—creates a unified brand experience regardless of the purchase point, preventing channel conflict and enhancing customer satisfaction. Crucially, designing recurring revenue models through subscription and bundled offerings enhances customer retention and predictability. By offering flexible payment plans or packaging relevant products/services together, you not only increase perceived value but also build long-term relationships and stable revenue streams. Neglecting an omnichannel sales approach can lead to fragmented customer experiences, missed sales opportunities in preferred channels, and a struggle to scale effectively. By championing integrated distribution strategies, you empower your organization to meet customers wherever they choose to buy, optimize sales efficiency, and ultimately drive the sustained revenue growth and market penetration of your product.

WHAT?	WHY?	HOW?	BEWARE!
E-COMMERCE AND MARKETPLACE INTEGRATION	**MAXIMIZED REACH**	**PLATFORM INTEGRATION**	**CHANNEL CONFLICT**
Optimizing product availability.	Increases product visibility and accessibility to a wider online audience.	Integrate inventory, pricing, and order management with e-commerce sites/marketplaces.	Managing potential conflicts or price disparities across multiple online channels.
RETAIL AND DIRECT SALES COORDINATION	**UNIFIED EXPERIENCE**	**OMNICHANNEL STRATEGY**	**OPERATIONAL COMPLEXITY**
Ensuring alignment between online and offline.	Provides a seamless customer journey regardless of channel.	Align pricing, promotions, and customer service across all channels.	Coordinating inventory and customer data across online/offline systems.
SUBSCRIPTION AND BUNDLED OFFERINGS	**RETENTION & VALUE**	**PRICING STRATEGY**	**CHURN MANAGEMENT**
Designing recurring revenue models.	Enhances customer retention and increases perceived value.	Develop subscription tiers and create product/service bundles.	Managing customer cancellations and ensuring subscription stickiness.

6. Post-launch analysis

For a business leader, meticulous post-launch analysis is a critical strategic imperative for evaluating product performance, gathering vital insights, and continuously refining strategies for long-term success. A launch is merely the beginning; ongoing assessment is crucial for sustained market presence and profitability. Central to this is rigorous user feedback and analytics monitoring. Continuously collecting insights through surveys, reviews, app analytics, and behavioral tracking allows you to iterate product improvements rapidly, addressing pain points and enhancing features based on real-world usage data. This data-driven approach ensures continuous product evolution.

Furthermore, optimizing customer support post-launch is crucial. Strengthening onboarding processes and providing robust technical assistance ensures seamless adoption and reduces early-stage frustration, directly impacting user satisfaction and retention. Proactive support turns new users into loyal customers. Crucially, conducting a comprehensive launch retrospective and roadmapping future iterations is vital. This involves assessing successes against predefined KPIs, identifying areas for enhancement, and translating these learnings into concrete plans for upcoming product iterations and strategic development. This closes the feedback loop, informing future innovation. Neglecting post-launch analysis can lead to unaddressed user issues, missed opportunities for improvement, high churn rates, and ultimately, a product that stagnates in the market. By championing a continuous cycle of evaluation, refinement, and strategic planning, you empower your organization to adapt quickly, continuously enhance product value, and ultimately secure the long-term success and competitive relevance of your product in a dynamic market.

WHAT?	WHY?	HOW?	BEWARE!
FEEDBACK ANALYTICS MONITORING	**PRODUCT ITERATION**	**DATA COLLECTION**	**DATA OVERLOAD**
Collecting insights for improvements.	Informs continuous product improvements based on real-world usage.	Implement analytics tools and feedback mechanisms (surveys, reviews).	Managing and interpreting vast amounts of feedback and usage data.
CUSTOMER SUPPORT OPTIMIZATION	**SEAMLESS ADOPTION**	**SUPPORT CHANNELS**	**SCALABILITY**
Strengthening onboarding and technical assistance.	Ensures positive initial user experience, boosting retention.	Optimize helpdesks, FAQs, chatbots, and training materials.	Scaling support effectively with rapid user growth.
LAUNCH RETROSPECTIVE AND FUTURE ROADMAPPING	**STRATEGIC LEARNING**	**REVIEW SESSIONS**	**BLAME AVOIDANCE**
Assessing successes and areas for enhancement.	Translates launch learnings into future product development plans.	Conduct post-mortem meetings to evaluate launch performance against goals.	Ensuring a blame-free environment for honest retrospective discussions.

Ai6x6™ Transformation Blueprint

MOMENTUM | Product Testing

"Move fast and break things. Unless you are breaking stuff, you are not moving fast enough." - Mark Zuckerberg.

Co-founder and CEO of Meta Platforms (formerly Facebook). He created Facebook from his Harvard dorm room, transforming it into the world's largest social media platform, leading its ambitious expansion into virtual reality and the metaverse.

The Momentum level is your Product Iteration, generated through the rapid improvement of the offering based on live feedback and testing. Multivariate loops allow the product to evolve at a speed that outpaces the competition.

The continuous process of refining and improving the product based on invaluable user feedback, evolving market trends, competitive analysis, and comprehensive performance data. This ongoing effort is essential for extending the product's lifespan, significantly enhancing the user experience, and maintaining a strong competitive edge in a dynamic market.

Testing involves iterative updates, bug fixes, feature enhancements, and performance improvements to ensure the product remains relevant, valuable, and continues to meet or exceed customer expectations over time.

The **"Product Testing"** chapter is paramount for business strategy maturity as it focuses on continuously improving the product based on user feedback and adapting to market needs. A mature business strategy involves ongoing optimization, including scaling, expanding reach and capabilities, and leveraging user feedback and analytics. This continuous refinement ensures the product remains competitive, relevant, and delivers increasing value to users. It's about proactively evolving the product in response to market dynamics, ensuring its long-term viability and sustained contribution to the organization's strategic goals. Here is how we approach our learning:

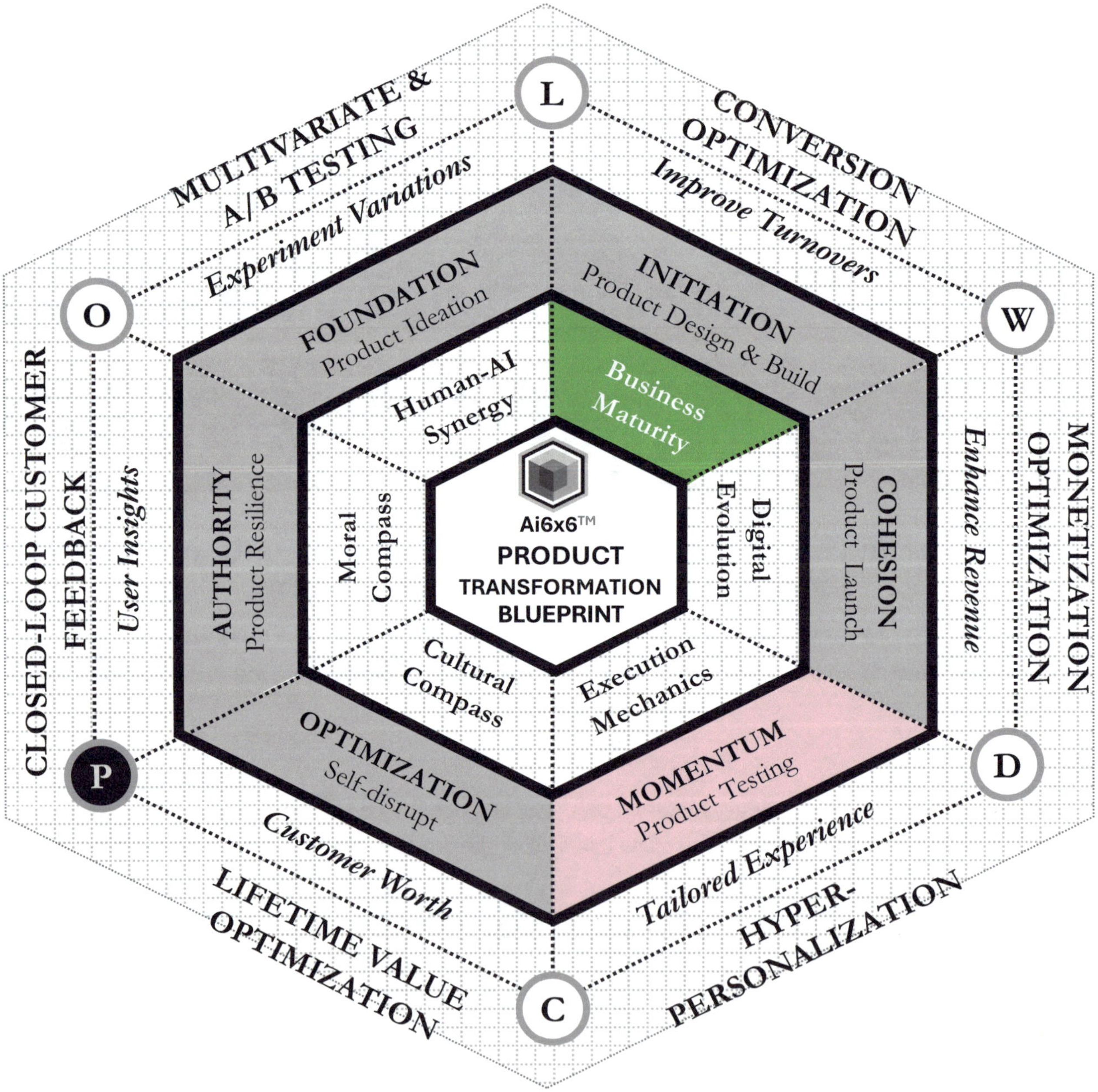

Ai6x6 BUSINESS MATURITY FOR OPTIMIZE | PRODUCT TRANSFORMATION

1. Multivariate & A/B testing

For a business leader, embracing Multivariate & A/B testing is a critical strategic imperative for continuously refining product performance through structured, data-driven experimentation. In a competitive digital landscape, intuition alone is insufficient for optimizing user experience and driving engagement; systematic testing provides empirical evidence for what truly works. Central to this is rigorous A/B testing methodologies. By comparing two variations of a web page, feature, or message to determine which performs better against a specific goal (e.g., conversion rate, engagement), you gain clear insights into user preferences and design effectiveness, leading to targeted improvements.

Furthermore, leveraging multivariate testing strategies allows for evaluating multiple elements simultaneously. This advanced testing method enables you to identify the optimal combination of interactions across various components, accelerating optimization efforts and pinpointing the most impactful design configurations. Crucially, establishing continuous iteration cycles is vital. Applying data-driven insights from testing to refine features, improve usability, and optimize user flows ensures that your product is constantly evolving based on real user behavior, leading to sustained enhancements and greater user satisfaction. Neglecting structured experimentation can lead to suboptimal product performance, inefficient resource allocation on unproven features, and a reactive approach to user engagement challenges. By championing multivariate and A/B testing, you empower your organization to make evidence-based product decisions, accelerate improvement cycles, and build a more user-centric and high-performing digital product, ultimately driving sustained success and competitive advantage.

WHAT?	WHY?	HOW?	BEWARE!
A/B TESTING METHODOLOGIES	**USER EXPERIENCE OPTIMIZATION**	**SPLIT TESTING**	**STATISTICAL SIGNIFICANCE**
Comparing variations for effectiveness.	Identifies best-performing elements for design and content.	Create two versions of a page/feature and show to different user groups.	Ensuring sufficient traffic and data for valid results.
MULTIVARIATE TESTING STRATEGIES	**OPTIMIZED INTERACTIONS**	**COMPLEX VARIANTS**	**TEST COMPLEXITY**
Evaluating multiple elements simultaneously.	Pinpoints the best combination of various elements for peak performance.	Test multiple combinations of changes on a single page.	High complexity and longer run times due to numerous variables.
CONTINUOUS ITERATION CYCLES	**SUSTAINED IMPROVEMENT**	**AGILE DEVELOPMENT**	**TEST FATIGUE**
Applying data-driven insights to refine features.	Ensures ongoing product enhancement based on user feedback.	Integrate testing results into agile sprint planning for rapid deployment.	Over-testing or users becoming accustomed to constant changes.

2. Conversion optimization

For a business leader, relentlessly pursuing conversion optimization is a strategic imperative for enhancing product effectiveness, directly driving higher engagement and increased sales. It's not enough to attract users; the true measure of success lies in converting their interest into desired actions, whether that's a purchase, a sign-up, or deeper feature adoption. Central to this is meticulous funnel optimization. By identifying key friction points and streamlining user flows within your product or website, you eliminate obstacles that deter users, guiding them seamlessly towards conversion goals and improving overall user experience.

Furthermore, rigorous call-to-action refinement is crucial. Improving the messaging, placement, and design of your calls-to-action (CTAs)—buttons, links, banners—ensures they are compelling, clear, and motivate users to take the next desired step, directly impacting conversion rates. Crucially, leveraging behavioral analytics is vital for uncovering optimization opportunities. By utilizing tools like heatmaps, click tracking, and user session recordings, you can gain deep insights into how users interact with your product, identify areas of confusion or abandonment, and pinpoint precise opportunities for improvement that lead to higher conversions. Neglecting conversion optimization can result in wasted marketing spend on traffic that doesn't convert, a stagnant user base, and missed revenue opportunities. By championing a continuous, data-driven approach to enhancing every step of the user journey, you empower your organization to unlock greater value from existing users, accelerate sales, and ultimately drive the sustained success and profitability of your product by maximizing its effectiveness in turning interest into action.

WHAT?	WHY?	HOW?	BEWARE!
FUNNEL OPTIMIZATION	**HIGHER CONVERSION**	**ANALYTICS & UX AUDITS**	**COMPLEXITY**
Identifying key friction points and streamlining user flows.	Guides users efficiently towards desired actions.	Map conversion funnels, identify drop-off points, and conduct UX audits.	Funnels with too many steps or external dependencies.
CALL-TO-ACTION REFINEMENT	**INCREASED CONVERSIONS**	**A/B TESTING**	**ATTENTION BLINDNESS**
Improving messaging and placement.	Directly influences user behavior and desired actions.	Test different CTA text, colors, sizes, and placements.	Users becoming desensitized to common CTA designs.
BEHAVIORAL ANALYTICS	**OPTIMIZATION OPPORTUNITIES**	**HEATMAP TOOLS**	**INTERPRETATION**
Leveraging heatmaps and user tracking.	Reveals how users interact, identifying areas for improvement.	Use tools like Hotjar or Crazy Egg to visualize user clicks, scrolls, and movement.	Translating visual data into actionable product changes.

For a business leader, relentless monetization optimization is a critical strategic imperative for maximizing revenue, enhancing profitability, and ensuring the long-term financial viability of your product. It's about continuously refining pricing and sales models to capture optimal value from your user base. Central to this is implementing dynamic pricing adjustments. By adapting cost structures in real-time based on factors like demand, competitive actions, and user segmentation, you can optimize revenue generation and market share, ensuring your product is priced optimally for various market conditions.

Furthermore, optimizing subscription and tiered pricing models is crucial. Creating scalable payment plans allows you to cater to diverse customer needs and usage levels, encouraging customer retention and building predictable recurring revenue streams. This ensures sustained profitability rather than one-off sales. Crucially, strategically deploying promotional strategies is vital for driving adoption and increasing perceived value. Implementing carefully timed seasonal offers, discounts, or loyalty incentives can attract new users, encourage upgrades, and reward long-term commitment without devaluing the product's core worth. Neglecting monetization optimization can lead to leaving significant revenue on the table, struggling with profitability, or losing customers due to rigid or uncompetitive pricing. By championing a continuous, data-driven approach to pricing and sales models, you empower your organization to maximize revenue capture, attract and retain a broad customer base, and ultimately drive the sustained financial success and scalability of your product in the marketplace. This financial acumen is fundamental to product leadership.

WHAT?	WHY?	HOW?	BEWARE!
DYNAMIC PRICING ADJUSTMENTS	**REVENUE MAXIMIZATION**	**ALGORITHM DEPLOYMENT**	**CUSTOMER PERCEPTION**
Adapting cost structures based on demand and segmentation.	Optimizes pricing for various market conditions and user segments.	Use AI/ML to adjust prices based on real-time data.	Users perceiving price changes as arbitrary or unfair.
SUBSCRIPTION AND TIERED MONETIZATION	**RETENTION & PREDICTABILITY**	**MODEL DESIGN**	**CHURN MANAGEMENT**
Creating scalable payment plans.	Enhances customer retention and provides recurring revenue.	Develop and test different subscription tiers (e.g., freemium, usage-based).	Effectively managing cancellations within subscription models.
PROMOTIONAL STRATEGIES	**ADOPTION & VALUE**	**CAMPAIGN PLANNING**	**DEVALUATION**
Implementing seasonal offers, discounts, and loyalty incentives.	Drives new user adoption and increases perceived value.	Design and execute seasonal sales, loyalty discounts, and bundle offers.	Risk of promotions eroding brand value or customer expectation of perpetual discounts.

4. Hyper-personalization

For a business leader, embracing hyper-personalization is a critical strategic imperative for customizing user experiences through advanced, AI-driven insights, deepening engagement, and fostering unparalleled customer loyalty. Beyond basic personalization, hyper-personalization tailors every interaction to the individual, creating a uniquely relevant and intuitive product experience. Central to this is enabling personalized content delivery. By leveraging AI to understand individual user preferences, behaviors, and past interactions, you can tailor recommendations, content feeds, product displays, and messaging in real-time, making every touchpoint feel uniquely relevant to that specific user.

Furthermore, developing adaptive user interfaces is crucial. This involves dynamically modifying layouts, navigation paths, and feature visibility to align with individual user habits, roles, or preferred workflows. The UI subtly shifts to meet their needs, enhancing ease of use and reducing friction. Crucially, implementing predictive engagement triggers allows you to anticipate customer needs and proactively deliver relevant information or calls to action. By using data-driven insights to forecast future behaviors, you can offer solutions or opportunities precisely when they are most receptive, creating a truly intuitive and proactive user journey. Neglecting hyper-personalization can lead to generic user experiences, lower engagement rates, and a struggle to differentiate your product in a crowded market where customers expect bespoke interactions. By championing advanced AI-driven personalization, you empower your organization to build highly engaging, intuitive, and differentiated products that delight users, drive deeper loyalty, and ultimately secure sustained competitive advantage and growth for your business.

WHAT?	WHY?	HOW?	BEWARE!
PERSONALIZED CONTENT DELIVERY	**ENHANCED RELEVANCE**	**RECOMMENDATION ENGINES**	**DATA PRIVACY**
Tailoring recommendations based on user preferences.	Maximizes user engagement by delivering highly tailored content.	Implement AI/ML-powered systems to personalize content feeds and product suggestions.	Balancing granular personalization with user privacy concerns.
ADAPTIVE USER INTERFACES	**IMPROVED USABILITY**	**DYNAMIC UI**	**COMPLEXITY**
Modifying layouts dynamically.	Optimizes the user experience by aligning with individual habits.	Develop AI-driven UI systems that adapt layouts and features based on user behavior.	Designing and maintaining highly dynamic and personalized interfaces.
PREDICTIVE ENGAGEMENT TRIGGERS	**PROACTIVE ENGAGEMENT**	**BEHAVIORAL FORECASTING**	**INTRUSION**
Anticipating customer needs.	Delivers timely information or offers when users are most receptive.	Use ML models to predict user intent and trigger personalized nudges.	Avoiding overly frequent or irrelevant triggers that annoy users.

For a business leader, relentlessly pursuing Lifetime Value Optimization is a critical strategic imperative for structuring long-term revenue growth and maximizing profitability from your existing customer base. It's about systematically increasing the value each customer brings to your business through intelligent strategies like upselling, cross-selling, and bundling. Central to this is implementing smart upsell strategies. By intelligently suggesting higher-tier products, premium versions, or upgraded services based on a customer's purchase patterns, usage data, and identified needs, you encourage them to move up the value chain, increasing their average revenue contribution.

Furthermore, developing effective cross-sell recommendations is crucial. Introducing complementary products or services that enhance a customer's initial purchase maximizes their overall spend and deepens their engagement with your brand, ensuring you meet all their related needs rather than allowing competitors to fill the gap. Crucially, employing strategic bundling tactics packages relevant offerings together. By combining multiple products or services into attractive bundles, you can increase perceived value, encourage larger initial purchases, and often boost retention by making the combined offering more compelling and convenient than individual purchases. Neglecting lifetime value optimization means leaving significant revenue on the table, constantly battling for new customer acquisition, and failing to fully leverage the profitability of your existing loyal base. By championing these revenue-expanding strategies, you empower your organization to drive sustainable growth, foster deeper customer relationships, and significantly enhance the long-term profitability of your product by maximizing the value of every customer interaction.

WHAT?	WHY?	HOW?	BEWARE!
SMART UPSELL STRATEGIES	**REVENUE GROWTH**	**RECOMMENDATION ENGINE**	**OVER-SELLING**
Suggesting higher-tier products.	Increases customer spending on premium offerings.	Use AI/ML to suggest upgrades based on usage and perceived need.	Aggressive upselling alienating customers.
CROSS-SELL RECOMMENDATIONS	**MAXIMIZED SPEND**	**DATA ANALYSIS**	**RELEVANCE**
Introducing complementary products.	Encourages customers to purchase additional, related products/services.	Identify product affinities and suggest complementary items.	Ensuring cross-sell suggestions are genuinely relevant and not random.
BUNDLING TACTICS	**PERCEIVED VALUE**	**PRODUCT COMBINATION**	**COMPLEXITY**
Packaging relevant offerings.	Increases average order value and enhances customer retention.	Create compelling bundles of related products/services.	Determining optimal bundle components and pricing.

6. Closed-loop customer feedback

For a business leader, establishing robust closed-loop customer feedback systems is a critical strategic imperative for continuously improving your product based on direct user insights, ensuring relevance, satisfaction, and long-term success. It's about systematically gathering, analyzing, and acting upon the voice of your customer. Central to this is implementing real-time feedback collection. By deploying immediate surveys, in-app polls, and accessible review mechanisms, you capture user insights precisely when they are interacting with your product, enabling immediate analysis and rapid response to emerging issues or opportunities.

Furthermore, integrating sentiment analysis is crucial for understanding the emotional landscape of your user base. By using AI to assess customer opinions, satisfaction levels, and emotional tones from text-based feedback (e.g., reviews, comments), you can quickly identify pain points and areas of delight, gaining a deeper qualitative understanding. Crucially, establishing iterative response mechanisms ensures that product features are continuously adjusted based on recurring feedback trends. This involves a disciplined process of prioritizing feedback, developing solutions, and releasing updates that directly address user needs, demonstrating responsiveness and building user trust. Neglecting closed-loop customer feedback can lead to products that become disconnected from user needs, escalating customer dissatisfaction, high churn rates, and a reactive approach to market shifts. By championing a continuous, data-driven approach to customer feedback, you empower your organization to build truly user-centric products, enhance customer satisfaction, and drive sustained innovation and loyalty, ultimately securing your business's long-term success in a competitive marketplace.

WHAT?	WHY?	HOW?	BEWARE!
REAL-TIME FEEDBACK COLLECTION	**IMMEDIATE INSIGHTS**	**IN-APP FEEDBACK**	**OVER-SURVEYING**
Implementing surveys, polls, and reviews.	Captures user sentiment and issues as they occur.	Deploy surveys, polls, and feedback widgets within the product.	Risk of annoying users with too many feedback requests.
SENTIMENT ANALYSIS INTEGRATION	**QUALITATIVE UNDERSTANDING**	**NLP & ML TOOLS**	**NUANCE**
Using AI to assess customer opinions.	Identifies emotional trends and satisfaction levels from text.	Integrate AI tools to analyze unstructured feedback data for sentiment.	AI potentially misinterpreting complex or ironic sentiment.
ITERATIVE RESPONSE MECHANISMS	**CONTINUOUS IMPROVEMENT**	**FEEDBACK TO ROADMAP**	**PRIORITIZATION**
Adjusting product features based on feedback.	Ensures product evolves to meet user needs and desires.	Establish a clear process for integrating feedback into product dev cycles.	Deciding which feedback to prioritize given resource constraints.

Ai6x6™ Transformation Blueprint

OPTIMIZATION | Self Disruption

"If you don't cannibalize yourself, someone else will." - Clayton Christensen.

An American academic and business consultant, best known for his theory of "disruptive innovation." His work explains how smaller companies can challenge established businesses, profoundly influencing business strategy, entrepreneurship, and innovation globally.

The Optimization level is your Product Disruption, involving the surgical decision to evolve past your own success to stay ahead. Managing obsolescence and disruption ensures the product remains at peak efficiency by self-correcting for the future.

The proactive and strategic decision to innovate or even replace your own successful products or business models before competitors have the chance to do so. This bold approach is crucial for long-term relevance, preventing market obsolescence, and ensuring the organization remains a pioneering market leader rather than a follower.

It requires a forward-thinking mindset, a willingness to cannibalize existing revenue streams for future growth, and the courage to embrace radical innovation to stay ahead in rapidly evolving industries.

"Self-Disruption" is critical for business strategy maturity because it focuses on the product's transition to a new version or its graceful phasing out in response to market changes. A mature business strategy embraces self-disruption, recognizing when to phase out, reinvent, or fundamentally change a product to adapt to evolving market changes or customer expectations. This strategic adaptability prevents stagnation, frees up resources for new innovations, and ensures the product portfolio remains aligned with future business opportunities. It's about making proactive, strategic choices to maintain market relevance. Here is how we approach our learning:

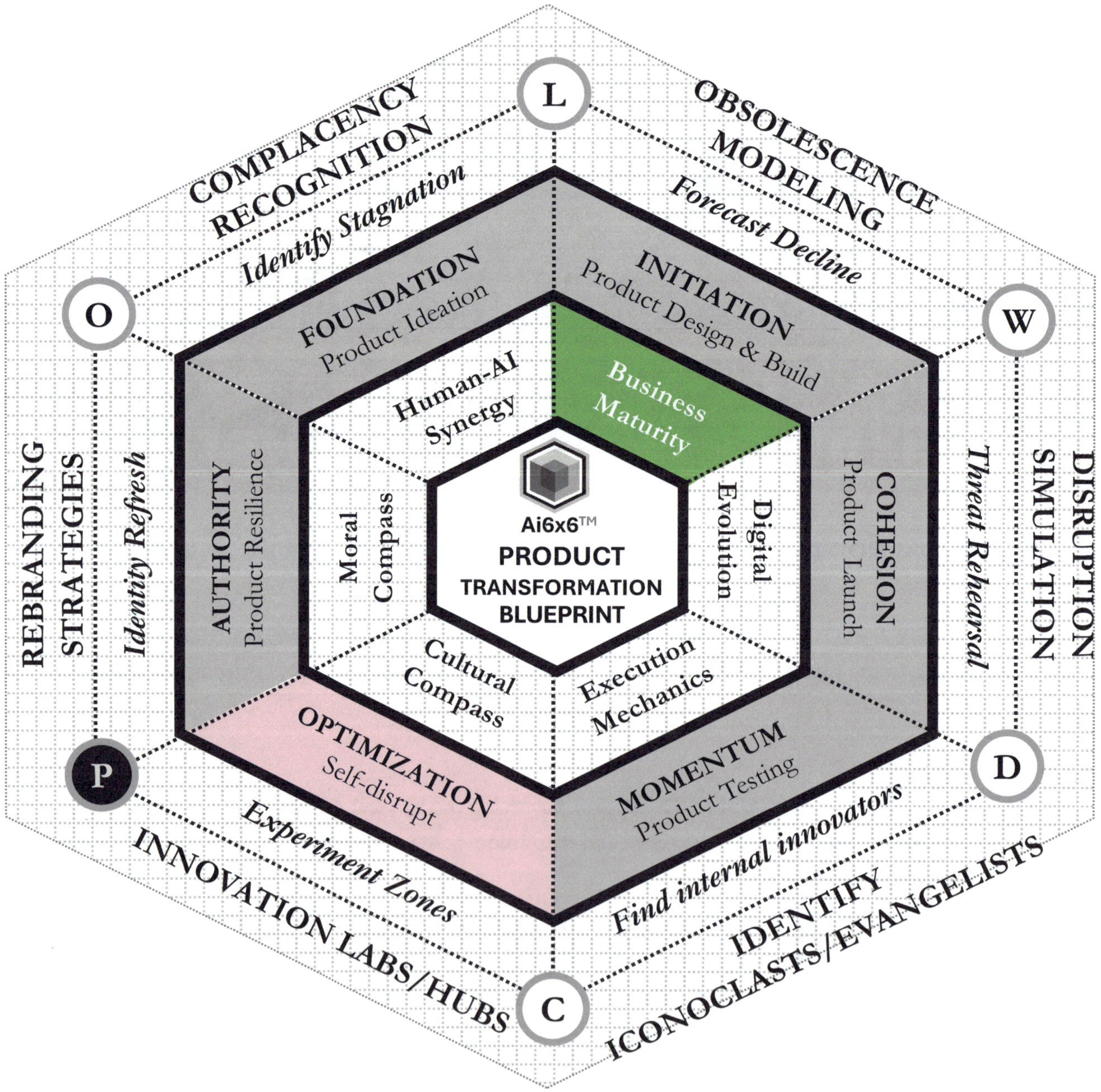

Ai6x6 BUSINESS MATURITY FOR SELF-DISRUPT | PRODUCT TRANSFORMATION

1. Complacency recognition

For a business leader, proactively engaging in complacency recognition is a critical strategic imperative for identifying early signs of stagnation within your organization and ensuring a continuous culture of innovation. In today's dynamic markets, resting on past successes is a recipe for obsolescence; continuous self-assessment is vital for sustained relevance. Central to this is rigorous performance benchmarking. Regularly assessing your achievements against both industry standards and internal historical performance helps identify any slowdowns in growth, declining market share, or a lack of differentiation compared to competitors. These metrics serve as early warning signals.

Furthermore, meticulous weakness identification is crucial. Analyzing signs of slowed growth, disengagement from market trends, or a perceived lack of unique value proposition reveals areas where your product or business may be becoming complacent. This requires honest, objective self-assessment, even when it's uncomfortable. Crucially, assessing your organizational culture ensures mindset shifts towards prioritizing adaptability and evolution. Leaders must foster an environment where challenging the status quo, embracing change, and continuous learning are celebrated, rather than resisting new ideas or clinging to outdated practices. Neglecting complacency recognition can lead to a gradual decline in competitiveness, missed market opportunities, and ultimately, forced, reactive disruption that is far more painful than proactive self-disruption. By championing continuous self-assessment and a culture of critical introspection, you empower your organization to stay agile, innovative, and proactive in a constantly changing environment, ultimately securing your business's long-term viability and leadership position.

WHAT?	WHY?	HOW?	BEWARE!
PERFORMANCE BENCHMARKING	**STAGNATION DETECTION**	**KPI ANALYSIS**	**DATA ACCESS**
Regularly assessing achievements against standards.	Identifies signs of slowed growth or declining performance.	Compare internal KPIs against industry benchmarks and historical data.	Difficulty obtaining reliable comparative data from competitors.
WEAKNESS IDENTIFICATION	**PROACTIVE INNOVATION**	**MARKET ANALYSIS**	**INTERNAL BIAS**
Analyzing signs of slowed growth or lack of differentiation.	Pinpoints areas needing strategic intervention and renewed focus.	Conduct SWOT analysis, gather customer feedback on perceived value.	Leadership reluctance to admit weaknesses or market disengagement.
ORGANIZATIONAL CULTURE ASSESSMENT	**CULTURAL ADAPTATION**	**SURVEYS & WORKSHOPS**	**RESISTANCE**
Ensuring mindset shifts for adaptability.	Fosters a mindset prioritizing continuous evolution and change.	Assess employee attitudes towards innovation, risk, and change.	Deep-seated cultural resistance to change or innovation.

2. Obsolescence modeling

For a business leader, proactively engaging in obsolescence modeling is a critical strategic imperative for predicting industry shifts, anticipating disruptive forces, and ensuring your products and business models stay ahead of the curve. In a world of accelerating technological change, traditional static planning is insufficient; foresight is key to sustained relevance. Central to this is rigorous trend forecasting. By utilizing AI-driven analytics to anticipate market changes, emerging technologies, and evolving consumer behaviors, you can identify nascent trends that could fundamentally disrupt your industry or render your current offerings obsolete. This proactive intelligence allows for early strategic adjustments.

Furthermore, conducting meticulous lifecycle analysis is crucial. Mapping the longevity of your product, service, or business against broader innovation cycles helps you understand where you are in the maturity curve and when new offerings or business models will be required to maintain competitiveness. This foresight informs R&D investments. Crucially, developing robust future-proofing strategies involves implementing agile transformation roadmaps for sustained relevance. This means continuously adapting your core capabilities, technologies, and business models to remain resilient against future disruptions, ensuring your organization can evolve before it's forced to. Neglecting obsolescence modeling can lead to reactive decision-making, being blindsided by new entrants or technologies, and a gradual erosion of market share. By championing a data-driven and forward-looking approach to anticipating industry shifts, you empower your organization to innovate proactively, adapt strategically, and ultimately secure your business's long-term viability and leadership position by staying ahead of the disruption curve.

WHAT?	WHY?	HOW?	BEWARE!
TREND FORECASTING	**PROACTIVE ADAPTATION**	**PREDICTIVE MODELS**	**ACCURACY**
Using AI-driven analytics to anticipate market changes.	Ensures the business stays ahead of emerging industry disruptions.	Deploy AI/ML to analyze market data, research, and patent filings.	Ensuring predictive models provide reliable forecasts of complex market shifts.
LIFECYCLE ANALYSIS	**STRATEGIC INVESTMENT**	**MATURITY CURVE**	**SUBJECTIVITY**
Mapping product, service, or business longevity.	Informs decisions on product refresh, reinvention, or retirement.	Assess products against industry lifecycle stages (growth, maturity, decline).	Difficulty objectively assessing a product's true position in its lifecycle.
FUTURE-PROOFING STRATEGIES	**SUSTAINED RELEVANCE**	**TRANSFORMATION PLANNING**	**RESISTANCE TO CHANGE**
Implementing agile transformation roadmaps.	Ensures the business remains resilient against future disruptions.	Develop agile roadmaps for technology and business model evolution.	Internal resistance to fundamental business or product transformation.

3. Disruption simulation

For a business leader, proactively engaging in disruption simulation is a critical strategic imperative for testing impact scenarios and robustly preparing your organization for competitive threats and unforeseen market shifts. In today's volatile business environment, simply planning for continuity is insufficient; you must actively anticipate and rehearse responses to potential disruptions to maintain agility and resilience. Central to this is conducting competitive role-playing exercises. Simulating the actions of new market entrants, the launch of disruptive technologies by rivals, or sudden shifts in consumer behavior allows your leadership team to anticipate threats, identify vulnerabilities, and develop proactive counter-strategies in a low-risk environment.

Furthermore, rigorous stress-testing internal processes is crucial. This involves evaluating operational vulnerabilities and pain points in the face of potential disruption events, ensuring your supply chains, IT infrastructure, and human resources are resilient enough to absorb shocks without critical failure. Crucially, developing adaptive response modeling designs frameworks to pivot and react effectively to changing market conditions. This involves creating flexible strategic playbooks and empowering teams with decision-making autonomy, enabling rapid, coordinated responses when real disruption occurs. Neglecting disruption simulation can leave your business vulnerable to being blindsided by new competitors, suffering prolonged operational downtime, and losing market share to more agile rivals. By championing this proactive, analytical, and rehearsal-based approach, you empower your organization to anticipate, prepare for, and ultimately thrive amidst market disruption, securing your business's long-term competitive advantage and resilience.

WHAT?	WHY?	HOW?	BEWARE!
COMPETITIVE ROLE-PLAYING	**THREAT ANTICIPATION**	**WAR-GAMING**	**REALISM**
Simulating market entrants or disruptive strategies.	Identifies potential threats and informs proactive counter-strategies.	Conduct tabletop exercises simulating competitor actions and market disruptions.	Designing simulations that accurately reflect complex market dynamics.
STRESS-TESTING INTERNAL PROCESSES	**OPERATIONAL RESILIENCE**	**VULNERABILITY AUDITS**	**DISRUPTION TO OPERATIONS**
Evaluating operational vulnerabilities.	Ensures internal systems and processes can withstand disruption.	Conduct comprehensive audits of supply chains, IT, and operational flows.	Performing stress tests on live systems can be disruptive.
ADAPTIVE RESPONSE MODELING	**AGILE REACTION**	**SCENARIO PLANNING**	**LACK OF FLEXIBILITY**
Designing frameworks to pivot and react effectively.	Enables quick, effective pivots to changing market conditions.	Develop flexible response frameworks for various disruption scenarios.	Frameworks becoming too rigid to adapt to unforeseen types of disruption.

4. Identify iconoclasts/evangelists

For a business leader, identifying and empowering iconoclasts and evangelists within your organization is a critical strategic imperative for driving meaningful change and fostering a culture of self-disruption. These unconventional thinkers and passionate advocates are often the catalysts for innovation, challenging the status quo and inspiring others to embrace transformation. Central to this is meticulous disruptor profiling. As a leader, you must actively identify individuals or teams within your organization who naturally challenge conventional norms, possess innovative mindsets, and are unafraid to question existing practices. These are often the hidden gems of transformative potential.

Furthermore, fostering internal thought leadership cultivation is crucial. Empowering these forward-thinking employees to champion transformation initiatives, share their ideas, and lead pilot projects creates internal momentum for change. Providing platforms for their voices to be heard encourages broader adoption of new ideas. Crucially, strategically engaging in external influencer partnerships can drive vision alignment and bring fresh perspectives. Collaborating with innovative industry leaders or external thought leaders can inject new thinking into your organization, inspire internal teams, and help validate emerging ideas against broader market trends. Neglecting to identify and support these internal change agents can lead to stagnation, missed opportunities for innovation, and a reactive approach to market shifts. By championing these iconoclasts and evangelists, you empower your organization to embrace discomfort, challenge assumptions, and proactively drive transformative change from within, ultimately securing your business's long-term adaptability and competitive edge.

WHAT?	WHY?	HOW?	BEWARE!
DISRUPTOR PROFILING	**INNOVATION CATALYST**	**OBSERVATION & ASSESSMENT**	**RESISTANCE**
Identifying individuals or teams that challenge norms.	Uncovers internal sources of transformative thinking.	Identify employees consistently offering unconventional ideas or challenging status quo.	Traditionalists or risk-averse leaders may resist recognizing disruptors.
INTERNAL THOUGHT LEADERSHIP	**CHANGE MOMENTUM**	**PLATFORM PROVISION**	**RESOURCE LIMITS**
Cultivating and empowering forward-thinking employees.	Creates internal advocacy and drives transformation initiatives.	Offer speaking opportunities, mentorship, and project leadership roles.	Limited resources for fostering internal thought leadership.
EXTERNAL INFLUENCER PARTNERSHIPS	**VISION ALIGNMENT**	**STRATEGIC ALLIANCES**	**INTEGRATION**
Collaborating with innovative industry leaders.	Brings fresh perspectives and validates internal ideas against external trends.	Form partnerships with external experts for workshops or advisory roles.	Effectively integrating external ideas without disrupting internal processes.

5. Innovation labs/hubs

For a business leader, establishing dedicated innovation labs or hubs is a critical strategic imperative for fostering experimentation, accelerating breakthrough ideas, and embedding a culture of proactive self-disruption within your organization. These dedicated environments provide the space and resources necessary to explore unconventional solutions away from daily operational pressures. Central to this is structuring dedicated innovation teams. These agile groups are empowered to test and prototype disruptive solutions rapidly, with a mandate to explore new technologies, business models, or experiences. Their focus on experimentation accelerates learning and validation.

Furthermore, fostering cross-disciplinary collaboration is crucial within these hubs. Encouraging diverse expertise—bringing together individuals from engineering, marketing, design, and even external partners—enhances ideation efforts, leading to more holistic and creative solutions that might not emerge from siloed teams. Crucially, implementing rapid prototyping and iteration frameworks reduces friction in experimentation. By creating a culture where failure is seen as a learning opportunity and quick iterations are encouraged, you accelerate the process of testing, refining, and scaling viable solutions, minimizing wasted resources on unproven concepts. Neglecting innovation labs or a dedicated focus on experimentation can lead to organizational inertia, a struggle to generate truly disruptive ideas, and a reactive approach to market changes. By championing these environments, you empower your organization to explore the unknown, incubate transformative ideas, and proactively drive the next wave of growth and relevance, ultimately securing your business's long-term competitive advantage through continuous innovation.

WHAT?	WHY?	HOW?	BEWARE!
DEDICATED INNOVATION TEAMS	**ACCELERATED EXPLORATION**	**AUTONOMOUS TEAMS**	**INTEGRATION**
Structuring agile groups for testing and prototyping.	Expedites testing and prototyping of disruptive solutions.	Form small, cross-functional teams with clear mandates for innovation.	Integrating successful innovations back into core business units.
CROSS-DISCIPLINARY COLLABORATION	**HOLISTIC SOLUTIONS**	**BRAINSTORMING SESSIONS**	**COMMUNICATION**
Encouraging diverse expertise.	Generates more comprehensive and creative ideas.	Organize workshops bringing together diverse skill sets and perspectives.	Bridging communication gaps between different disciplines.
RAPID PROTOTYPING AND ITERATION FRAMEWORKS	**EFFICIENT VALIDATION**	**AGILE TOOLS**	**RESOURCE ALLOCATION**
Reducing friction in experimentation.	Speeds up testing and refinement of new concepts.	Provide tools and processes for quick build-test-learn cycles.	Ensuring sufficient resources (time, budget) for rapid iteration.

For a business leader, strategically deploying rebranding strategies is a critical imperative for reinventing identity, repositioning for market resurgence, and adapting to significant market changes or evolving customer expectations. In a dynamic business landscape, a brand can become stale, misaligned, or simply lose relevance, necessitating a bold transformation to capture new opportunities. Central to this is managing brand narrative evolution. This involves refreshing your messaging and market perception to align with emerging audience expectations, crafting a new compelling story that resonates with current and future customers while honoring your core values.

Furthermore, meticulous market repositioning is crucial. This entails redefining your competitive stance and strategic objectives to differentiate your brand effectively in new or shifting landscapes. It's about consciously shaping how your brand is perceived relative to competitors in a renewed context, attracting new segments. Crucially, implementing comprehensive product and service reinvention is vital. Adapting core offerings to fit evolving consumer demands and industry trends ensures that your brand's promise is backed by tangible, relevant solutions. This might involve significant product overhauls, new service lines, or a complete reimagining of your value delivery. Neglecting a necessary rebranding can lead to brand erosion, loss of market share, and an inability to connect with new generations of customers. By championing a bold and thoughtful rebranding strategy, you empower your organization to revitalize its market presence, attract new audiences, and ultimately secure sustained relevance and growth for your business through a powerful act of strategic reinvention.

WHAT?	WHY?	HOW?	BEWARE!
BRAND NARRATIVE EVOLUTION	**RELEVANCE & CONNECTION**	**STORYTELLING WORKSHOPS**	**INCONSISTENCY**
Refreshing messaging and market perception.	Ensures the brand resonates with current and future audiences.	Develop new narratives and messaging aligned with strategic goals.	Maintaining a cohesive narrative across all touchpoints during a rebrand.
MARKET REPOSITIONING	**COMPETITIVE ADVANTAGE**	**POSITIONING STATEMENTS**	**PERCEPTION SHIFT**
Redefining competitive stance and strategic objectives.	Differentiates the brand in new or evolving markets.	Research market gaps and redefine competitive positioning.	Difficulty changing long-held customer perceptions of the brand.
PRODUCT AND SERVICE REINVENTION	**CUSTOMER ADOPTION**	**R&D INVESTMENT**	**RESOURCE INTENSITY**
Adapting core offerings.	Ensures offerings align with evolving consumer demands.	Invest in R&D to adapt or create new products/services.	Significant investment of time, money, and effort for reinvention.

Ai6x6™ Transformation Blueprint

AUTHORITY | Product Resilience

"Failure is simply the opportunity to begin again, this time more intelligently." - Henry Ford.

Founder of the Ford Motor Company and pioneer of the assembly line. He revolutionized the automobile industry by making cars affordable through efficient manufacturing, transforming transportation and impacting industrial manufacturing worldwide.

The Authority level is your Product Resilience, attained when the product ecosystem becomes a permanent fixture of the industry economy. Circular economy and resilience strategies ensure the product remains an enduring industry standard.

The inherent capacity of a product and its entire supporting ecosystem to effectively adapt, swiftly recover, and ultimately thrive amidst unpredictable market changes, technological shifts, or significant disruptions. This ensures sustained value delivery, maintains market presence, and allows the product to endure and succeed even in turbulent or rapidly evolving environments.

Building resilience into a product involves robust architecture, flexible design, and a continuous feedback loop that enables rapid iteration and adaptation to unforeseen challenges, ensuring its longevity and continued market viability.

The **"Product Resilience"** chapter is fundamental for business strategy maturity as it focuses on ensuring the long-term viability of products through continuous improvement and strategic foresight. A mature business strategy builds product resilience by embedding mechanisms for ongoing adaptation, security, and performance enhancement. This commitment ensures products can withstand market shifts, technological advancements, and competitive pressures. It's about sustaining the product's relevance and innovating for future growth, ensuring that the product portfolio consistently supports the organization's strategic objectives and maintains its competitive edge over time. Here is how we approach our learning:

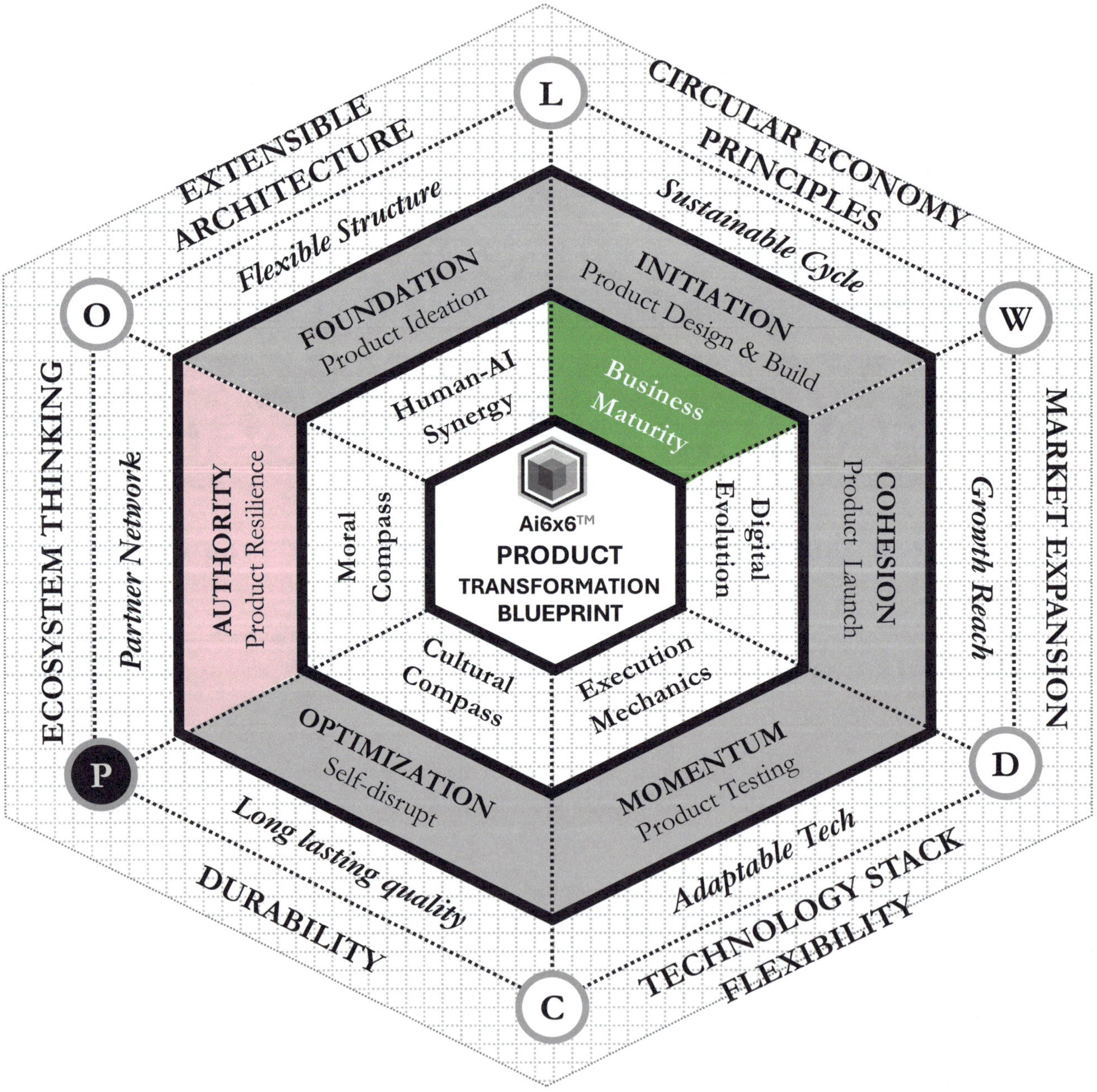

Ai6x6 BUSINESS MATURITY FOR RESILIENCE | PRODUCT TRANSFORMATION

1. Extensible architecture

For a business leader, championing an extensible architecture is a critical strategic imperative for designing adaptable systems that can evolve seamlessly with rapidly changing technological and business needs, ensuring long-term product resilience and competitive advantage. In today's dynamic digital landscape, rigid, monolithic systems are a liability; flexibility and future-proofing are paramount. Central to this is embracing a microservices framework. By enabling modular development where product components are built as independent, self-contained services, you allow for independent scalability, rapid updates, and isolated deployments, making your system inherently more resilient and easier to adapt without disrupting the entire product.

Furthermore, an API-first design approach is crucial. Prioritizing interoperability by designing every component with well-defined APIs ensures seamless integrations with third-party services, internal systems, and future functionalities. This standardization fosters flexibility and reduces integration friction. Crucially, building a plug-and-play infrastructure allows for effortless expansion and functionality enhancements without major overhauls. This modularity means new features or technologies can be added or swapped out with minimal disruption, accelerating innovation and reducing technical debt. Neglecting extensible architecture can lead to rigid systems that are slow to update, costly to maintain, and unable to adapt to new market demands or technological advancements. By championing this forward-thinking architectural approach, you empower your organization to build future-ready products, accelerate innovation cycles, and maintain agility, ultimately securing your business's long-term competitive edge and sustained success in an ever-evolving technological landscape.

WHAT?	WHY?	HOW?	BEWARE!
MICROSERVICES FRAMEWORK	**SCALABILITY & RESILIENCE**	**SERVICE DECOMPOSITION**	**COMPLEXITY**
Enabling modular development.	Allows independent scaling and rapid updates for components.	Break down monolithic applications into smaller, independent services.	Increased operational complexity in managing distributed microservices.
API-FIRST DESIGN	**SEAMLESS INTEGRATIONS**	**API STANDARDS**	**SECURITY**
Prioritizing interoperability.	Supports fluid interactions with internal and external services.	Design and publish well-documented APIs for all product components.	Ensuring robust API security and access control.
PLUG-AND-PLAY INFRASTRUCTURE	**AGILITY & FLEXIBILITY**	**MODULAR COMPONENTS**	**COMPATIBILITY**
Allowing effortless expansion.	Enables quick addition of new features without major overhauls.	Design for interchangeable components and clear interfaces.	Ensuring new modules are always compatible with the existing core.

2. Circular economy principles

For a business leader, integrating circular economy principles into your product strategy is a critical imperative for implementing sustainable practices, extending product lifecycle, and enhancing your brand's environmental stewardship. Beyond merely regulatory compliance, a circular approach drives innovation, reduces costs, and resonates with environmentally conscious consumers. Central to this is adopting resource-efficient production methods. This involves optimizing material usage, minimizing waste generation, and reducing energy consumption throughout the manufacturing process to minimize your environmental impact and enhance operational efficiency.

Furthermore, designing for reusability and refurbishment is crucial. By creating products with durable, modular components that can be easily repaired, reused, or refurbished, you extend their utility and reduce the need for new raw materials, contributing to a more sustainable consumption model. Crucially, establishing closed-loop systems involves mechanisms for reclaiming, repurposing, and significantly reducing waste at the end of a product's life. This might include take-back programs, recycling initiatives, or partnerships that ensure materials re-enter the production cycle, minimizing landfill waste. Neglecting circular economy principles can lead to increased material costs, reputational damage from unsustainable practices, and a failure to meet evolving consumer and regulatory expectations for environmental responsibility. By championing a circular approach, you empower your organization to reduce its ecological footprint, innovate in sustainable design, attract environmentally conscious customers, and ultimately secure your business's viability and brand leadership in a world increasingly valuing sustainability.

WHAT?	WHY?	HOW?	BEWARE!
RESOURCE-EFFICIENT PRODUCTION	**ENVIRONMENTAL IMPACT**	**LEAN MANUFACTURING**	**SUPPLY CHAIN COMPLEXITY**
Optimizing material usage.	Minimizes ecological footprint and enhances operational efficiency.	Implement lean processes to reduce waste and optimize material use.	Difficulty in sourcing sustainable materials globally.
REUSABILITY AND REFURBISHMENT	**EXTENDED UTILITY**	**MODULAR DESIGN**	**COST**
Designing products with recyclable components.	Extends product lifespan and reduces demand for new resources.	Design products with easily replaceable or upgradeable components.	Initial higher design and manufacturing costs for durable, repairable products.
CLOSED-LOOP SYSTEMS	**WASTE REDUCTION**	**RECYCLING PROGRAMS**	**LOGISTICS**
Establishing mechanisms for reclaiming and repurposing waste.	Minimizes landfill waste and promotes resource recovery.	Implement take-back programs and partnerships for material recycling.	Complexities in collecting, sorting, and processing returned products for reuse.

3. Market expansion

For a business leader, strategically pursuing market expansion is a critical imperative for ensuring product adaptability for growth, maximizing revenue, and significantly broadening your audience reach. In a competitive global landscape, continuous growth often requires looking beyond existing markets and customer segments. Your ability to scale and diversify is paramount. Central to this is mastering vertical scaling. This involves expanding product lines to cater to premium segments or niche markets within your existing industry. By offering specialized versions or higher-value propositions, you can capture more revenue from current markets and deepen your competitive moat.

Furthermore, meticulously planning horizontal scaling is crucial. This entails entering new geographic markets or diversifying offerings across different industries where your product's core value proposition can be applied. This broadens your customer base and reduces reliance on a single market. Crucially, demand-driven adaptations are vital for success in new territories. Refining product features, adjusting pricing strategies, and localizing messaging based on emerging consumer trends and cultural nuances in new markets ensures relevance and successful adoption. Neglecting market expansion can lead to stagnant growth, over-reliance on a single market, and a vulnerability to competitive saturation. By championing a thoughtful and adaptive market expansion strategy, you empower your organization to unlock new revenue streams, reach diverse audiences, and ultimately secure the sustained growth and long-term competitive advantage of your product in an ever-widening global arena.

WHAT?	WHY?	HOW?	BEWARE!
VERTICAL SCALING	**REVENUE MAXIMIZATION**	**PREMIUM OFFERINGS**	**MARKET ACCEPTANCE**
Expanding product lines to cater to premium segments.	Captures higher value from existing market segments.	Develop high-end versions or specialized products for niche segments.	Ensuring premium offerings are perceived as valuable.
HORIZONTAL SCALING	**AUDIENCE REACH**	**INTERNATIONALIZATION**	**CULTURAL DIFFERENCES**
Entering new geographic markets or diversifying offerings.	Broadens customer base and diversifies revenue streams.	Localize products, marketing, and distribution for new regions.	Adapting to diverse cultural norms and consumer behaviors.
DEMAND-DRIVEN ADAPTATIONS	**MARKET RELEVANCE**	**LOCALIZED RESEARCH**	**RESOURCE INTENSITY**
Refining features and pricing based on trends.	Ensures product resonates with new market demands.	Conduct market research in new territories to inform adaptations.	Significant investment for product localization and marketing.

4. Technology stack flexibility

For a business leader, actively ensuring technology stack flexibility is a critical strategic imperative for guaranteeing product adaptability, reducing technical debt, and fostering long-term resilience against technological shifts. In a rapidly evolving digital landscape, proprietary lock-in and rigid architectures are liabilities; agility and interoperability are paramount. Central to this is embracing vendor neutrality. Avoiding reliance on single-vendor proprietary solutions maintains long-term technological independence, allowing you to choose the best tools for each specific need, preventing costly migrations, and ensuring your product can integrate with a wider ecosystem.

Furthermore, adopting containerization strategies is crucial. Deploying modular instances (containers) of your applications improves portability across different environments (cloud, on-premise) and enhances scalability, making your product easier to deploy, manage, and update without conflicts. Crucially, leveraging open-source collaboration provides access to community-driven innovations for continuous enhancements. Contributing to and utilizing open-source projects allows you to benefit from collective intelligence, accelerate development, and reduce licensing costs, ensuring your product's technology stack remains cutting-edge and robust. Neglecting technology stack flexibility can lead to high operational costs, slow development cycles, an inability to adapt to new technologies, and a significant competitive disadvantage. By championing open-source, vendor-agnostic, and containerized solutions, you empower your organization to build highly adaptable, cost-efficient, and future-proof products, accelerating innovation and ultimately securing your business's long-term resilience and market leadership in a dynamic tech landscape.

WHAT?	WHY?	HOW?	BEWARE!
VENDOR NEUTRALITY	**TECHNOLOGICAL INDEPENDENCE**	**OPEN STANDARDS**	**SUPPORT**
Avoiding proprietary lock-in.	Enables flexibility and reduces reliance on single vendors.	Prioritize open-source tools and technologies with open APIs.	Potential lack of dedicated enterprise support compared to proprietary solutions.
CONTAINERIZATION STRATEGIES	**IMPROVED PORTABILITY**	**DOCKER/KUBERNETES**	**COMPLEXITY**
Deploying modular instances for portability and scalability.	Facilitates easy deployment across different environments.	Package applications into containers and use orchestration tools.	Managing and orchestrating a large number of containers.
OPEN-SOURCE COLLABORATION	**CONTINUOUS ENHANCEMENTS**	**COMMUNITY ENGAGEMENT**	**SECURITY**
Leveraging community-driven innovations.	Benefits from global developer community contributions.	Contribute to and utilize relevant open-source projects.	Ensuring the security and reliability of open-source components.

5. Durability

For a business leader, consciously enhancing product durability is a critical strategic imperative for extending product longevity, reducing environmental impact, and fostering customer trust. In an era of rapid consumption, designing for durability differentiates your brand and aligns with growing consumer demand for sustainable, long-lasting products. Central to this is integrating repairability into product design. Developing products with accessible, standardized, and interchangeable components ensures they can be easily repaired by users or service providers, reducing the need for costly replacements and extending their useful life.

Furthermore, ensuring reverse-compatibility is crucial. Designing newer versions of your product to remain functional and interoperable with legacy systems and technologies minimizes the need for forced upgrades, provides value to long-term customers, and reduces electronic waste. Crucially, taking responsibility for end-of-life practices is vital. Establishing comprehensive recycling programs and secure disposal strategies minimizes environmental harm, ensuring that products are either reclaimed for materials or responsibly decommissioned when they truly reach the end of their lifecycle. Neglecting product durability can lead to frustrated customers due to planned obsolescence, increased waste, reputational damage, and a failure to meet growing environmental expectations. By championing a design philosophy focused on longevity, repairability, and responsible disposal, you empower your organization to build high-quality, sustainable products, enhance customer satisfaction, and secure your business's long-term brand reputation and competitive advantage through genuine environmental stewardship.

WHAT?	WHY?	HOW?	BEWARE!
REPAIRABILITY INTEGRATION	**EXTENDED LIFESPAN**	**MODULAR DESIGN**	**COST**
Developing products with accessible components.	Reduces waste and offers cost savings for customers.	Design products with easily replaceable and standard components.	Higher initial manufacturing costs for repairable designs.
REVERSE-COMPATIBILITY	**CUSTOMER VALUE**	**API/PROTOCOL DESIGN**	**DEVELOPMENT COMPLEXITY**
Ensuring newer versions function with legacy systems.	Extends product utility for existing users, reducing upgrade pressure.	Maintain backward compatibility in software updates and hardware interfaces.	Increased complexity in maintaining compatibility with older systems.
END-OF-LIFE RESPONSIBILITY	**ENVIRONMENTAL STEWARDSHIP**	**RECYCLING PROGRAMS**	**COST & LOGISTICS**
Establishing recycling programs and disposal strategies.	Minimizes environmental harm and enhances brand reputation.	Implement take-back schemes and partnerships for product recycling.	High costs and logistical challenges of implementing recycling programs.

6. Ecosystem thinking

For a business leader, embracing ecosystem thinking is a critical strategic imperative for fostering product resilience through robust partnerships, industry standardization, and seamless interoperability. In today's interconnected digital landscape, no product exists in a vacuum; its strength is increasingly defined by its ability to integrate and thrive within a broader network of complementary technologies, services, and partners. Central to this is rigorous industry standards alignment. Complying with widely accepted regulations and technical frameworks (e.g., API standards, data formats, communication protocols) ensures that your product can seamlessly connect and communicate with other components in the ecosystem, reducing friction and enhancing market adoption.

Furthermore, designing for cross-platform compatibility is crucial. Creating solutions that function seamlessly across different environments—operating systems, cloud providers, device types—maximizes your product's reach and utility, preventing fragmentation and enhancing user experience. Crucially, cultivating strategic partnerships is vital for shared innovation and stability. Collaborating with other ecosystem stakeholders—from technology providers to service integrators and distribution partners—allows you to co-create value, leverage complementary strengths, and build a more robust and resilient offering that meets diverse customer needs. Neglecting ecosystem thinking can lead to isolated products, limited market reach, increased integration challenges, and a vulnerability to more integrated competitors. By championing partnerships, standardization, and interoperability, you empower your organization to build products that are resilient, widely accessible, and future-proof, ultimately securing your business's long-term competitive advantage and sustained success within a collaborative digital world.

WHAT?	WHY?	HOW?	BEWARE!
INDUSTRY STANDARDS ALIGNMENT	**SEAMLESS INTEGRATION**	**STANDARD ADOPTION**	**EVOLVING STANDAGES**
Complying with widely accepted regulations.	Facilitates easier integration with other ecosystem components.	Adhere to open standards (e.g., OpenAPI, OAuth) in product development.	Keeping up with rapidly evolving industry standards.
CROSS-PLATFORM COMPATIBILITY	**MAXIMIZED REACH**	**PLATFORM-AGNOSTIC DEVELOPMENT**	**TESTING COMPLEXITY**
Designing solutions that function seamlessly across environments.	Ensures product accessibility across diverse operating systems and devices.	Build products using frameworks that support multiple platforms.	Extensive testing required across numerous platforms and devices.
STRATEGIC PARTNERSHIPS	**SHARED INNOVATION**	**ECOSYSTEM MAPPING**	**PARTNER ALIGNMENT**
Collaborating with ecosystem stakeholders.	Leverages external strengths for product development and market reach.	Identify key partners (tech providers, integrators) for collaboration.	Ensuring clear objectives and trust with external partners.

BLUEPRINT SERIES

SECTION D

Digital Evolution

Ai6x6™ Transformation Blueprint

Story 1 — The Birth of COBOL and Business Computing (1959–1965): Early enterprises struggled to translate business logic into machine-readable form, limiting the usefulness of emerging computers. COBOL introduced a standardized language that mirrored human reasoning, enabling organizations to encode processes consistently. This shift allowed systems to scale beyond isolated experiments. Human analysts defined rules while machines executed them with precision. The partnership created the first bridge between business intent and digital execution.

Story 2 — ARPANET and Packet Switching (1969–1972): Communication networks were fragile, centralized, and prone to failure. Packet switching introduced a decentralized architecture that allowed information to flow even when parts of the network failed. Humans designed protocols while machines routed data autonomously. This resilience transformed digital communication from a technical novelty into a strategic infrastructure. ARPANET demonstrated how distributed systems could scale without collapsing. It laid the foundation for the modern internet.

Story 3 — Xerox PARC and the Graphical User Interface (1973–1983): Computers were powerful but inaccessible to most users, requiring specialized knowledge to operate. Xerox PARC introduced the graphical user interface, enabling humans to interact with machines visually rather than through code. This shift democratized computing and redefined usability. Machines handled complexity behind the scenes while humans engaged through intuitive metaphors. The interface became a catalyst for mass digital adoption. It showed that digital evolution accelerates when technology adapts to human cognition.

Story 4 — The Rise of Enterprise Resource Planning (1980s–1990s): Organizations operated in silos, with fragmented systems and inconsistent data. ERP platforms unified finance, operations, and supply chains into integrated digital environments. Machines enforced standardization while humans redesigned processes around shared information. This integration improved visibility, coordination, and control. ERP became the backbone of global enterprises. It demonstrated how digital evolution transforms not just tools but entire operating models.

Story 5 — The Commercial Internet and E-Commerce (1994–2005): The internet shifted from academic infrastructure to commercial ecosystem, enabling new forms of interaction, distribution, and value creation. Websites became digital storefronts, and payment systems matured into secure transactional platforms. Machines automated logistics while humans reimagined customer experience. The digital marketplace expanded global reach and lowered barriers to entry. It marked a turning point where digital capability became essential for competitiveness.

Story 6 — Cloud Computing and Elastic Infrastructure (2006–2015): Organizations struggled with rigid, capital-intensive infrastructure that limited experimentation and scale. Cloud computing introduced elastic resources that could expand or contract instantly. Machines automated provisioning while humans focused on architecture and innovation. This shift accelerated deployment cycles, reduced cost, and enabled global reach. Cloud became the foundation for modern digital ecosystems. It demonstrated how digital evolution thrives when infrastructure becomes fluid, on-demand, and programmable.

LESSONS FROM A PRODUCT TRANSFORMATION PERSPECTIVE

Products evolve from static artifacts to dynamic digital systems shaped by integration, usability, and continuous iteration. Early digital products encoded business logic, while later innovations introduced intuitive interfaces, interconnected platforms, and cloud-based adaptability. As digital capability expanded, products became more modular, scalable, and responsive to user behavior. Machines enabled rapid testing and deployment, while humans guided design, ethics, and experience. Product maturity emerges when offerings can evolve without losing coherence—improving through feedback, data, and shared architectures. Digital evolution transforms products into living systems that grow alongside customer needs and organizational capability.

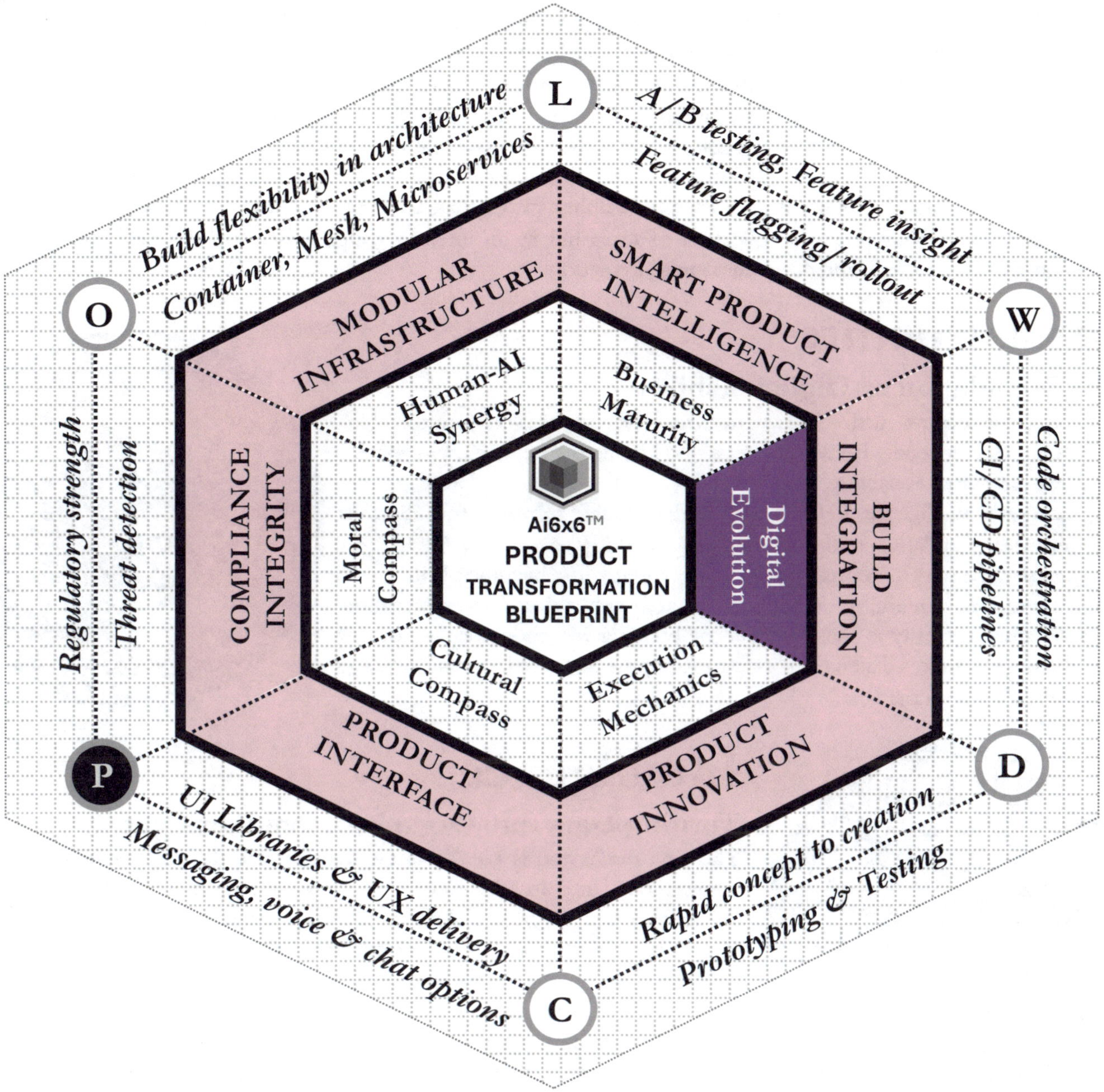

The **Ai6x6 Research Lab** chose this lens to solve the "personality gap" that often leaves users feeling disconnected from the technology they use. We believe a product must possess **Digital Gravitas**—the quiet confidence of being "Modular by Nature" yet singular in purpose. This isn't about the complexity of the code; it is about the mental shift toward creating "The Thinking Product" that communicates its intent clearly. It is vital because a leader cannot "Inspire" if the product's interface is a source of friction rather than a "Hello" of belonging.

By mastering this evolution, the leader moves from a state of technical anxiety to a state of absolute strategic command over the product's digital soul. This lens ensures that the **Synced to Create** integration is built on a foundation of **Integrity**, where every line of code is "Built with Purpose." When a leader understands the "Interface" of their creation, they gain the freedom to lead with an undistracted focus on **Invention**. This lens provides the psychological peace of mind that comes from knowing the product's backbone is secure, reliable, and designed for human synergy.

Digital Evolution is no longer a technical initiative. It is the structural rewiring of how an organization thinks, operates, and competes. For years, leaders treated digital transformation as a project—something with a start date, a roadmap, and a finish line. But the world has moved beyond that framing. Digital Evolution is not a destination; it is a permanent state of adaptation. It is the recognition that technology is no longer an enabler of strategy but a **shaper** of it. The organizations that thrive are those that treat digital capability as a living system—continuously upgraded, continuously integrated, and continuously aligned with the realities of the market. This lens reframes Digital Evolution as a leadership discipline rather than a technology program. It challenges leaders to see digital capability not as a collection of tools but as an interconnected ecosystem that influences every decision, every workflow, and every customer interaction.

Infrastructure

The Foundation of Digital Stability

The journey begins with **Infrastructure**, the foundation upon which everything else depends. Infrastructure is not just hardware or cloud capacity; it is the stability, scalability, and reliability of the systems that carry the organization's intelligence. Leaders who underestimate infrastructure discover that innovation collapses under its own weight. Without a strong foundation, every new capability becomes fragile, every integration becomes brittle, and every attempt at modernization becomes a patch rather than a transformation. Infrastructure maturity is the discipline of building for endurance—ensuring that the organization's digital backbone can support the speed, complexity, and volatility of modern markets.

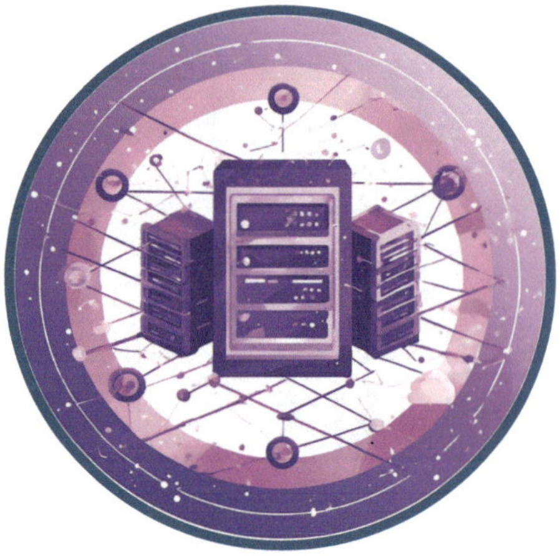

Intelligence

Turning Data Into Organizational Foresight

The next dimension is **Intelligence**, representing the organization's ability to convert data into insight and insight into action. Intelligence is not about dashboards or reports; it is about building a system that learns. It requires data that flows freely, models that are trained responsibly, and decisions that are informed by evidence rather than assumption. Leaders who embrace intelligence understand that it is not a technical asset but a strategic one. Intelligence shapes priorities, exposes risks, and reveals opportunities that would otherwise remain invisible. It is the cognitive engine of Digital Evolution—the mechanism through which the organization perceives, interprets, and responds to its environment.

Integration

Making the Enterprise Whole

Integration is where Digital Evolution becomes real. Organizations often accumulate technologies faster than they can connect them. Systems become islands, data becomes siloed, and teams become fragmented. Integration is the discipline that ensures the enterprise operates as a coherent whole. It is the work of connecting platforms, harmonizing data, and designing workflows that move across boundaries. Without integration, digital investments become isolated experiments rather than enterprise capabilities. Integration maturity is the art of coherence—ensuring that the organization's digital landscape functions as a single, unified system rather than a patchwork of disconnected tools.

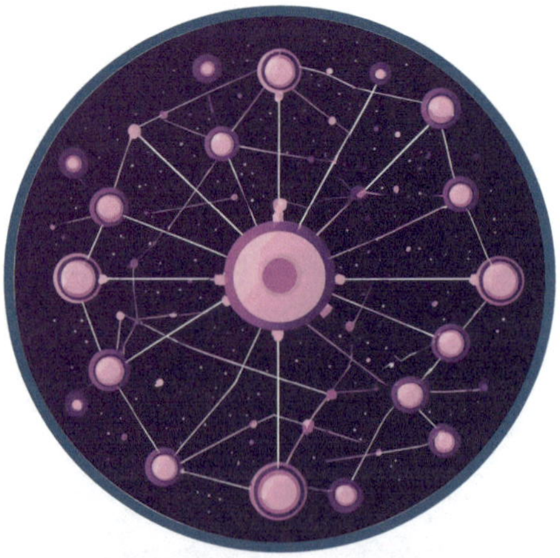

Innovation

Expanding the Field of Possibility

The fourth dimension, **Innovation**, expands the field of possibility. Innovation is not about chasing trends or adopting technology for its own sake. It is about using digital capability to reimagine how value is created, delivered, and experienced. Innovation demands that leaders challenge assumptions, explore alternatives, and test ideas quickly. It requires a culture where experimentation is encouraged, failure is reframed as learning, and progress is measured by insight rather than perfection. Innovation maturity is the discipline of curiosity—using digital tools not just to optimize the present but to invent the future.

Interface

Where Technology Meets Human Behavior

Interface represents the human experience of Digital Evolution. It is the point where technology meets behavior, where systems meet expectations, and where design meets trust. Leaders who ignore interface discover that even the most powerful systems fail if people cannot or will not use them. Interface is not cosmetic; it is strategic. It shapes adoption, accelerates learning, and determines whether digital capability becomes embedded or abandoned. Interface maturity is the discipline of empathy—designing digital experiences that are intuitive, accessible, and aligned with human needs.

Integrity

The Guardrail of Responsible Evolution

The final dimension, **Integrity**, grounds Digital Evolution in responsibility. Integrity ensures that systems are secure, data is protected, and decisions are ethical. It prevents speed from becoming recklessness and innovation from becoming harm. Leaders who prioritize integrity build organizations that are trusted, resilient, and prepared for the scrutiny that accompanies digital capability. Integrity maturity is the discipline of stewardship—ensuring that digital evolution strengthens the organization's moral and operational foundations rather than compromising them.

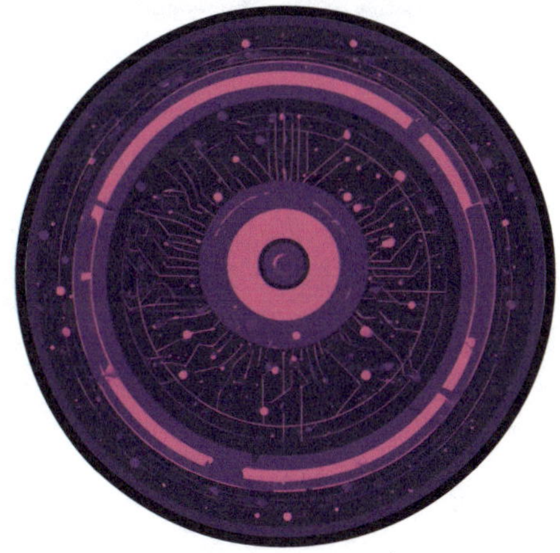

Digital Evolution is not a technology story. It is a leadership story. It is the recognition that the future belongs to organizations that treat digital capability as a core competency, not a side project. It is the shift from building tools to building systems, from adopting technology to evolving with it, and from reacting to change to shaping it. In this future, Digital Evolution is not a department, a project, or a budget line. It is the operating philosophy of the enterprise. It shapes how decisions are made, how systems are built, how teams collaborate, and how value is created. It turns technology from a source of disruption into a source of stability. It transforms the organization from a follower of change into a shaper of it.

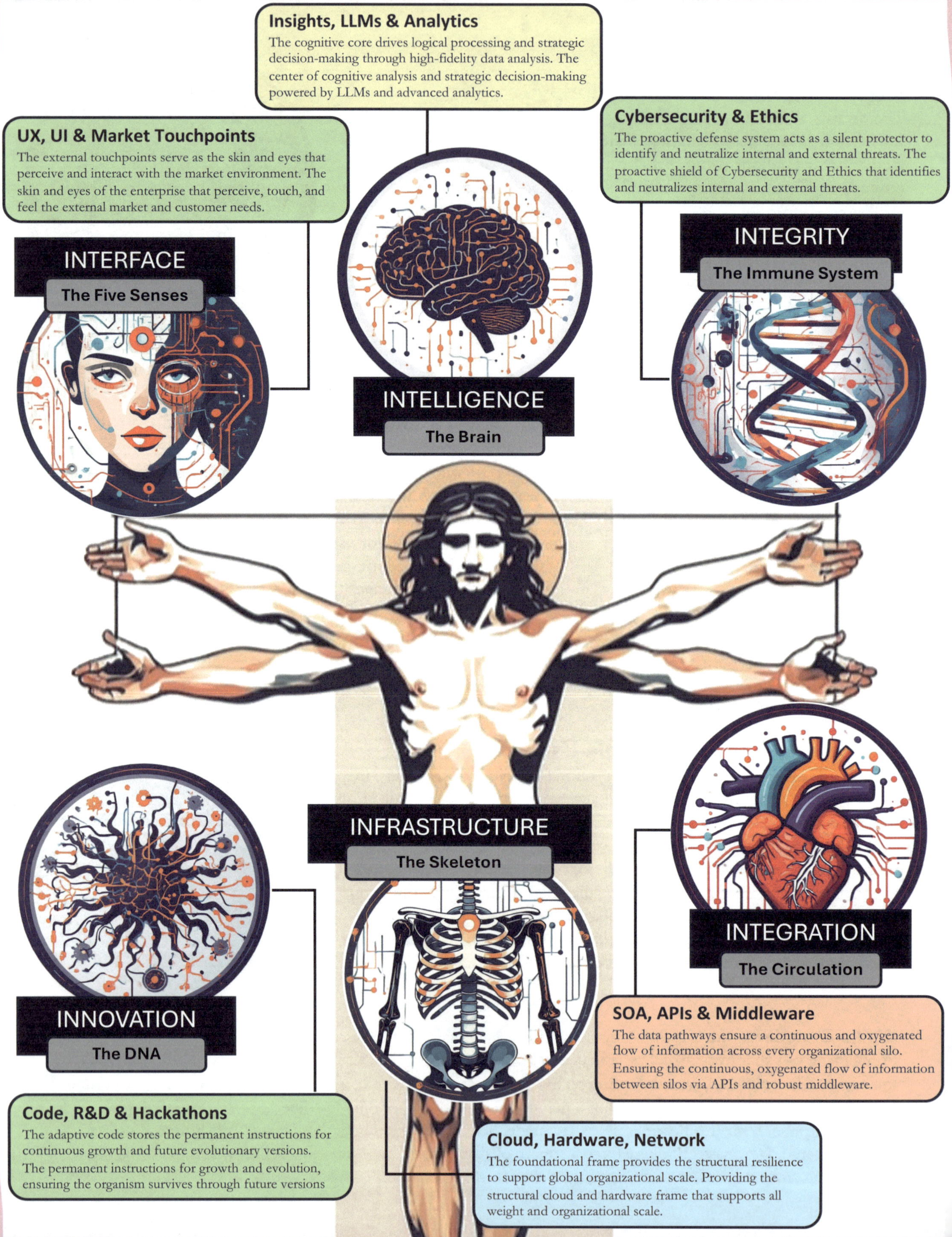
Insights, LLMs & Analytics
The cognitive core drives logical processing and strategic decision-making through high-fidelity data analysis. The center of cognitive analysis and strategic decision-making powered by LLMs and advanced analytics.
UX, UI & Market Touchpoints
The external touchpoints serve as the skin and eyes that perceive and interact with the market environment. The skin and eyes of the enterprise that perceive, touch, and feel the external market and customer needs.
Cybersecurity & Ethics
The proactive defense system acts as a silent protector to identify and neutralize internal and external threats. The proactive shield of Cybersecurity and Ethics that identifies and neutralizes internal and external threats.
INTERFACE
The Five Senses
INTELLIGENCE
The Brain
INTEGRITY
The Immune System
INFRASTRUCTURE
The Skeleton
INTEGRATION
The Circulation
INNOVATION
The DNA
SOA, APIs & Middleware
The data pathways ensure a continuous and oxygenated flow of information across every organizational silo. Ensuring the continuous, oxygenated flow of information between silos via APIs and robust middleware.
Code, R&D & Hackathons
The adaptive code stores the permanent instructions for continuous growth and future evolutionary versions. The permanent instructions for growth and evolution, ensuring the organism survives through future versions
Cloud, Hardware, Network
The foundational frame provides the structural resilience to support global organizational scale. Providing the structural cloud and hardware frame that supports all weight and organizational scale.

Ai6x6™ Transformation Blueprint

Modular 'Infrastructure'

"You should always build with change in mind." — Martin Fowler

Software development leader and author; key voice in modular architecture and refactoring strategy.

Modularity isn't a technical choice—it's a transformation commitment. PLM systems, microservices, and container orchestration signal maturity where product adaptability becomes a native trait. "Modular Infrastructure" means not bolting on flexibility—it means embedding it from inception. In transformation, change doesn't disrupt—it reroutes. Systems built for composability absorb market shifts without architecture fatigue. Enterprises with infrastructure agility accelerate innovation across cycles—not because they're lucky, but because they're structured to stretch. Modularity is not about simplicity—it's about sustaining strategy through volatility.

For product transformation, the "Modular Infrastructure" chapter is foundational for digital maturity because it lays the architectural groundwork for building agile, scalable, and resilient digital products. A digitally mature organization adopts modular approaches like microservices and container orchestration, coupled with robust PLM systems and DevOps toolchains. This infrastructure enables rapid development, continuous delivery, and efficient management of product lifecycles. It allows the organization to quickly adapt to market demands, iterate on features, and deliver innovative solutions with speed and quality, ensuring products remain competitive and future-proof in the digital age.

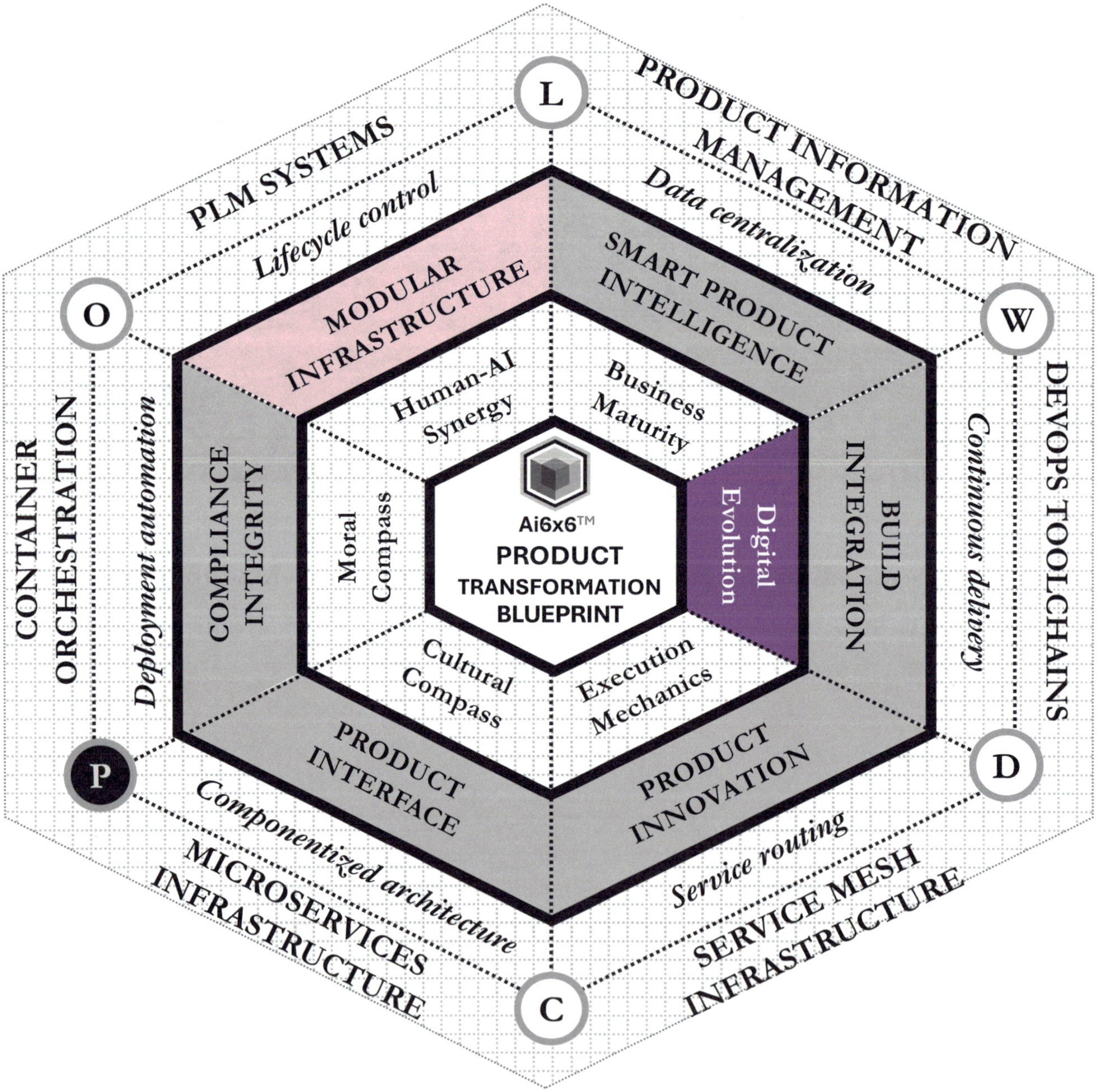

Ai6x6 DIGITAL EVOLUTION FOR INFRASTRUCTURE | PRODUCT TRANSFORMATION

1. PLM SYSTEMS

Product Lifecycle Management (PLM) Systems are comprehensive software solutions that manage a product's entire lifecycle from conception, design, and manufacturing to service and disposal. They centralize product data, facilitate collaboration among cross-functional teams, and manage workflows related to design iterations, bill of materials (BOM), and regulatory compliance. By providing a single source of truth for all product-related information, PLM systems streamline product development processes, reduce time-to-market, improve product quality, and ensure compliance with industry standards. This holistic approach supports innovation, minimizes errors, and optimizes resource utilization across the product's lifespan, crucial for complex product development in industries like manufacturing, aerospace, and consumer goods.

Without PLM Systems, organizations often suffer from fragmented product data spread across disparate systems, leading to inefficiencies, errors, and significant delays in product development. Design changes might not propagate consistently, bill of materials can be inaccurate, and collaboration among engineering, manufacturing, and supply chain teams becomes chaotic. The absence of a centralized PLM solution results in higher development costs, increased risks of product defects, challenges in meeting compliance requirements, and slower time-to-market, ultimately hindering innovation and competitiveness by making product management reactive and prone to costly mistakes.

Implementation	Challenges
* Define product data model & workflows * Select appropriate PLM software * Migrate existing product data * Integrate with CAD/ERP/CRM systems * Train engineering & cross-functional teams * Plan for continuous data governance	* Complex data migration from legacy systems * Integrating with diverse engineering tools * User adoption & change management * Data security & intellectual property protection * Customization complexity & maintenance * Measuring clear ROI on process improvements

Products	Platform Cost	Project Cost	Duration
PTC Windchill	$20-100K	$150-700K	6-18 m
Siemens Teamcenter	$25-120K	$200-900K	8-24 m
Dassault Systèmes ENOVIA	$20-100K	$150-700K	6-18 m
Autodesk Fusion 360 Manage	$5-30K	$30-150K	3-9 m
Aras Innovator (Open Source Core)	$10-50K	$60-300K	4-12 m
Oracle Agile PLM	$15-80K	$100-500K	5-16 m

Product Information Management (PIM) systems are centralized platforms that collect, manage, and enrich all product-related data from various sources, ensuring accuracy, consistency, and completeness across all sales and marketing channels. This includes attributes like product descriptions, specifications, images, videos, pricing, and localized content. By acting as a single source of truth for product data, PIM systems enable organizations to deliver rich, consistent, and accurate product content to e-commerce websites, mobile apps, print catalogs, and marketplaces. This improves customer experience, accelerates time-to-market for new products, reduces manual errors, and supports global expansion by efficiently managing localized product information at scale.

Without a dedicated PIM system, organizations often face fragmented and inconsistent product data spread across spreadsheets, disparate databases, and various departmental silos. This leads to manual data entry errors, outdated product information, and significant delays in updating content across multiple sales channels. The absence of a single source of truth for product information results in poor customer experiences due to inaccurate listings, increased support queries, higher product return rates, and slowed market entry for new products, ultimately hindering sales, scalability, and brand reputation in a multi-channel retail environment.

Implementation	Challenges
* Audit existing product data sources * Select a PIM platform * Define data model & attributes * Migrate & cleanse product data * Integrate with e-commerce, ERP & DAM * Establish data governance & enrichment workflows	* Complex data migration & cleansing * Integrating with diverse legacy systems * User adoption & data ownership issues * Managing product data variations (SKUs, locales) * Ensuring data quality & consistency * Measuring clear ROI on content accuracy

Products	Platform Cost	Project Cost	Duration
Akeneo PIM	$5-30K	$30-150K	3-9 m
Salsify	$10-50K	$60-300K	4-12 m
Contentserv	$10-40K	$50-250K	3-10 m
PIMcore (Open Source Core)	$2-15K	$10-70K	2-6 m
Informatica Product 360	$15-70K	$80-400K	4-12 m
Stibo Systems STEP PIM	$20-80K	$100-500K	5-16 m

DevOps Toolchains are integrated sets of tools that automate and streamline the various phases of software development and operations, encompassing planning, coding, building, testing, releasing, deploying, operating, and monitoring. These toolchains facilitate continuous integration (CI), continuous delivery (CD), and continuous monitoring (CM), fostering a culture of collaboration, automation, and rapid feedback loops between development and operations teams. By automating repetitive tasks, improving communication, and enabling faster, more reliable software releases, DevOps toolchains significantly reduce lead time from commit to deploy, decrease deployment failures, and enhance system stability. This accelerates innovation, improves product quality, and ensures faster delivery of value to customers through efficient software lifecycle management.

Without a well-defined DevOps Toolchain, organizations experience significant friction and inefficiencies in their software development and deployment processes. Manual handoffs between development, testing, and operations teams lead to errors, delays, and inconsistent deployments. The absence of automation in builds, tests, and releases results in slower time-to-market for new features, higher rates of production defects, and increased operational overhead. This ultimately hinders the ability to respond rapidly to market demands, deliver continuous value to customers, and maintain a competitive edge through agile and reliable software delivery.

Implementation	Challenges
* Define SDLC stages & desired automation * Select & integrate tools (Git, Jenkins, Docker) * Establish CI/CD pipelines * Implement infrastructure as code * Set up monitoring & alerting * Foster a DevOps culture & training	* Tool sprawl & complex integrations * Ensuring security throughout the pipeline * Managing infrastructure as code at scale * Overcoming organizational silos & resistance * Maintaining pipeline stability & performance * Measuring true ROI on automation & efficiency

Products	Platform Cost	Project Cost	Duration
GitLab (Complete DevOps Platform)	$0.5-10K	$5-50K	1-4 m
Azure DevOps	$0.5-8K	$5-40K	1-3 m
Jenkins (Open Source)	N/A	$5-30K	1-3 m
Harness (CD Platform)	$2-15K	$10-70K	1-4 m
CircleCI	$0.5-5K	$2-20K	1-2 m
Atlassian Jira/Bitbucket/Confluence	$0.2-5K	$1-20K	0.5-2 m

Service Mesh Infrastructure is a dedicated infrastructure layer that handles service-to-service communication within a microservices architecture. It typically uses a "sidecar proxy" deployed alongside each service instance, intercepting all network traffic. This layer provides capabilities like traffic management (routing, load balancing), observability (metrics, tracing, logging), and security (encryption, authentication, authorization) without requiring changes to the application code itself. By abstracting network concerns from application logic, Service Mesh Infrastructure simplifies the development and operation of complex distributed systems, improves reliability, enhances security, and provides deep insights into service interactions, crucial for scalable and resilient product delivery in cloud-native environments.

Without Service Mesh Infrastructure, managing service-to-service communication in a microservices environment becomes highly complex and prone to errors. Developers must embed network resilience, observability, and security logic directly into each application, leading to duplicated effort, inconsistent implementations, and increased code complexity. The absence of a dedicated communication layer makes it difficult to diagnose network issues, enforce consistent policies, and implement advanced traffic management strategies, ultimately hindering the scalability, reliability, and security of distributed applications and slowing down product innovation.

Implementation	Challenges
* Define microservices architecture & needs	* Significant operational complexity & learning curve
* Select a service mesh (e.g., Istio, Linkerd)	* Performance overhead of sidecar proxies
* Deploy sidecar proxies alongside services	* Integrating with existing application codebases
* Configure traffic management rules	* Debugging issues across services & mesh
* Implement observability (metrics, tracing)	* Managing configuration & policy updates
* Plan for security policies & access control	* Ensuring consistent security policies at scale

Products	Platform Cost	Project Cost	Duration
Istio (Open Source)	N/A	$20-100K	3-9 m
Linkerd (Open Source)	N/A	$15-70K	2-8 m
AWS App Mesh	$0.1-5K	$5-30K	1-3 m
Consul Connect (HashiCorp)	$1-10K	$10-50K	2-4 m
Kong Konnect (API Gateway & Mesh)	$5-30K	$20-100K	2-6 m
Red Hat OpenShift Service Mesh	$10-50K	$60-250K	3-9 m

Microservices Infrastructure refers to the foundational set of technologies and practices that support the development, deployment, and management of applications composed of small, independent, and loosely coupled services. This includes elements like containerization (e.g., Docker), container orchestration (e.g., Kubernetes), API gateways, and specialized databases for each service. By enabling teams to develop, deploy, and scale services independently, Microservices Infrastructure promotes agility, resilience, and modularity in software development. This architectural approach allows for faster innovation cycles, easier maintenance, and enhanced scalability, crucial for building complex, evolving digital products that can quickly adapt to changing market demands.

Without Microservices Infrastructure, organizations often rely on monolithic application architectures, which become difficult to scale, update, and maintain as they grow in complexity. Changes in one part of the application can necessitate redeploying the entire system, leading to slow release cycles, increased risk of outages, and challenges in adopting new technologies. The absence of a modular infrastructure hinders agility, makes it difficult to isolate failures, and prevents independent scaling of components, ultimately impacting the speed of product delivery, system reliability, and the ability to innovate rapidly in response to market needs.

Implementation	Challenges
* Define service boundaries & APIs * Adopt containerization (Docker) * Implement container orchestration (Kubernetes) * Set up API gateways & service discovery * Design distributed data management * Plan for distributed logging & monitoring	* High initial setup complexity & learning curve * Managing distributed data consistency * Increased operational overhead * Debugging issues across multiple services * Ensuring robust inter-service communication * Organizational shift required for development teams

Products	Platform Cost	Project Cost	Duration
Docker (Containerization)	$0-5K	$2-20K	1-2 m
Kubernetes (Container Orchestration)	N/A	$20-100K	3-9 m
AWS ECS / EKS	$0.5-10K	$5-50K	1-4 m
Google Kubernetes Engine (GKE)	$0.5-10K	$5-50K	1-4 m
Azure Kubernetes Service (AKS)	$0.5-10K	$5-50K	1-4 m
HashiCorp Nomad (Scheduler)	$1-10K	$5-30K	1-3 m

Container Orchestration is the automated management of containerized applications, including tasks such as deployment, scaling, networking, and load balancing. Tools like Kubernetes, Docker Swarm, and Apache Mesos automate the process of deploying, scaling, and managing containers across clusters of hosts. They ensure that containers are running efficiently, are highly available, and can communicate with each other, abstracting away the underlying infrastructure. By automating the complexities of managing distributed containerized applications, Container Orchestration significantly improves operational efficiency, enhances application scalability and resilience, and enables rapid and consistent deployments. This is a cornerstone for modern cloud-native product development, allowing organizations to deliver new features and updates with unprecedented speed and reliability.

Without Container Orchestration, managing containerized applications at scale becomes a highly manual, complex, and error-prone process. Deploying, scaling, and networking individual containers across multiple hosts would require significant human effort, leading to inconsistencies, operational bottlenecks, and increased risk of downtime. The absence of automated orchestration hinders the ability to achieve high availability, rapid scaling, and efficient resource utilization in a containerized environment, ultimately slowing down product development cycles, impacting application reliability, and preventing organizations from fully leveraging the benefits of container technology for modern product delivery.

Implementation	Challenges
* Containerize applications (Docker) * Select an orchestration tool (e.g., Kubernetes) * Design container deployment strategies * Implement networking & service discovery * Configure persistent storage for stateful apps * Plan for logging, monitoring & security	* High learning curve & operational complexity * Resource management & cost optimization * Networking configuration for complex deployments * Managing persistent storage for stateful workloads * Security considerations in a distributed environment * Ensuring high availability & disaster recovery

Products	Platform Cost	Project Cost	Duration
Kubernetes (Open Source)	N/A	$20-100K	3-9 m
Amazon EKS (Managed Kubernetes)	$0.5-10K	$5-50K	1-4 m
Google Kubernetes Engine (GKE)	$0.5-10K	$5-50K	1-4 m
Azure Kubernetes Service (AKS)	$0.5-10K	$5-50K	1-4 m
Docker Swarm (Built into Docker)	N/A	$2-15K	1-2 m
Red Hat OpenShift	$10-60K	$50-250K	3-9 m

CHAPTER
D2

Ai6x6™ Transformation Blueprint

Product 'Intelligence'

"Build products that tell you what works." — Leah Culver

Co-founder of Breaker and respected in the developer community for creating data-driven user experiences.

Products that think don't guess—they observe. Usage analytics, heatmaps, and A/B testing reflect a maturity where insight isn't retrospective—it's real-time. "Product Intelligence" builds intelligence into its own interface, behavior, and flow. Crash reports and user session recordings speak strategy before customers do. Transformation emerges when feedback loops are systematized, not solicited. Intelligent products detect friction, measure delight, and evolve autonomously. Teams don't just iterate—they listen passively and act deliberately. Intelligence isn't added—it's architected into every tap, toggle, and touchpoint.

The "Product Intelligence" chapter is crucial for product transformation's digital maturity as it focuses on embedding data-driven insights directly into product development and optimization. A digitally mature organization leverages advanced analytics, A/B testing tools, and usage analytics platforms to deeply understand product usage, identify pain points, and predict user behavior. This intelligence, complemented by crash analytics and user session recording, ensures that product decisions are informed by real-world data, leading to continuous improvement, enhanced user experiences, and products that truly resonate with the market, driving adoption and satisfaction.

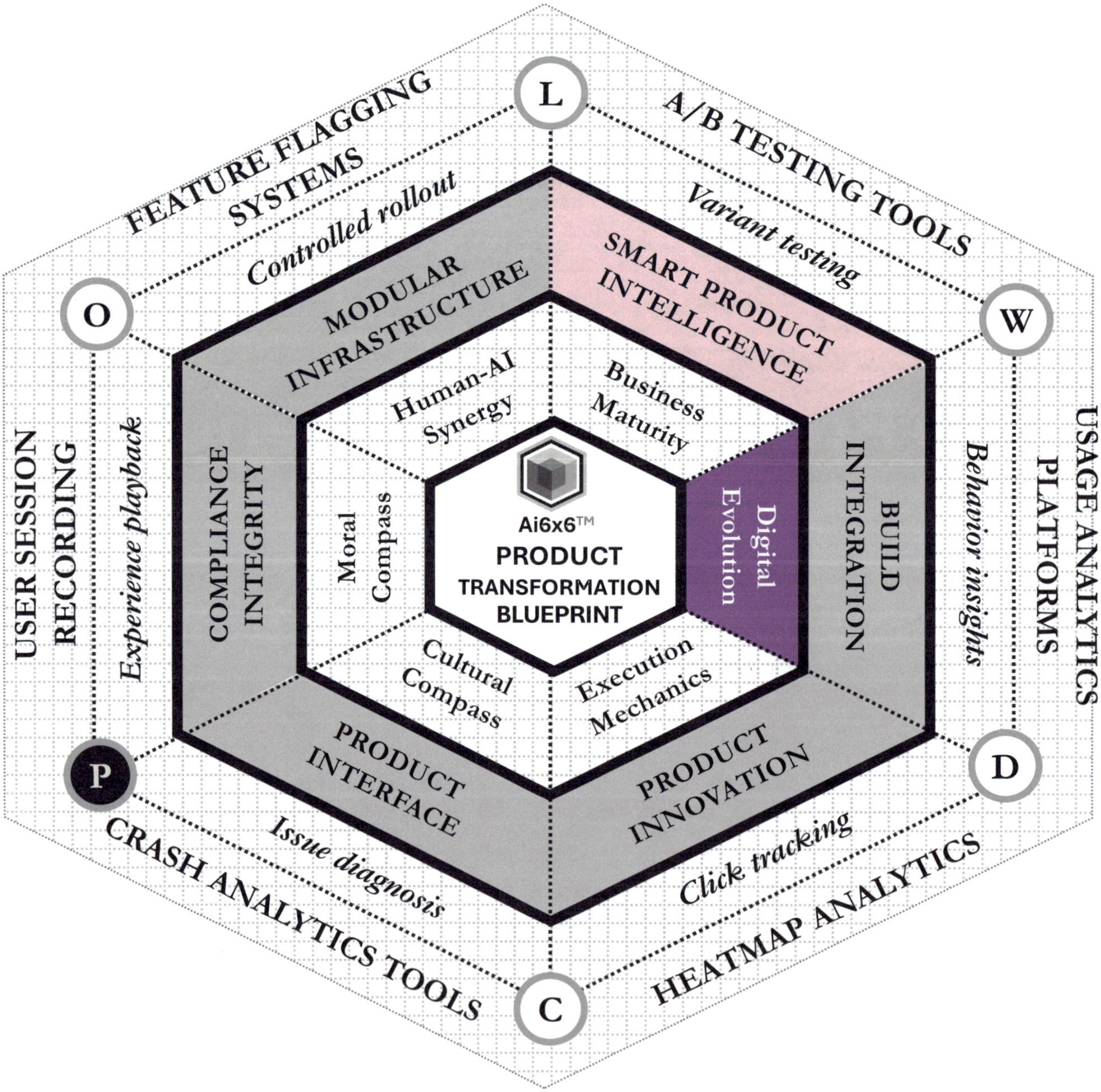

Ai6x6 DIGITAL EVOLUTION FOR INTELLIGENCE | PRODUCT TRANSFORMATION

1. FEATURE FLAGGING SYSTEMS

Feature Flagging Systems, also known as feature toggles, are software tools that allow development teams to remotely enable or disable features in a product without deploying new code.1 This capability is achieved by wrapping new features in conditional code blocks that can be turned on or off via a centralized dashboard.2 By decoupling feature release from code deployment, feature flagging systems enable safe continuous delivery, A/B testing, gradual rollouts (canary releases), and targeted feature releases to specific user segments.3 This significantly reduces deployment risk, allows for rapid experimentation, facilitates personalized user experiences, and provides a powerful mechanism for controlling the release of new product capabilities, crucial for agile product development and iteration.4

Without Feature Flagging Systems, organizations must perform full code deployments every time they want to enable, disable, or modify a product feature. This leads to slower release cycles, increased deployment risk, and a cumbersome process for A/B testing or rolling out features incrementally. The absence of dynamic control over features makes it difficult to conduct controlled experiments, respond quickly to production issues by toggling problematic features off, or target specific user groups with new functionalities. This ultimately hinders rapid iteration, limits experimentation capabilities, and slows down the delivery of value to users, impacting competitive agility in product development.

Implementation	Challenges
* Define features to be controlled by flags * Select a feature flagging platform * Integrate SDKs into application code * Design flag naming conventions & lifecycle * Implement flag management workflows * Plan for A/B testing & rollout strategies	* Managing flag proliferation & technical debt * Ensuring consistent flag states across systems * Debugging issues related to flag interactions * Performance overhead from flag evaluations * Securing flag access & preventing unauthorized changes * Educating teams on effective flag usage

Products	Platform Cost	Project Cost	Duration
LaunchDarkly	$2-15K	$10-70K	1-4 m
Optimizely Full Stack	$5-30K	$20-100K	2-6 m
Split.io	$5-25K	$20-90K	2-5 m
Flagsmith	$0.5-5K	$2-20K	0.5-2 m
Unleash (Open Source)	N/A	$5-30K	1-3 m
ConfigCat	$0.1-2K	$0.5-10K	0.5-1.5 m

A/B Testing Tools are software platforms that enable product and marketing teams to compare two or more versions of a webpage, app feature, email, or other digital content to determine which one performs better. These tools randomly split user traffic between the different versions and track key metrics (e.g., conversion rates, click-through rates, engagement) to identify statistical winners.5 By providing data-driven insights into user preferences and behavior, A/B testing tools facilitate continuous optimization of product features, user interfaces, and marketing messages.6 This leads to improved user experience, higher conversion rates, and better product market fit by systematically validating hypotheses and making informed decisions based on real user interactions.

Without A/B Testing Tools, organizations must rely on intuition, anecdotal evidence, or slow, qualitative research to make product and design decisions. This leads to suboptimal user experiences, lower conversion rates, and a high risk of investing in features or designs that do not resonate with users. The absence of a systematic way to test hypotheses with real user data hinders continuous optimization, prevents rapid learning, and makes it difficult to quantify the impact of changes. This ultimately slows down product iteration, increases the cost of poor decisions, and limits an organization's ability to drive measurable improvements in product performance and user engagement.

Implementation	Challenges
* Define test hypotheses & success metrics * Select an A/B testing platform * Implement test variations & tracking code * Configure traffic segmentation * Analyze results & identify statistical significance * Plan for iterative testing & optimization	* Ensuring statistical validity of results * Managing multiple concurrent tests * Preventing "sample pollution" across tests * Integrating with analytics & data platforms * Overcoming organizational resistance to data * Ensuring consistent user experience during tests

Products	Platform Cost	Project Cost	Duration
Google Optimize (Free/Paid)	$0-5K	$2-20K	1-3 m
Optimizely Web / Experimentation	$5-30K	$20-100K	2-6 m
VWO (Visual Website Optimizer)	$2-15K	$10-70K	1-4 m
Split.io (Experimentation Focus)	$5-25K	$20-90K	2-5 m
Adobe Target	$10-50K	$60-250K	3-9 m
AB Tasty	$5-20K	$20-80K	2-5 m

3. USAGE ANALYTICS PLATFORMS

Usage Analytics Platforms are tools that collect, process, and visualize data on how users interact with a digital product or service (e.g., website, mobile app, software). They track metrics such as page views, clicks, session duration, feature adoption, conversion funnels, and user journeys. By providing deep insights into user behavior, feature popularity, and common user flows, these platforms enable product managers, designers, and marketers to understand actual product engagement, identify areas for improvement, and make data-driven decisions.7 This leads to more intuitive product design, optimized user flows, increased feature adoption, and a better overall user experience, ultimately driving product growth and user retention by revealing how customers truly engage with the product.

Without Usage Analytics Platforms, organizations operate with limited visibility into how users actually interact with their products. Decisions about feature development, UI/UX improvements, and marketing strategies are often based on assumptions rather than concrete data, leading to suboptimal product experiences and wasted development efforts. The absence of granular insights into user behavior makes it difficult to identify pain points, measure feature adoption, understand conversion bottlenecks, or personalize user journeys effectively. This ultimately hinders continuous product optimization, slows down innovation, and limits the ability to deliver a truly user-centric product that resonates with its audience.

Implementation	Challenges
* Define key product metrics & KPIs * Select a usage analytics platform * Implement tracking code (SDKs, tags) * Configure events & user properties * Build dashboards & reports * Plan for data governance & privacy	* Ensuring accurate & comprehensive data collection * Managing data volume & processing at scale * Interpreting complex user behavior patterns * Integrating with other data sources (CRM, marketing) * Data privacy & compliance (GDPR, CCPA) * User adoption within product & marketing teams

Products	Platform Cost	Project Cost	Duration
Google Analytics 4 (GA4)	$0-5K	$5-30K	1-3 m
Mixpanel	$2-15K	$10-70K	1-4 m
Amplitude	$5-30K	$20-100K	2-6 m
Heap	$5-25K	$20-90K	2-5 m
Productboard (with analytics integrations)	$1-10K	$5-30K	1-3 m
PostHog (Open Source)	N/A	$2-15K	0.5-2 m

Heatmap Analytics are visual tools that provide graphical representations of user interaction on web pages or mobile app screens, using color gradients to indicate areas of high (hot) and low (cold) engagement. Common types include click maps (showing where users click), scroll maps (showing how far users scroll), and attention maps (showing where users focus their gaze). By revealing exactly how users interact with content, where they click, where they hesitate, or where they ignore elements, heatmap analytics offer invaluable qualitative insights into user behavior and interface usability.8 This helps product designers and marketers optimize layouts, improve content placement, identify friction points, and enhance conversion rates by making data-driven decisions about visual design and user flow.9

Without Heatmap Analytics, organizations rely on quantitative data alone (like page views or clicks) which doesn't fully explain why users behave a certain way on a page. This leaves product and design teams guessing about optimal content placement, effective call-to-actions, or areas of user confusion. The absence of visual insights into user interaction makes it difficult to diagnose usability issues, optimize visual hierarchy, or understand engagement patterns on a granular level. This ultimately leads to suboptimal UI/UX design, missed opportunities for conversion optimization, and a less intuitive product experience driven by assumptions rather than direct observation of user interaction.

Implementation	Challenges
* Define pages/screens for heatmap analysis * Select a heatmap analytics tool * Install tracking code on target pages/apps * Configure session recording (optional but recommended) * Analyze heatmaps & identify patterns * Plan for A/B testing insights from heatmaps	* Ensuring accurate data collection across devices * Performance impact of tracking scripts * Interpreting complex visual data patterns * Data privacy implications (masking sensitive data) * Managing high volume of session recordings * Differentiating between intentional & accidental clicks

Products	Platform Cost	Project Cost	Duration
Hotjar	$0.1-5K	$0.5-10K	0.5-1.5 m
Crazy Egg	$0.1-2K	$0.5-5K	0.5-1.5 m
Contentsquare	$10-50K	$60-250K	3-9 m
FullStory (with heatmaps)	$5-25K	$20-90K	2-5 m
Smartlook	$0.5-5K	$2-20K	1-2 m
Mouseflow	$0.5-2K	$0.5-10K	0.5-1.5 m

5. CRASH ANALYTICS TOOLS

Crash Analytics Tools are software solutions that automatically detect, collect, and report on application crashes and errors in real-time, providing developers with detailed diagnostic information such as stack traces, device context, and user breadcrumbs leading up to the crash. These tools integrate directly into mobile apps, web applications, or desktop software, providing immediate visibility into critical issues that impact user experience. By swiftly identifying the root cause of crashes, Crash Analytics Tools enable development teams to prioritize and resolve bugs faster, improve application stability, and minimize negative user impact.10 This is essential for maintaining product quality, enhancing user satisfaction, and ensuring a reliable and robust user experience.

Without Crash Analytics Tools, organizations are largely reliant on manual user reports or general server logs to identify and diagnose application crashes, leading to delayed issue detection, vague problem descriptions, and prolonged troubleshooting. This results in a frustrating user experience due to persistent bugs, increased customer support burden, and significant delays in delivering stable product updates. The absence of automated, detailed crash reporting hinders the ability to proactively identify critical bugs, efficiently allocate development resources, and maintain a high-quality product, ultimately impacting user retention, brand reputation, and the overall reliability of the software.

Implementation	Challenges
* Integrate SDK into application codebase * Configure symbolication for readable stack traces * Set up error reporting & alerts * Define crash reporting policies (privacy) * Integrate with issue tracking (Jira, GitHub) * Plan for continuous monitoring & prioritization	* Ensuring comprehensive crash coverage across devices * Managing data volume of crash reports * Debugging complex crashes (memory, threading) * Handling data privacy for crash reports * Prioritizing critical crashes vs. minor errors * Integrating with diverse development environments

Products	Platform Cost	Project Cost	Duration
Sentry	$0.1-5K	$0.5-10K	0.5-1.5 m
Firebase Crashlytics (Google)	N/A	$0.5-5K	0.5-1 m
Instabug	$0.1-5K	$0.5-10K	0.5-1.5 m
Bugsnag	$0.1-5K	$0.5-10K	0.5-1.5 m
Raygun	$0.1-2K	$0.5-5K	0.5-1.5 m
AppCenter (Microsoft)	N/A	$0.5-5K	0.5-1 m

User Session Recording tools capture and replay actual user interactions on a website or mobile application, providing a video-like recording of a user's mouse movements, clicks, scrolls, taps, and form interactions. These recordings offer unparalleled qualitative insights into how individual users navigate, engage with, and potentially struggle with a product's interface. By visually demonstrating user journeys and identifying points of friction, confusion, or abandonment, User Session Recording helps product managers and UX designers diagnose usability issues, validate design assumptions, and uncover hidden pain points.11 This leads to more intuitive product design, improved conversion funnels, and a deeper understanding of real-world user behavior, vital for continuous product improvement and optimization.

Without User Session Recording, organizations lack the ability to observe real user behavior directly, missing critical qualitative insights into why users get stuck, abandon tasks, or experience frustration.12 Relying solely on quantitative data means product and design teams are left to hypothesize about user difficulties, leading to less effective design changes and longer troubleshooting times for usability issues. The absence of visual replays of user sessions makes it difficult to pinpoint exact points of friction, understand user context, or validate assumptions about user flows, ultimately hindering the ability to create truly intuitive and frictionless product experiences that drive user satisfaction and conversion.

Implementation	Challenges
* Define pages/flows for recording * Select a session recording tool * Install tracking code on target pages/apps * Implement privacy masking for sensitive data * Analyze recordings for patterns & issues * Plan for continuous monitoring & issue prioritization	* Managing large volumes of recording data * Ensuring data privacy (masking sensitive fields) * Performance impact of recording scripts * Identifying key sessions from many recordings * Integrating insights with quantitative analytics * User adoption within product & UX teams

Products	Platform Cost	Project Cost	Duration
Hotjar	$0.1-5K	$0.5-10K	0.5-1.5 m
FullStory	$5-25K	$20-90K	2-5 m
Smartlook	$0.5-5K	$2-20K	1-2 m
Mouseflow	$0.5-2K	$0.5-10K	0.5-1.5 m
Contentsquare (Enterprise focus)	$10-50K	$60-250K	3-9 m
Clarity (Microsoft - Free)	N/A	$0.5-2K	0.5-1 m

Ai6x6™ Transformation Blueprint

Build 'Integration'

"Continuous delivery is about building software so it's always ready to release." — Jez Humble , DevOps thought leader and co-author of "Continuous Delivery"; influential in pipeline automation.

Integration at the product level isn't just technical—it's temporal. CI/CD pipelines, artifact repositories, and webhook-driven frameworks enable velocity with integrity. "Build integration" means teams don't work sequentially—they work symphonically. In transformation, speed must meet quality—and integration delivers both. Mature organizations reduce deployment anxiety because their pipelines are expressive, monitored, and modular. Product strategy isn't delayed by environment setup—it's expressed through it. Integration doesn't just reduce error—it reinforces excellence through rhythm, not rigidity.

"Build Integration" is vital for product transformation's digital maturity because it ensures seamless collaboration and efficient workflows across the entire product development lifecycle. A digitally mature organization prioritizes robust CI/CD pipelines, artifact repositories, and package managers to synchronize code flow, build processes, and testing frameworks. This deep integration, supported by webhooks for dev tools and integration testing, eliminates silos between development, operations, and other teams. It accelerates time-to-market and fosters a cohesive environment where diverse teams can collaborate effectively to deliver high-quality digital products with speed and reliability.

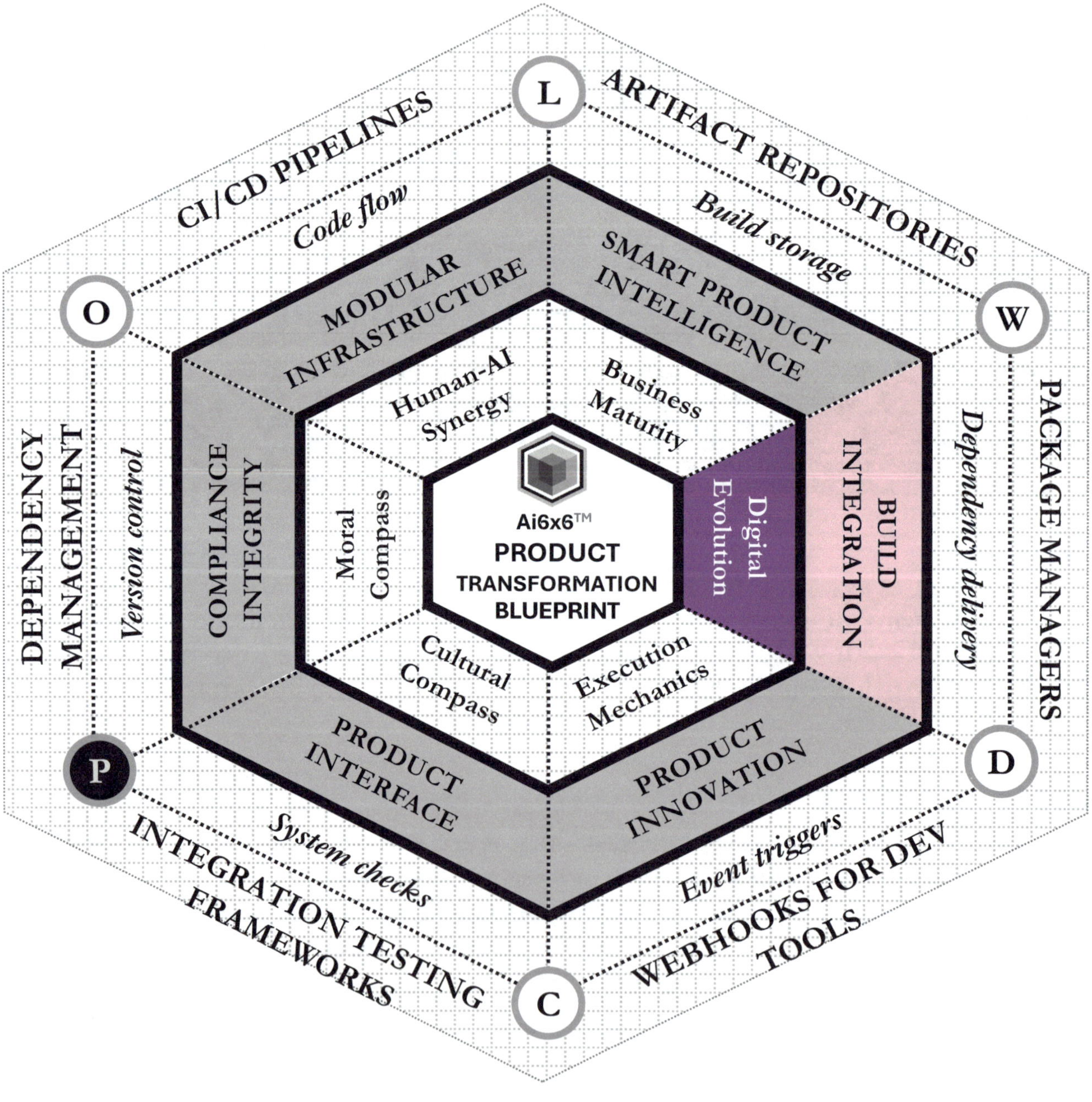

Ai6x6 DIGITAL EVOLUTION FOR INTEGRATION | PRODUCT TRANSFORMATION

1. CI/CD PIPELINES

CI/CD Pipelines (Continuous Integration/Continuous Delivery or Deployment Pipelines) are automated workflows that standardize and streamline the process of building, testing, and deploying software.1 Continuous Integration (CI) involves frequently merging code changes into a central repository, followed by automated builds and tests to detect integration errors early.2 Continuous Delivery (CD) extends this by automatically preparing validated code for release, while Continuous Deployment automatically releases it to production.3 By automating these critical steps, CI/CD pipelines significantly reduce manual errors, accelerate software delivery cycles, and ensure that a product is always in a deployable state.4 This enables rapid iteration, faster delivery of features to users, and improved product quality by providing continuous feedback and minimizing friction in the software development lifecycle.5

Without CI/CD Pipelines, organizations face slow, manual, and error-prone software release processes.6 Code integration conflicts become frequent, testing is often delayed and incomplete, and deployments are high-risk, infrequent events. The absence of automation in these critical steps leads to extended time-to-market for new features, accumulation of technical debt, increased risk of production bugs, and significant operational overhead.7 This ultimately hinders an organization's agility, slows down innovation, and reduces its ability to respond quickly to market demands, impacting product competitiveness and customer satisfaction due to inconsistent and delayed software releases.

Implementation	Challenges
* Version control code (e.g., Git) * Select a CI/CD platform * Automate build & test processes * Configure deployment stages & environments * Integrate with monitoring & alerting * Plan for pipeline security & governance	* Complex setup & maintenance for large projects * Managing diverse build & test environments * Ensuring pipeline security & access control * Integrating with disparate tools & systems * Debugging pipeline failures efficiently * Overcoming cultural resistance to automation

Products	Platform Cost	Project Cost	Duration
Jenkins (Open Source)	N/A	$5-30K	1-3 m
GitLab CI/CD	$0.5-10K	$5-50K	1-4 m
CircleCI	$0.5-5K	$2-20K	1-2 m
GitHub Actions	$0.5-5K	$2-20K	1-2 m
Azure DevOps Pipelines	$0.5-8K	$5-40K	1-3 m
Harness (Enterprise CD)	$2-15K	$10-70K	1-4 m

2. ARTIFACT REPOSITORIES

Artifact Repositories are centralized storage systems for managing and versioning binary artifacts (such as compiled code, libraries, images, and packages) generated during the software development process.8 They serve as a single source of truth for all build artifacts, ensuring consistency and traceability across different environments (development, testing, production).9 By providing robust version control, access control, and metadata management for these binary assets, Artifact Repositories streamline CI/CD pipelines, prevent "dependency hell," and improve the reliability of software builds and deployments.10 This critical infrastructure component ensures that the right version of every component is used consistently, accelerating product delivery, reducing build failures, and enhancing the overall integrity and security of the software supply chain.

Without Artifact Repositories, organizations risk fragmented storage of build artifacts across various local machines or unmanaged network drives, leading to versioning inconsistencies, "dependency hell," and difficulties in reproducing builds.11 The absence of a centralized, version-controlled repository makes it challenging to ensure that the correct binary components are used in deployments, increasing the risk of production issues and security vulnerabilities. This ultimately slows down software delivery, hinders traceability, complicates rollbacks, and reduces the overall reliability and integrity of the product development lifecycle, leading to inefficiencies and increased operational risk.

Implementation	Challenges
* Define artifact types & naming conventions * Select an artifact repository solution * Integrate with CI/CD pipelines * Establish access control & security policies * Plan for retention policies & storage scaling * Educate teams on artifact management best practices	* Migrating existing artifacts & dependencies * Managing storage costs for large volumes * Ensuring high availability & disaster recovery * Security of stored binaries & access * Integrating with diverse build tools & package managers * Enforcing consistent versioning & metadata

Products	Platform Cost	Project Cost	Duration
JFrog Artifactory	$5-30K	$20-100K	2-6 m
Sonatype Nexus Repository	$2-15K	$10-70K	1-4 m
GitLab Package Registry	$0.5-10K	$5-30K	1-3 m
GitHub Packages	$0.5-5K	$2-20K	0.5-2 m
AWS CodeArtifact	$0.1-2K	$0.5-10K	0.5-1.5 m
Azure Artifacts	$0.1-2K	$0.5-10K	0.5-1.5 m

Package Managers are software tools that automate the process of installing, updating, configuring, and removing software packages (libraries, frameworks, and applications) within a development environment or operating system. They handle dependencies between packages, ensuring that all necessary components are present and compatible. Examples include npm for Node.js, Maven for Java, pip for Python, and NuGet for .NET. By standardizing dependency management and providing a robust way to include external code, Package Managers significantly simplify development workflows, reduce "dependency hell," and ensure consistent build environments.12 This accelerates product development by making it easier to leverage open-source components, manage external libraries securely, and maintain reliable software builds.13

Without Package Managers, developers would manually download, configure, and manage external libraries and dependencies, leading to significant inefficiencies, version conflicts, and "dependency hell."14 This results in inconsistent build environments across development machines, increased risk of production bugs due to mismatched versions, and substantial time wasted on resolving dependency issues. The absence of automated package management hinders reproducibility of builds, complicates security patching, and slows down development, ultimately impacting product quality, delivery speed, and the overall maintainability of the software product.

Implementation	Challenges
* Identify languages & ecosystems used * Select appropriate package managers * Configure private/internal registries (if needed) * Establish dependency versioning policies * Integrate into CI/CD pipelines * Plan for security scanning of dependencies	* Managing complex dependency trees * Resolving version conflicts (dependency hell) * Ensuring security of third-party packages * Maintaining private package registries * Performance overhead of package resolution * Keeping up with frequent package updates

Products	Platform Cost	Project Cost	Duration
npm (Node.js)	N/A	$0.5-5K	0.5-1 m
Maven/Gradle (Java)	N/A	$1-10K	0.5-2 m
pip (Python)	N/A	$0.5-5K	0.5-1 m
NuGet (.NET)	N/A	$0.5-5K	0.5-1 m
Yarn (Node.js Alternative)	N/A	$0.5-5K	0.5-1 m
Conan (C/C++)	N/A	$1-10K	0.5-2 m

4. WEBHOOKS FOR DEV TOOLS

Webhooks for Dev Tools are automated, real-time notifications sent from one software application to another when a specific event occurs. In the context of development tools, webhooks enable seamless integration and automation across different systems, such as version control (e.g., GitHub, GitLab), CI/CD platforms (e.g., Jenkins), project management (e.g., Jira), and communication tools (e.g., Slack).15 For example, a webhook can trigger a CI build when code is pushed to a repository, or send a notification when a bug is created.16 By providing instant communication between disparate tools, webhooks automate workflows, improve team collaboration, and accelerate feedback loops, essential for agile product development and continuous delivery.17

Without Webhooks for Dev Tools, organizations rely on manual triggers, polling mechanisms, or complex custom integrations to synchronize information between various development tools. This leads to delayed notifications, out-of-sync data, and significant manual overhead for coordinating tasks across different systems. The absence of real-time event-driven communication hinders automation, slows down feedback loops between development stages, and increases the potential for human error. This ultimately impacts development efficiency, collaboration, and the speed at which product features can be developed, tested, and deployed, creating bottlenecks in the software development lifecycle.

Implementation	Challenges
* Identify tools to integrate * Configure webhooks in source tool * Set up endpoints to receive webhooks * Implement logic to process webhook payloads * Secure webhook endpoints (signatures, TLS) * Plan for error handling & retries	* Managing security of webhook endpoints * Handling high volume of webhook events * Ensuring reliable delivery & processing * Debugging integration issues across tools * Parsing & interpreting diverse payload formats * Preventing infinite loops or cascades of events

Products	Platform Cost	Project Cost	Duration
GitHub Webhooks	N/A	$0.5-5K	0.5-1 m
GitLab Webhooks	N/A	$0.5-5K	0.5-1 m
Jira Webhooks	N/A	$0.5-5K	0.5-1 m
Slack Incoming Webhooks	N/A	$0.5-5K	0.5-1 m
Zapier / Make (Automation)	$0.1-2K	$0.5-5K	0.5-1.5 m
Custom Webhook Handler (Node.js, Python)	N/A	$2-15K	1-2 m

Integration Testing Frameworks are software tools and methodologies used to test the interactions and data flow between different modules, components, or services within a software product, or between a product and external systems (like databases, APIs, or third-party services). Unlike unit tests that focus on individual code units, integration tests verify that these units work correctly when combined.18 By simulating real-world interactions and ensuring seamless communication between integrated parts, these frameworks detect interface defects, data mismatches, and compatibility issues early in the development cycle.19 This significantly improves product reliability, reduces the risk of production errors, and ensures that complex systems function as a cohesive whole, critical for delivering robust and dependable software products.

Without Integration Testing Frameworks, organizations risk undetected defects in the interfaces and interactions between different parts of their software product, or between their product and external systems.20 This leads to a higher likelihood of critical bugs appearing in production environments, causing system failures, data corruption, and poor user experiences. The absence of systematic integration testing results in costly rework, delayed releases, and a reactive approach to quality assurance.21 This ultimately impacts product reliability, increases support burden, and erodes customer trust due to frequent and unpredictable software malfunctions stemming from unverified inter-component communications.

Implementation	Challenges
* Define integration points & test scenarios * Select an appropriate testing framework * Write integration test cases * Set up isolated test environments * Integrate tests into CI/CD pipelines * Plan for test data management & cleanup	* Complex environment setup & teardown * Managing external service dependencies * Ensuring test isolation & reproducibility * Long test execution times * Debugging failures across multiple components * Maintaining test suites as systems evolve

Products	Platform Cost	Project Cost	Duration
Selenium (Web UI)	N/A	$5-30K	1-3 m
Cypress (Web UI)	N/A	$5-25K	1-3 m
Postman (API Testing)	$0-2K	$1-10K	0.5-2 m
Playwright (Web UI)	N/A	$5-20K	1-3 m
Jest (JavaScript Unit/Integration)	N/A	$0.5-5K	0.5-1 m
TestContainers (Docker based)	N/A	$2-10K	1-2 m

6. DEPENDENCY MANAGEMENT

Dependency Management in software development is the practice of identifying, tracking, and resolving the external libraries, modules, and components that a software product relies on to function. This involves using tools and processes to automatically fetch required dependencies, ensure version compatibility, and manage transitive dependencies (dependencies of dependencies).22 By standardizing how external code is incorporated and managed, Dependency Management ensures consistent builds, reduces conflicts between libraries, and simplifies security patching.23 This is crucial for maintaining the stability, security, and reproducibility of software products, allowing development teams to efficiently leverage the vast ecosystem of open-source and proprietary components while mitigating risks associated with external code.24

Without proper Dependency Management, software development teams face severe challenges with "dependency hell," where incompatible versions of libraries or missing components lead to build failures, runtime errors, and inconsistent behavior across different environments.25 Manually tracking and resolving these dependencies is a time-consuming, error-prone, and unsustainable task, especially in large and complex projects. The absence of automated dependency management hinders reproducibility of builds, complicates security vulnerability remediation, and increases the technical debt, ultimately impacting product stability, development efficiency, and the overall integrity of the software product.

Implementation	Challenges
* Identify all external dependencies * Select a dependency management tool * Configure dependency resolution & caching * Establish versioning & update policies * Integrate with build systems & CI/CD * Plan for dependency security scanning	* Resolving complex transitive dependencies * Preventing version conflicts (dependency hell) * Ensuring security of third-party libraries * Managing licensing compliance for open source * Performance overhead of dependency resolution * Keeping dependencies updated without breaking changes

Products	Platform Cost	Project Cost	Duration
Maven/Gradle (Java)	N/A	$1-10K	0.5-2 m
npm/Yarn (Node.js)	N/A	$0.5-5K	0.5-1 m
pip/Poetry (Python)	N/A	$0.5-5K	0.5-1 m
NuGet (.NET)	N/A	$0.5-5K	0.5-1 m
Go Modules (Go)	N/A	$0.5-5K	0.5-1 m
Bundler (Ruby)	N/A	$0.5-5K	0.5-1 m

Ai6x6™ Transformation Blueprint

Product 'Innovation'

"Move fast and make things. You'll learn by shipping." — Julie Zhuo
Former VP of Product Design at Facebook; shaped iterative design culture through prototyping.

Invention isn't about novelty—it's about iteration that inspires. Hackathon platforms, MVP frameworks, and crowd testing channels aren't events—they're habitats. "Product Innovation" create a cadence where bold ideas move quickly but meaningfully. Mature transformation turns prototyping into culture—measuring not just utility but resonance. When Figma drafts meet beta feedback and idea platforms collect internal wisdom, products no longer launch—they evolve. Studios act like sentient spaces—listening to customer signals, internal creativity, and business logic. The result isn't speed—it's strategic spontaneity.

The "Product Innovation:" chapter is paramount for product transformation's digital maturity as it focuses on cultivating a culture of rapid experimentation, creativity, and continuous improvement in product development. A digitally mature organization employs tools and methodologies like hackathon platforms, rapid MVP platforms, and beta testing engines to quickly test ideas, gather feedback, and iterate on product concepts. This commitment to agile innovation, supported by design tools like Figma and crowd testing, ensures the organization can consistently deliver groundbreaking products that meet evolving customer needs and stay ahead of the competition, fostering a spirit of continuous invention.

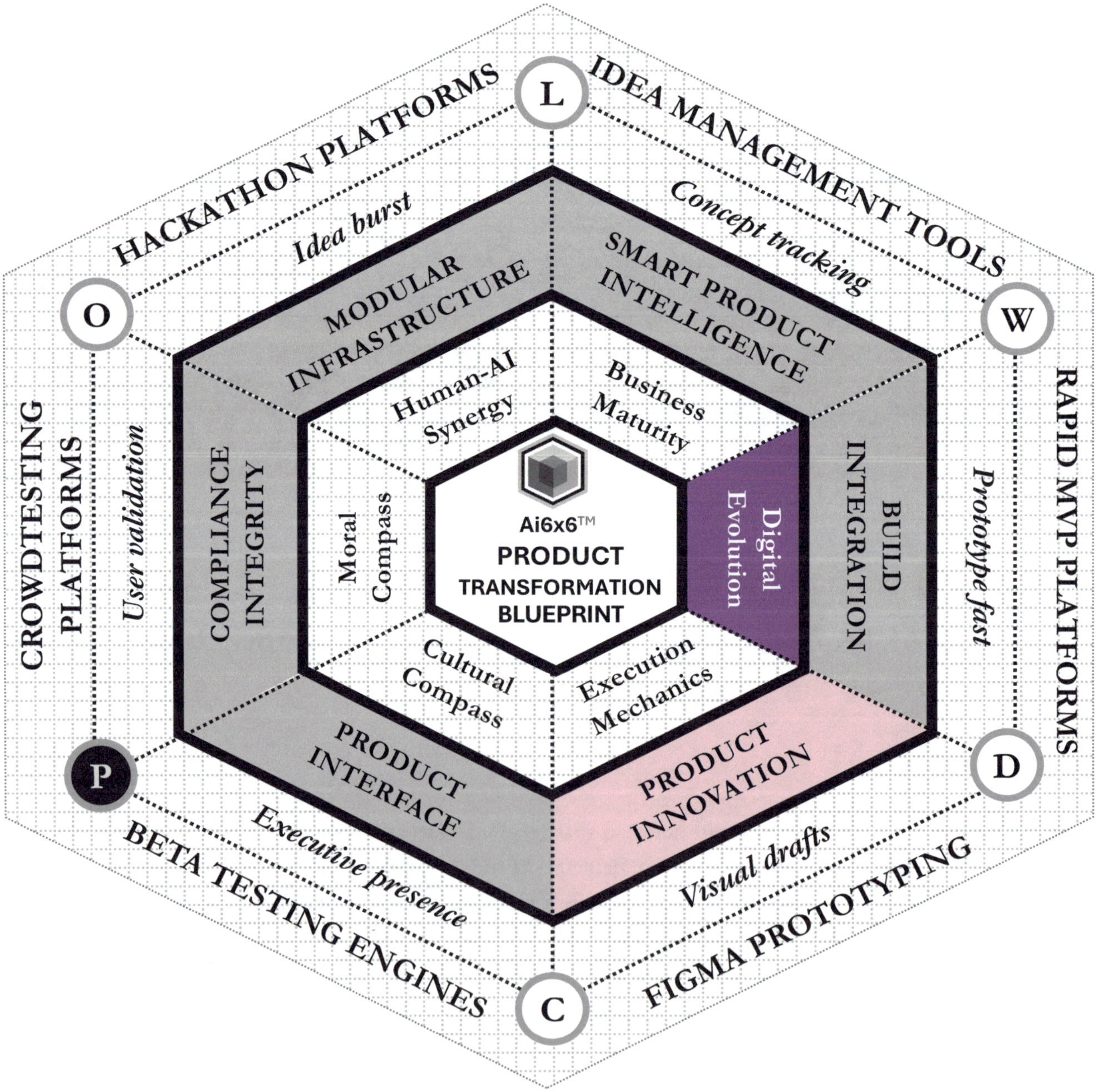

Ai6x6 DIGITAL EVOLUTION FOR INNOVATION | PRODUCT TRANSFORMATION

1. HACKATHON PLATFORMS

Hackathon Platforms are specialized online or hybrid platforms that facilitate the organization and execution of hackathons, innovation challenges, and idea generation events. These platforms provide tools for team formation, idea submission, project management, mentor matching, judging, and prize distribution. By offering a structured and engaging environment, Hackathon Platforms enable organizations to quickly generate a high volume of innovative ideas, foster cross-functional collaboration, identify new talent, and rapidly prototype solutions to specific business challenges or market opportunities. This accelerates internal innovation, enhances employee engagement, and provides a fertile ground for discovering groundbreaking product concepts that might otherwise go unexplored, driving a culture of continuous invention.

Without dedicated Hackathon Platforms, organizations attempting to run innovation challenges often face significant logistical hurdles, including inefficient team formation, disorganized idea submissions, cumbersome judging processes, and limited visibility into project progress. This results in reduced participation, less impactful outcomes, and a fragmented experience for participants and organizers alike. The absence of a streamlined platform hinders the ability to scale innovation efforts, effectively capture and nurture new ideas, and consistently leverage hackathons as a strategic tool for product development and talent discovery, ultimately stifling organic innovation and collaboration within the organization.

Implementation	Challenges
* Define hackathon theme & goals * Select a hackathon platform * Promote event & recruit participants * Provide resources (APIs, data, mentors) * Establish judging criteria & process * Plan for post-hackathon idea nurturing	* Attracting & retaining participant engagement * Ensuring fair judging & clear criteria * Managing diverse project types & technologies * Integrating with internal systems (e.g., HR) * Post-hackathon idea follow-through & funding * Measuring clear ROI on innovation outcomes

Products	Platform Cost	Project Cost	Duration
Devpost	$5-20K	$2-10K	0.5-2 m
HackerEarth	$5-25K	$5-20K	1-3 m
BeMyApp	$10-30K	$10-50K	1-3 m
Topcoder (Crowdsourcing)	$15-50K	$20-100K	2-4 m
ChallengePost (now Devpost)	N/A	N/A	N/A
Custom Internal Portal	N/A	$30-150K	2-7 m

Idea Management Tools are software platforms designed to capture, organize, evaluate, and prioritize new ideas from employees, customers, or partners. These tools typically include features for idea submission, collaborative brainstorming, voting, commenting, and tracking an idea's progress through an innovation funnel. By providing a structured and transparent process for collecting and assessing ideas, Idea Management Tools foster a culture of continuous innovation, harness collective intelligence, and ensure that valuable insights are not lost. This helps organizations identify promising product enhancements, new features, or entirely new product concepts, accelerating the innovation pipeline and ensuring that product development is driven by a diverse range of perspectives and market needs.

Without dedicated Idea Management Tools, organizations struggle to systematically collect and evaluate innovative ideas from their employees, customers, or partners. Ideas often reside in disparate emails, documents, or informal conversations, leading to lost opportunities, duplicated efforts, and a lack of transparency in the innovation process. The absence of a structured system for idea capture and evaluation hinders the ability to prioritize promising concepts, track their development, and provide feedback to contributors. This ultimately stifles organic innovation, reduces employee engagement in problem-solving, and slows down the pipeline for new product features and strategic initiatives.

Implementation	Challenges
* Define idea submission process & criteria * Select an idea management platform * Promote system to target audience * Establish moderation & evaluation workflows * Integrate with project management tools * Plan for feedback loops & recognition	* Encouraging consistent idea submission * Managing idea volume & quality * Objectively evaluating & prioritizing ideas * Overcoming organizational silos for input * Ensuring transparency & feedback to contributors * Measuring impact of ideas implemented

Products	Platform Cost	Project Cost	Duration
Aha! Ideas	$1-10K	$5-30K	1-3 m
Ideascale	$5-30K	$20-100K	2-6 m
Brightidea	$10-50K	$60-250K	3-9 m
UserVoice (Feedback & Ideas)	$1-8K	$5-25K	1-3 m
Spigit (Muzo - Part of Planview)	$10-40K	$50-200K	3-8 m
Microsoft Teams/SharePoint (Custom)	N/A	$5-30K	1-3 m

Rapid MVP (Minimum Viable Product) Platforms are development environments or tools designed to quickly build and launch a basic version of a product with just enough features to satisfy early adopters and gather validated learning. These platforms often leverage low-code/no-code capabilities, pre-built components, and simplified development workflows to accelerate the prototyping and deployment process. By enabling quick iteration and real-world testing of core product hypotheses, Rapid MVP Platforms significantly reduce the time and cost associated with initial product development. This allows organizations to validate market demand, gather early user feedback, and pivot or persevere based on real data, accelerating the innovation cycle and minimizing risk in new product ventures.

Without Rapid MVP Platforms, organizations face extended development cycles and higher costs for building even the most basic version of a new product. This leads to delays in market entry, reduced opportunities for early user feedback, and increased risk of building features that no one wants or needs. The absence of tools that facilitate quick prototyping and deployment hinders the ability to rapidly validate market assumptions, test core value propositions, and iterate based on real-world user interaction. This ultimately slows down innovation, increases investment risk in unproven ideas, and limits an organization's agility in exploring new product opportunities.

Implementation	Challenges
* Define core value proposition & MVP features * Select a rapid MVP development platform * Utilize pre-built components & templates * Design for quick user feedback loops * Plan for scalability beyond MVP * Focus on speed over perfection	* Striking balance between "minimum" and "viable" * Overcoming limitations of low-code/no-code platforms * Ensuring scalability & maintainability for growth * Integrating with complex existing systems * Data security & compliance in rapid builds * Avoiding feature creep during MVP phase

Products	Platform Cost	Project Cost	Duration
Bubble (No-Code App Builder)	$0.1-2K	$0.5-10K	0.5-2 m
Adalo (No-Code Mobile App)	$0.1-2K	$0.5-10K	0.5-2 m
Webflow (No-Code Web Design)	$0.1-1K	$0.5-5K	0.5-1.5 m
Glide (No-Code Apps from Sheets)	$0.1-1K	$0.5-5K	0.5-1 m
Firebase (Backend as a Service)	$0-1K	$1-10K	0.5-2 m
OutSystems (Low-Code Dev)	$5-30K	$20-100K	2-6 m

Figma Prototyping refers to the capabilities within Figma, a collaborative cloud-based design tool, that allow designers to create interactive prototypes of user interfaces. These prototypes simulate the user experience by linking design screens with interactive elements (clicks, hovers, scrolls), creating realistic flows that mimic a working application. By enabling rapid prototyping and iterative design, Figma Prototyping facilitates early user testing, gathers valuable feedback on usability and flow, and allows stakeholders to visualize and experience product concepts before development begins. This significantly reduces development rework, accelerates the design-to-development handoff, and ensures that the final product meets user needs effectively, fostering a highly collaborative and efficient product design process.

Without Figma Prototyping or similar interactive design tools, product teams rely on static mockups or wireframes, which fail to convey the dynamic user experience and interactive flows of a product. This leads to misunderstandings between designers, developers, and stakeholders, resulting in costly rework during development when usability issues are discovered late. The absence of interactive prototypes hinders effective user testing, delays feedback loops, and makes it difficult to validate design assumptions before committing significant engineering resources. This ultimately slows down the product development cycle, increases design-to-development friction, and raises the risk of building a product with suboptimal user experience.

Implementation	Challenges
* Create UI designs in Figma * Define interactions & user flows * Link screens to create navigable prototype * Share prototypes for feedback & testing * Integrate with user testing platforms * Plan for version control & design system updates	* Managing complex interaction logic in prototypes * Performance of large prototypes * Ensuring realistic feel for advanced interactions * Maintaining design consistency across large projects * Handoff to development teams (specs, assets) * User adoption within design & product teams

Products	Platform Cost	Project Cost	Duration
Figma (Prototyping capabilities)	$0-0.5K	$0.1-2K	0.1-0.5 m
Adobe XD	$0-0.5K	$0.1-2K	0.1-0.5 m
Sketch (with InVision/Marvel integration)	$0-0.5K	$0.1-2K	0.1-0.5 m
InVision (Prototyping & Collaboration)	$0.1-1K	$0.5-5K	0.5-1 m
Framer (Code-based Prototyping)	$0.1-1K	$0.5-5K	0.5-1 m
Axure RP	$0.1-0.5K	$0.5-2K	0.5-1 m

5. BETA TESTING ENGINES

Beta Testing Engines are platforms and systems designed to manage the process of testing pre-release versions of a product with a select group of real users (beta testers) outside of the core development team. These engines facilitate tester recruitment, feedback collection (bugs, usability issues, feature requests), bug reporting, and communication between product teams and testers. By providing a structured environment for real-world testing, Beta Testing Engines enable organizations to uncover critical bugs, validate product market fit, gather qualitative feedback on usability, and build anticipation before a full public launch. This helps refine the product based on actual user experiences, reduces post-launch issues, and ensures a higher quality, more user-centric product release, crucial for successful market entry.

Without dedicated Beta Testing Engines, organizations typically rely on informal, manual processes for gathering feedback on pre-release products. This leads to disorganized bug reports, inefficient communication with testers, difficulties in tracking feedback, and a lack of systematic insight into real-world product usage. The absence of a structured beta testing process hinders the ability to effectively uncover critical bugs before launch, validate usability with a representative user base, and gather actionable feedback at scale. This ultimately increases the risk of launching a product with significant defects or usability issues, leading to poor user adoption, negative reviews, and costly post-launch fixes.

Implementation	Challenges
* Define beta testing goals & scope * Select a beta testing platform * Recruit & onboard beta testers * Set up feedback channels & bug reporting * Establish communication plan with testers * Plan for data analysis & iteration based on feedback	* Recruiting diverse & engaged beta testers * Managing bug reports & feature requests * Maintaining tester engagement over time * Ensuring data privacy for tester feedback * Integrating feedback into product roadmap * Dealing with conflicting or irrelevant feedback

Products	Platform Cost	Project Cost	Duration
Centercode	$5-30K	$20-100K	2-6 m
TestFairy (Mobile Beta Testing)	$1-10K	$5-30K	1-3 m
TestFlight (Apple - iOS apps)	N/A	$0.5-2K	0.5-1 m
Google Play Console (Android apps)	N/A	$0.5-2K	0.5-1 m
UserTesting (for moderated tests)	$2-15K	$10-50K	1-3 m
BetaFamily	$0.5-5K	$2-10K	0.5-2 m

Crowdtesting Platforms leverage a global community of professional testers to perform various types of software testing (e.g., functional, usability, localization, performance) on a product. Unlike traditional in-house testing or beta testing with a selected group, crowdtesting offers access to a diverse pool of testers using a wide range of devices, operating systems, and real-world environments. By tapping into this scalable and geographically distributed workforce, Crowdtesting Platforms enable organizations to achieve broader test coverage, uncover edge-case bugs, and gain insights from diverse user perspectives quickly and cost-effectively. This accelerates quality assurance, ensures a higher quality product across varied user contexts, and provides valuable feedback on localization and real-world usability, crucial for launching robust and globally ready products.

Without Crowdtesting Platforms, organizations typically rely on smaller, in-house QA teams or limited beta tester pools, which often struggle to achieve comprehensive test coverage across a multitude of devices, operating systems, and real-world user scenarios. This leads to missed bugs, suboptimal user experiences in specific environments, and challenges in validating product readiness for diverse global markets. The absence of scalable, on-demand testing from a wide range of users hinders the ability to uncover obscure defects, gain authentic user insights from varied demographics, and ensure consistent product quality across all potential customer contexts, ultimately impacting market adoption and brand reputation.

Implementation	Challenges
* Define test scope & objectives * Select a crowdtesting platform * Provide clear test instructions & access * Set up bug reporting & feedback channels * Review & filter incoming test results * Plan for iterative testing cycles	* Ensuring quality & relevance of test reports * Managing communication with a distributed crowd * Data security & confidentiality with external testers * Filtering noise from valuable feedback * Integrating with internal bug tracking systems * Cost management for ongoing testing needs

Products	Platform Cost	Project Cost	Duration
Applause	$10-50K	$20-100K	2-6 m
Testlio	$10-40K	$20-80K	2-5 m
uTest (Applause)	$5-30K	$10-70K	1-4 m
UserTesting (Crowd feedback)	$2-15K	$5-50K	1-3 m
Global App Testing	$5-20K	$10-70K	1-4 m
Test IO	$5-25K	$15-80K	2-5 m

CHAPTER
D5

BLUEPRINT SERIES

Ai6x6™ Transformation Blueprint

Product 'Interface'

"People ignore design that ignores people." — Frank Chimero , Designer and author; widely quoted for human-centered design insights that shape modern UX principles.

Interfaces aren't visual—they're visceral. In-app messaging, voice control, and customizable UI libraries convert software from service into self-awareness. "Product Interface" isn't cute—it's strategic. Transformation maturity makes interaction intuitive before it's taught. Accessible design ensures products don't exclude; embedded SDKs ensure dashboards speak fluently. When users sense empathy in architecture, engagement deepens. Interfaces must respect cognition, not challenge it. UX isn't decoration—it's emotional symmetry. Products declare their personality not in features—but in how they greet, guide, and grow with users.

"Product Interface" is critical for product transformation's digital maturity because it emphasizes designing intuitive, accessible, and engaging user experiences that make the product feel human. A digitally mature organization understands that the product's interface is its primary voice, and invests in in-app messaging, customizable UI libraries, and accessibility testing suites. This attention to interface design, including interactive prototypes and voice control integration, ensures products are not only functional but also delightful, easy to use, and inclusive, driving adoption, satisfaction, and long-term customer loyalty by creating seamless and positive interactions.

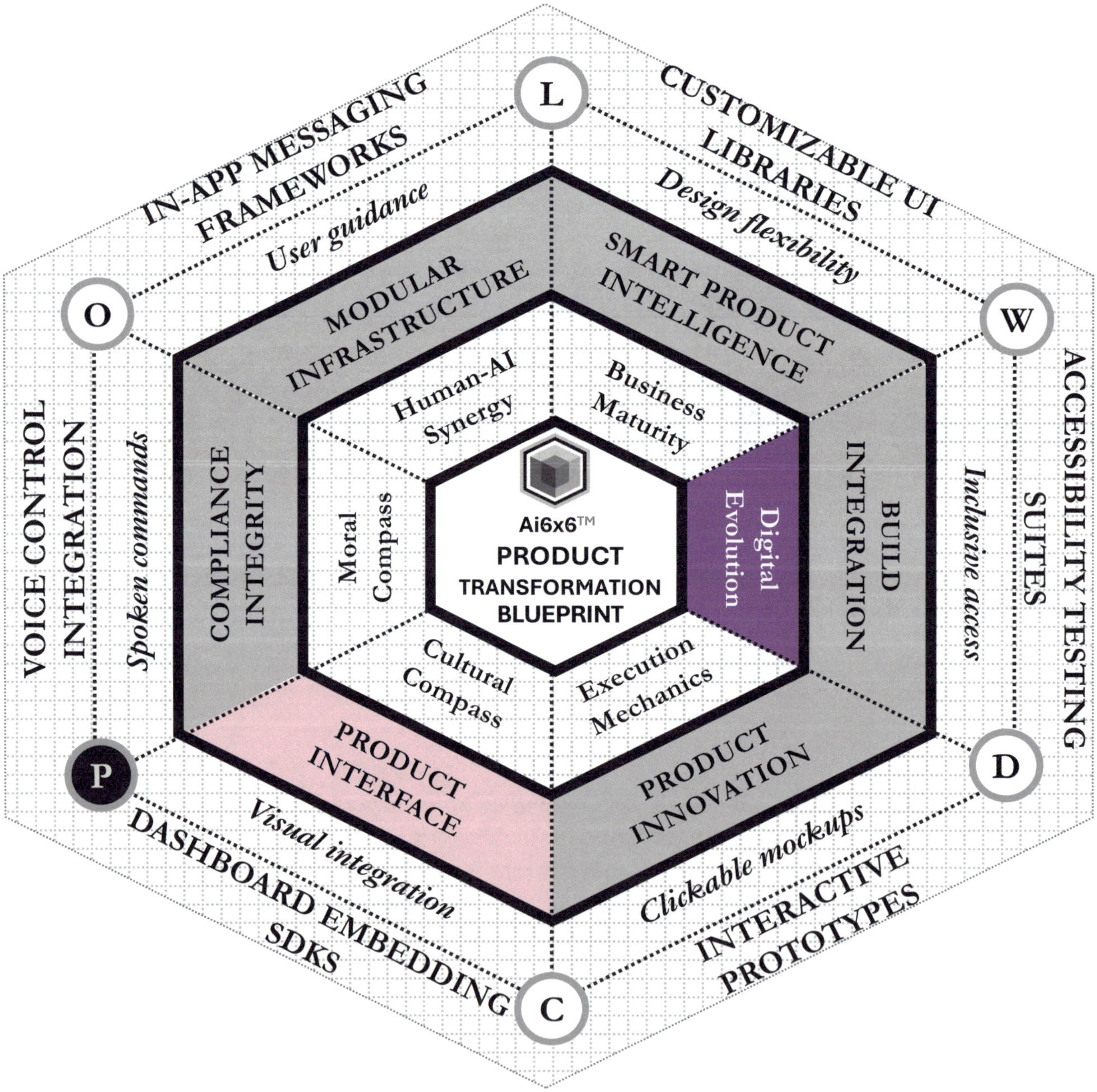

Ai6x6 DIGITAL EVOLUTION FOR INTERFACE | PRODUCT TRANSFORMATION

1. IN-APP MESSAGING FRAMEWORKS

In-App Messaging Frameworks are software development kits (SDKs) and platforms that enable product teams to send targeted, contextual messages directly within a mobile or web application. These messages can include onboarding guides, feature announcements, personalized offers, urgent alerts, or feedback requests, appearing as banners, pop-ups, or full-screen overlays. By delivering relevant communications at key moments in the user journey, In-App Messaging Frameworks enhance user engagement, improve feature adoption, guide users through complex workflows, and reduce churn. This allows for real-time interaction, personalized experiences, and efficient communication without requiring users to leave the application, ultimately driving product activation, retention, and overall user satisfaction.

Without In-App Messaging Frameworks, organizations are limited to generic external communication channels (e.g., email, push notifications) or static in-app elements that cannot adapt to individual user behavior. This results in less effective user onboarding, lower feature discovery rates, and missed opportunities to provide timely assistance or personalized guidance. The absence of dynamic, contextual in-app messaging hinders the ability to engage users proactively at critical moments, deliver targeted information, or collect direct feedback within the application. This ultimately impacts user retention, product adoption, and the overall stickiness and user-friendliness of the product experience.

Implementation	Challenges
* Define messaging goals (onboarding, feature) * Select an in-app messaging platform * Integrate SDK into application code * Design message templates & content * Define user segments & triggers * Plan for A/B testing & performance monitoring	* Designing non-intrusive yet effective messages * Managing message fatigue & user annoyance * Ensuring precise targeting & segmentation * Integrating with user analytics for triggers * A/B testing different message strategies * Maintaining performance on low-end devices

Products	Platform Cost	Project Cost	Duration
Braze	$5-30K	$20-100K	2-6 m
Intercom	$2-15K	$10-70K	1-4 m
Leanplum	$5-25K	$20-90K	2-5 m
Localytics	$2-10K	$10-50K	1-3 m
Iterable	$5-30K	$20-100K	2-6 m
Firebase In-App Messaging (Google)	N/A	$0.5-5K	0.5-1.5 m

2. CUSTOMIZABLE UI LIBRARIES

Customizable UI Libraries are collections of pre-built user interface components (e.g., buttons, forms, navigation menus, modals) that are designed to be highly flexible and adaptable to a product's specific branding, design system, and functional requirements. These libraries often provide extensive styling options, theme support, and clear API documentation, allowing developers to easily integrate and modify components without building everything from scratch. By offering a robust foundation for consistent and branded interfaces, Customizable UI Libraries accelerate frontend development, ensure design consistency across the product, and reduce technical debt. This empowers development teams to build beautiful, intuitive, and unique product experiences efficiently, while maintaining a cohesive visual language across the application.

Without Customizable UI Libraries, development teams often resort to building every UI component from scratch or using rigid, unadaptable libraries. This leads to inconsistent user interfaces across different parts of a product, slower development cycles due to repetitive coding, and difficulty in maintaining brand consistency. The absence of a reusable, flexible set of UI components results in increased technical debt, higher development costs, and a fragmented user experience. This ultimately impacts product quality, designer-developer collaboration, and the ability to rapidly iterate on the user interface while maintaining a cohesive and polished look and feel.

Implementation	Challenges
* Define product's design system & branding * Select a suitable UI library * Customize components to match design specs * Integrate into application frontend * Establish component usage guidelines * Plan for ongoing library maintenance & updates	* Ensuring full adherence to design system * Overcoming customization limitations of library * Managing performance of complex components * Keeping up with library updates & breaking changes * Training developers on library usage & best practices * Balancing custom code with library standards

Products	Platform Cost	Project Cost	Duration
Material-UI (React)	N/A	$1-10K	0.5-2 m
Ant Design (React)	N/A	$1-10K	0.5-2 m
Bootstrap (CSS Framework)	N/A	$0.5-5K	0.5-1 m
Chakra UI (React)	N/A	$0.5-5K	0.5-1 m
Vue Material (Vue.js)	N/A	$0.5-5K	0.5-1 m
Kendo UI (Various Frameworks)	$0.5-5K	$2-20K	1-2 m

Accessibility Testing Suites are software tools and platforms designed to identify and report accessibility issues in digital products (websites, mobile apps, software) to ensure they are usable by people with disabilities. These suites perform automated scans for common accessibility violations (e.g., missing alt text, insufficient color contrast, improper keyboard navigation) and often provide guided manual testing features. By integrating accessibility checks into the development and QA process, Accessibility Testing Suites help organizations meet compliance standards (e.g., WCAG, Section 508), avoid legal risks, and expand their user base to include individuals with diverse abilities. This ensures a more inclusive product, enhances user experience for all, and demonstrates a commitment to ethical and responsible product design.

Without Accessibility Testing Suites, organizations risk launching digital products that are unusable or difficult to navigate for people with disabilities, leading to exclusion of a significant user segment, potential legal liabilities, and damage to brand reputation. Relying on manual, ad-hoc accessibility checks is inefficient, prone to error, and often misses critical violations of compliance standards (e.g., WCAG). The absence of systematic accessibility testing hinders the ability to identify and remediate barriers for disabled users proactively, making the product less inclusive and potentially exposing the organization to discrimination lawsuits. This ultimately impacts market reach, user satisfaction, and the ethical standing of the product.

Implementation	Challenges
* Define target accessibility standards (WCAG) * Select an accessibility testing suite * Integrate into CI/CD pipelines & QA workflow * Conduct automated & manual testing * Train developers & QAs on accessibility best practices * Plan for continuous monitoring & remediation	* Interpreting complex accessibility guidelines * Balancing automated vs. manual testing efforts * Prioritizing & remediating discovered issues * Integrating testing into agile development cycles * Ensuring accessibility for dynamic content * User adoption & awareness within dev teams

Products	Platform Cost	Project Cost	Duration
axe DevTools (Deque Systems)	$0-5K	$0.5-10K	0.5-2 m
Lighthouse (Google - built-in Chrome)	N/A	$0.1-1K	0.1-0.5 m
Accessibility Insights (Microsoft)	N/A	$0.1-1K	0.1-0.5 m
WAVE (WebAIM)	N/A	$0.1-1K	0.1-0.5 m
SortSite (PowerMapper)	$0.1-1K	$0.5-2K	0.5-1 m
Pa11y (Open Source)	N/A	$0.5-5K	0.5-1 m

Interactive Prototypes are high-fidelity, clickable simulations of a digital product's user interface and experience. They allow users and stakeholders to interact with a product concept as if it were a live application, without any underlying code development. These prototypes go beyond static mockups by incorporating transitions, animations, and responsive behaviors, providing a realistic feel of the final product. By enabling early and extensive user testing, gathering actionable feedback on usability and flow, and visually communicating product vision to stakeholders, Interactive Prototypes significantly reduce development rework, validate design decisions before coding, and accelerate the iterative design process. This ensures the final product is intuitive, user-friendly, and aligned with user expectations, driving successful product adoption.

Without Interactive Prototypes, product design and development teams often rely on static wireframes or low-fidelity mockups, which fail to convey the dynamic user experience and interactive nuances of a product. This leads to misunderstandings, misinterpretations, and costly design changes discovered late in the development cycle. The absence of a realistic, interactive simulation hinders effective user testing, delays critical feedback, and makes it difficult for stakeholders to truly visualize and validate the product concept. This ultimately results in slower design iterations, increased development costs due to rework, and a higher risk of launching a product with a suboptimal or confusing user interface.

Implementation	Challenges
* Define user flows & key interactions * Select a prototyping tool (Figma, Adobe XD) * Create high-fidelity UI designs * Add interactive elements (clicks, hovers, transitions) * Conduct user testing & gather feedback * Iterate based on user insights	* Managing complexity of large, detailed prototypes * Ensuring realistic feel without actual code * Keeping prototypes updated with design changes * Handoff and communication with development * Performance issues with very complex prototypes * User engagement during testing sessions

Products	Platform Cost	Project Cost	Duration
Figma (Prototyping)	$0-0.5K	$0.1-2K	0.1-0.5 m
Adobe XD	$0-0.5K	$0.1-2K	0.1-0.5 m
Sketch (with plugins like InVision)	$0-0.5K	$0.1-2K	0.1-0.5 m
Axure RP	$0.1-0.5K	$0.5-2K	0.5-1 m
InVision	$0.1-1K	$0.5-5K	0.5-1 m
Framer	$0.1-1K	$0.5-5K	0.5-1 m

5. DASHBOARD EMBEDDING SDKS

Dashboard Embedding SDKs (Software Development Kits) are toolkits that allow developers to seamlessly integrate interactive dashboards and data visualizations from business intelligence (BI) platforms directly into their own applications, websites, or portals. These SDKs provide APIs and components that enable secure authentication, dynamic filtering, and customization of the embedded content, maintaining the look and feel of the host application. By bringing powerful analytics and reporting capabilities directly into the user's workflow, Dashboard Embedding SDKs enhance product value, provide real-time insights, and create a more data-driven user experience. This empowers users with self-service analytics without leaving the application, driving deeper engagement, informed decision-making, and increased product stickiness.

Without Dashboard Embedding SDKs, organizations would typically require users to switch between their core product and a separate BI or analytics platform to access data insights. This creates a fragmented and inconvenient user experience, reducing the likelihood of users leveraging valuable data. The absence of seamless embedding capabilities hinders the ability to provide real-time, in-context analytics, limits data-driven decision-making directly within the product, and reduces the perceived value of integrated reporting. This ultimately impacts user engagement, slows down insight-to-action cycles, and makes the product less compelling for users who require immediate access to performance metrics and visual data.

Implementation	Challenges
* Select a BI platform with embedding SDK * Configure secure authentication (SSO, JWT) * Integrate SDK into product frontend * Design placement & context for dashboards * Implement dynamic filtering & customization * Plan for performance optimization & scaling	* Ensuring seamless user authentication & authorization * Maintaining consistent UI/UX with host app * Performance of embedded dashboards * Managing data security & access controls * Customizing embedded dashboards effectively * Licensing costs for embedding analytics

Products	Platform Cost	Project Cost	Duration
Tableau Embedded Analytics	$5-30K	$20-100K	2-6 m
Power BI Embedded (Microsoft)	$5-25K	$20-90K	2-5 m
Looker Embedded Analytics (Google)	$10-50K	$60-250K	3-9 m
Sisense Fusion (Embedded BI)	$10-40K	$50-200K	3-8 m
Metabase (Open Source / Self-Hosted)	N/A	$5-30K	1-3 m
Qlik Sense Embedded	$5-30K	$20-100K	2-6 m

6. VOICE CONTROL INTEGRATION

Voice Control Integration involves embedding voice recognition and natural language processing (NLP) capabilities directly into a product, allowing users to interact with it using spoken commands. This can range from simple voice commands for navigation or data entry to more complex conversational interfaces for task completion. By leveraging advanced speech-to-text and NLU technologies, Voice Control Integration provides a hands-free, intuitive, and highly accessible mode of interaction, particularly beneficial for users with disabilities, multitasking scenarios, or environments where touch/visual interaction is impractical. This enhances product usability, speeds up task completion, and offers a more natural, engaging user experience, positioning the product at the forefront of evolving human-computer interaction paradigms.

Without Voice Control Integration, products rely solely on traditional visual or touch-based interfaces, limiting their accessibility and convenience in certain contexts (e.g., in-car systems, smart home devices, or for users with motor impairments). This means users are unable to interact with the product hands-free, leading to less intuitive experiences, slower task completion in specific scenarios, and exclusion of users who prefer or require voice-based interaction. The absence of voice control hinders the ability to provide an inclusive, convenient, and cutting-edge user experience, ultimately impacting market appeal, user satisfaction, and the competitive positioning of the product in an increasingly voice-enabled world.

Implementation	Challenges
* Define voice commands & user flows * Select a voice AI/NLP SDK * Integrate SDK into product interface * Train voice models for specific vocabulary * Design conversational UI & error handling * Plan for continuous improvement & testing	* Ensuring high accuracy in speech recognition * Handling diverse accents, dialects & background noise * Designing natural & forgiving voice commands * Integrating with product's core logic * Performance impact & latency * User adoption & privacy concerns

Products	Platform Cost	Project Cost	Duration
Google Cloud Speech-to-Text / Dialogflow	$0.5-5K	$5-30K	1-3 m
Amazon Alexa Voice Service (AVS)	N/A	$5-25K	1-3 m
Microsoft Azure AI Speech	$0.5-5K	$5-30K	1-3 m
IBM Watson Assistant (Voice Integration)	$0.5-5K	$5-25K	1-3 m
SiriKit (Apple - iOS apps)	N/A	$2-15K	1-2 m
Custom Voice Recognition Engine	N/A	$50-250K	3-9 m

Ai6x6™ Transformation Blueprint

Compliance 'Integrity'

"The best compliance is proactive. Bake it into your build." — Brian Krebs
Cybersecurity journalist; widely followed for exposing gaps and promoting secure-by-design approaches.

Compliance is not legal insulation—it's strategic architecture. Security scans, vulnerability monitoring, and recall readiness position products to scale without suspicion. "Compliance Integrity" means compliance isn't added post-launch—it's embedded pre-code. In transformation, resilience is a design principle. Mature organizations don't fear regulation—they rehearse for it. Licensing, documentation, and safety standards are no longer burdens—they're blueprints. When compliance becomes intuitive, trust becomes repeatable. Product excellence isn't just what it does—it's what it's ready for when it matters most.

The "Compliance Integrity" chapter is fundamental for product transformation's digital maturity as it highlights the non-negotiable importance of security, compliance, and ethical considerations embedded in product development. A digitally mature organization builds integrity by design, utilizing security scanning tools, vulnerability management, and license compliance to ensure products are robust, trustworthy, and adhere to all legal and safety standards. This commitment, extending to regulatory submission platforms and recall management, protects users, maintains brand reputation, and underpins the long-term success and sustainability of digital products in an increasingly regulated and security-conscious world.

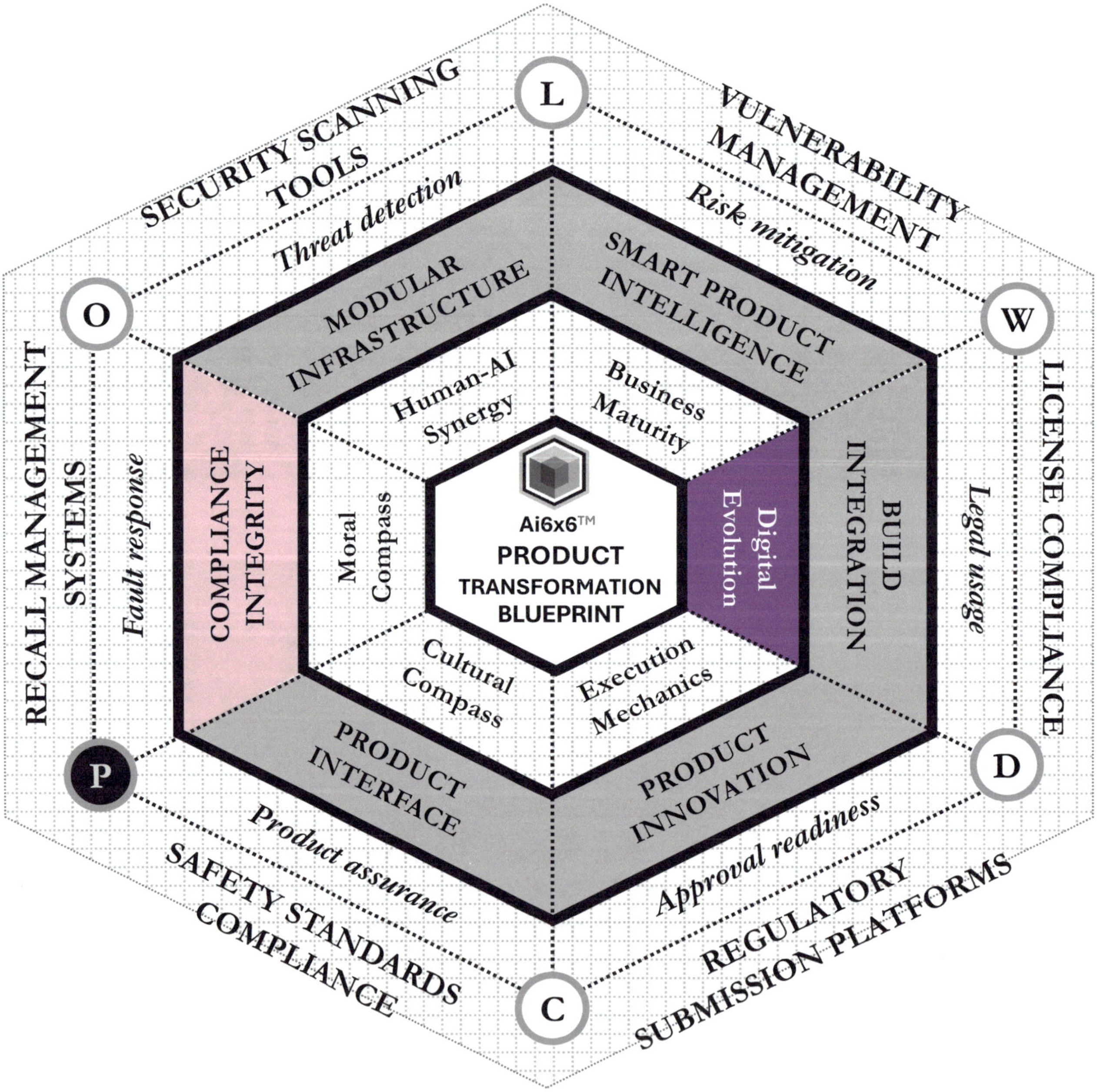

Ai6x6 DIGITAL EVOLUTION FOR INTEGRITY | PRODUCT TRANSFORMATION

1. SECURITY SCANNING TOOLS

Security Scanning Tools are software applications designed to identify vulnerabilities and weaknesses in a product's code, dependencies, and deployed environment. These tools range from Static Application Security Testing (SAST) for source code analysis, Dynamic Application Security Testing (DAST) for runtime vulnerabilities, to Software Composition Analysis (SCA) for open-source component vulnerabilities. By automatically scanning for known exploits, misconfigurations, and security flaws throughout the development lifecycle, Security Scanning Tools enable development teams to proactively identify and remediate risks before product deployment. This significantly enhances product security, reduces the attack surface, ensures compliance with security standards, and protects sensitive data, crucial for building trustworthy and resilient software products.

Without Security Scanning Tools, organizations are left to manually inspect code and environments for vulnerabilities, a process that is highly inefficient, error-prone, and cannot keep pace with modern development speeds or the evolving threat landscape. This leads to a higher probability of releasing products with critical security flaws, making them susceptible to cyberattacks, data breaches, and reputational damage. The absence of automated security checks results in a reactive approach to security, increased costs for remediation post-launch, and potential non-compliance with industry regulations, ultimately undermining product integrity and customer trust.

Implementation	Challenges
* Define security policies & standards * Select appropriate scanning tools (SAST, DAST, SCA) * Integrate scans into CI/CD pipelines * Automate vulnerability detection & reporting * Establish remediation workflows & ownership * Plan for continuous monitoring & re-scans	* High volume of false positives requiring triage * Integrating tools across diverse tech stacks * Performance impact on CI/CD pipelines * Prioritizing & remediating critical vulnerabilities * Ensuring comprehensive scan coverage * Keeping up with evolving threat landscape

Products	Platform Cost	Project Cost	Duration
Snyk (SCA, SAST, Container)	$2-15K	$10-70K	1-4 m
Veracode (SAST, DAST, SCA)	$5-30K	$20-100K	2-6 m
Checkmarx (SAST, SCA, DAST)	$5-25K	$20-90K	2-5 m
GitLab Ultimate (Built-in SAST/DAST/SCA)	$1-10K	$5-50K	1-4 m
OWASP ZAP (DAST - Open Source)	N/A	$0.5-5K	0.5-1.5 m
Qualys Web Application Scanning	$2-10K	$5-30K	1-3 m

Vulnerability Management is a systematic process of identifying, evaluating, treating, and reporting on security vulnerabilities within an organization's products, systems, and infrastructure. It involves continuous monitoring, scanning, and analysis to discover security weaknesses, followed by prioritization based on risk level and remediation efforts. Unlike one-off security scans, vulnerability management is an ongoing, cyclical process that aims to reduce the overall attack surface and improve the security posture over time. By centralizing vulnerability data, facilitating remediation workflows, and providing clear reporting, Vulnerability Management helps organizations proactively reduce security risks, ensure compliance, and maintain the integrity and resilience of their products against evolving cyber threats.

Without a robust Vulnerability Management process, organizations face a reactive and chaotic approach to addressing security weaknesses, leading to an increasing backlog of unaddressed vulnerabilities and a higher risk of successful cyberattacks. Disparate scan results, a lack of centralized tracking, and unclear ownership for remediation efforts create a "whack-a-mole" scenario where critical vulnerabilities are missed or addressed too slowly. The absence of systematic vulnerability management results in higher operational costs for incident response, potential data breaches, and non-compliance with industry regulations, ultimately eroding trust in the product and compromising its security integrity.

Implementation	Challenges
* Define asset inventory & risk categories	* Managing vast number of reported vulnerabilities
* Select a vulnerability management platform	* Prioritizing based on true business risk
* Integrate with security scanners & ITSM	* Coordinating remediation across teams
* Establish prioritization & remediation workflows	* False positives & noise in scan results
* Implement continuous monitoring & reporting	* Keeping up with new vulnerabilities & patches
* Plan for regular security assessments & audits	* Demonstrating continuous improvement in posture

Products	Platform Cost	Project Cost	Duration
Tenable.io / Nessus	$5-30K	$20-100K	2-6 m
Qualys Vulnerability Management	$5-25K	$20-90K	2-5 m
Rapid7 InsightVM	$5-20K	$20-80K	2-5 m
CrowdStrike Falcon Spotlight	$10-40K	$50-200K	3-8 m
Wiz (Cloud Native VM)	$10-50K	$60-250K	3-9 m
OpenVAS (Open Source)	N/A	$5-30K	1-3 m

License Compliance in product development refers to the process of ensuring that all open-source software (OSS) and third-party libraries used within a product adhere to their respective licensing terms and conditions. This involves identifying all external components, understanding their licenses (e.g., MIT, GPL, Apache), and verifying that their use, modification, and distribution comply with legal obligations. Tools for License Compliance automate the scanning of codebases to detect and report on OSS licenses, track dependencies, and manage potential conflicts or obligations. By ensuring legal adherence to OSS licenses, organizations avoid costly legal disputes, protect their intellectual property, and maintain ethical sourcing practices, crucial for building sustainable and legally sound software products.

Without systematic License Compliance practices and tools, organizations risk inadvertently violating open-source software (OSS) licenses, which can lead to costly legal disputes, intellectual property claims, forced disclosure of proprietary code, and severe reputational damage. Manually tracking and verifying licenses for every third-party component is an impossible task in modern software development, leading to blind spots and unknown legal risks. The absence of automated license scanning and management results in uncontrolled legal exposure, hinders the ability to distribute the product securely, and undermines the ethical and legal foundation of the product, ultimately impacting market viability and investor confidence.

Implementation	Challenges
* Define OSS policy & risk appetite * Select a license compliance tool * Integrate scans into CI/CD pipelines * Establish license review & approval workflows * Maintain accurate inventory of dependencies * Plan for legal review & policy updates	* Interpreting complex & ambiguous license terms * Managing transitive dependencies & their licenses * Dealing with license conflicts & incompatible licenses * High volume of components in modern software * Ensuring accuracy of automated detection * Educating developers on license obligations

Products	Platform Cost	Project Cost	Duration
Black Duck Software (Synopsys)	$5-30K	$20-100K	2-6 m
Snyk (Open Source Security & License)	$2-15K	$10-70K	1-4 m
WhiteSource (Mend)	$5-25K	$20-90K	2-5 m
FOSSA	$2-10K	$10-50K	1-3 m
SPDX Tools (Open Source Standards)	N/A	$1-10K	0.5-2 m
DepShield (GitHub)	N/A	$0.5-5K	0.5-1 m

4. REGULATORY SUBMISSION PLATFORMS

Regulatory Submission Platforms are specialized software solutions that streamline and automate the process of preparing, compiling, and submitting product-related documentation to regulatory bodies (e.g., FDA for medical devices, FCC for electronics, regional data privacy authorities). These platforms often include features for document management, workflow automation, audit trails, and compliance checks against specific regulatory standards. By centralizing the complex and highly scrutinized submission process, Regulatory Submission Platforms reduce manual errors, accelerate time-to-market for regulated products, and ensure meticulous adherence to legal and industry requirements. This is critical for products in sectors like healthcare, finance, or highly regulated consumer goods, where compliance is paramount for market access and avoiding severe penalties.

Without dedicated Regulatory Submission Platforms, organizations in regulated industries face an extremely complex, manual, and error-prone process for preparing and submitting product documentation. This leads to significant delays in market entry, increased risk of non-compliance, and potential rejections from regulatory bodies due to missing or incorrect information. The absence of a centralized, auditable system for managing submission artifacts hinders traceability, complicates version control, and makes it difficult to demonstrate adherence to strict regulatory guidelines. This ultimately results in higher operational costs, prolonged time-to-market, and a heightened risk of legal and financial penalties, jeopardizing the product's ability to reach its intended market.

Implementation	Challenges
* Identify applicable regulations & standards	* Interpreting complex & evolving regulations
* Select a regulatory submission platform	* Ensuring data integrity & audit trails
* Map internal processes to platform workflows	* Managing large volumes of documentation
* Migrate existing compliance documentation	* Integrating with diverse internal systems
* Integrate with QMS/PLM systems	* Staying updated with regulatory changes
* Train compliance & product teams	* High cost of validation & qualification

Products	Platform Cost	Project Cost	Duration
Veeva Vault RIM	$20-100K	$150-700K	6-18 m
MasterControl QMS (with submissions)	$15-70K	$100-500K	5-16 m
Sparta Systems TrackWise	$10-50K	$60-300K	4-12 m
Dassault Systèmes BIOVIA	$20-90K	$120-600K	5-18 m
ArisGlobal LifeSphere Regulatory	$15-60K	$80-400K	4-14 m
Oracle Argus (Pharmacovigilance)	$10-50K	$70-350K	4-12 m

Safety Standards Compliance in product development refers to the adherence to specific industry- or government-mandated safety regulations and certifications, ensuring that a product does not pose undue risks to users or the environment. This involves incorporating safety by design principles, conducting rigorous testing (e.g., functional safety, electrical safety, material safety), and maintaining meticulous documentation to demonstrate conformity with standards like ISO 26262 (automotive functional safety), IEC 61508 (industrial functional safety), or UL/CE markings (product safety). By embedding safety compliance throughout the product lifecycle, organizations protect consumers, avoid legal liabilities, and build trust in their products' reliability. This is critical for preventing accidents, reducing recalls, and ensuring market acceptance in safety-critical domains.

Without a systematic approach to Safety Standards Compliance, organizations risk launching products that are inherently unsafe, leading to severe user injuries or fatalities, costly product recalls, massive legal liabilities, and irreparable damage to brand reputation. Ad-hoc or informal safety checks are insufficient to meet stringent regulatory requirements and can result in significant design flaws being overlooked. The absence of integrated safety by design principles and rigorous validation processes hinders the ability to demonstrate due diligence, obtain necessary certifications, and ensure market access for safety-critical products. This ultimately jeopardizes public trust, incurs enormous financial penalties, and can lead to criminal charges, severely impacting the long-term viability of the product and the company.

Implementation	Challenges
* Identify applicable safety standards * Integrate safety requirements into design * Conduct formal safety analyses (e.g., FMEA) * Implement rigorous safety testing & validation * Maintain comprehensive safety documentation * Plan for ongoing audits & certification renewals	* Interpreting complex technical safety standards * Balancing safety requirements with cost/performance * Managing functional safety in complex systems * Conducting exhaustive validation & verification * Adapting to evolving safety regulations * Securing and demonstrating traceability of safety evidence

Products	Platform Cost	Project Cost	Duration
PTC Windchill (Safety Analysis modules)	$15-70K	$100-500K	6-18 m
Siemens Teamcenter (Safety & Compliance)	$20-90K	$120-600K	6-18 m
Jama Connect (Requirements & Traceability)	$5-30K	$20-100K	2-6 m
Polarion ALM (Siemens - for traceability)	$5-25K	$20-90K	2-5 m
MedAccred Software (Medical Device Safety)	$10-40K	$50-200K	3-9 m
Custom FMEA/HAZOP Software	N/A	$30-150K	3-9 m

6. RECALL MANAGEMENT SYSTEMS

Recall Management Systems are specialized software solutions designed to plan, execute, and track product recalls efficiently and compliantly. These systems provide capabilities for managing recall notices, contacting affected customers, tracking returned products, coordinating remediation (repair, replacement, refund), and generating regulatory reports. By centralizing and automating the complex recall process, Recall Management Systems enable organizations to respond swiftly and effectively to safety or quality issues, minimize public risk, protect brand reputation, and ensure full compliance with regulatory requirements. This is a critical component for risk mitigation and maintaining customer trust, ensuring that a company can handle adverse product events with transparency and accountability.

Without dedicated Recall Management Systems, organizations attempting to manage a product recall face a chaotic, manual, and highly inefficient process, leading to slow response times, ineffective communication with affected customers, and potential non-compliance with regulatory deadlines. This results in prolonged public safety risks, severe damage to brand reputation, increased legal exposure, and significant financial losses due to disorganized logistics and unrecovered products. The absence of a structured system hinders the ability to quickly identify affected batches, trace products, notify customers effectively, and report accurately to authorities, ultimately exacerbating the impact of a recall and undermining public trust in the product and the company.

Implementation	Challenges
* Define recall protocols & communication plan * Select a recall management platform * Integrate with CRM, ERP & supply chain * Establish clear roles & responsibilities * Implement customer notification & tracking * Plan for reporting & post-recall analysis	* Swiftly identifying affected product batches * Effectively reaching all affected customers * Managing logistics of product return/replacement * Ensuring regulatory reporting accuracy * Protecting brand image during crisis * Data privacy for customer contact information

Products	Platform Cost	Project Cost	Duration
TraceLink (Recall & Patient Solutions)	$15-70K	$80-400K	4-12 m
MasterControl QMS (Recall Management)	$10-50K	$60-300K	4-12 m
SAP S/4HANA (Quality Management)	$20-100K	$150-700K	6-18 m
Oracle Agile PLM (Quality & Compliance)	$15-80K	$100-500K	5-16 m
Compliant Recall (SaaS)	$5-30K	$20-100K	2-6 m
Custom ERP/PLM Module	N/A	$50-250K	4-10 m

BLUEPRINT SERIES

SECTION E

Execution Mechanics

Ai6x6™ Transformation Blueprint

Story 1 — The Manhattan Project Coordination Model (1942–1946): The Manhattan Project united scientists, engineers, and military leaders across dispersed sites under extreme secrecy and urgency. Workstreams were interdependent, requiring precise coordination without full visibility. Structured reporting, controlled interfaces, and disciplined compartmentalization enabled progress despite complexity. Human judgment guided trade-offs while machines and labs executed specialized tasks. The project demonstrated how execution succeeds when governance, clarity, and cadence align. It became a template for managing high-stakes, multi-disciplinary programs.

Story 2 — The Berlin Airlift Logistics System (1948–1949): With land routes blocked, Allied forces had to supply an entire city solely by air. Aircraft schedules, loading patterns, and turnaround times were optimized to the minute. Machines provided lift and speed, but humans orchestrated sequencing, prioritization, and contingency planning. Continuous feedback loops refined operations daily. The airlift proved that disciplined execution can overcome structural constraints. It showed how precision, rhythm, and adaptability create operational breakthroughs.

Story 3 — The Interstate Highway Construction Program (1956–1970s): The United States embarked on building a nationwide highway network requiring coordination across states, contractors, and engineering disciplines. Standardized designs ensured consistency while local teams adapted to terrain and conditions. Machines handled heavy construction, but humans managed sequencing, safety, and quality. The program relied on predictable funding, clear specifications, and phased execution. It demonstrated how large-scale delivery succeeds when structure and flexibility coexist. The highways became a symbol of disciplined national execution.

Story 4 — Toyota's Just-in-Time Logistics (1960s–1980s): As Toyota refined its production system, logistics became a critical execution layer. Parts arrived exactly when needed, minimizing inventory and waste. Machines tracked flow, but humans monitored signals, resolved disruptions, and maintained rhythm. Kanban created transparency across the value chain. The system required precision, trust, and continuous calibration. It showed how execution excellence emerges when every component of the workflow reinforces timing, quality, and flow.

Story 5 — FedEx's Overnight Delivery Network (1973–1990s): FedEx built a hub-and-spoke system capable of moving packages across the country overnight. Aircraft, sorting machines, and routing algorithms provided speed, but humans managed exceptions, weather disruptions, and customer commitments. The network operated on tight cycles that demanded flawless coordination. Data informed decisions while people ensured reliability. The system demonstrated how execution mechanics scale when tech and human oversight operate in synchronized cadence.

Story 6 — The Rise of Agile Software Delivery (2001–2015): Software teams struggled with long cycles, rigid plans, and late-stage failures. Agile introduced iterative delivery, cross-functional collaboration, and continuous feedback. Machines automated testing and deployment, while humans prioritized work, clarified requirements, and adapted to change. Execution shifted from linear phases to dynamic cycles. The approach improved predictability and responsiveness. It became a global standard for modern execution, proving that disciplined iteration outperforms rigid planning.

LESSONS FROM A PRODUCT TRANSFORMATION PERSPECTIVE

Product excellence depends on execution mechanics that translate intent into consistent outcomes. As organizations matured, products evolved from handcrafted outputs to systematically delivered solutions shaped by standards, iteration, and feedback. Machines automate repeatable steps, while humans refine design, resolve ambiguity, and ensure alignment with user needs. Execution maturity ensures that products are built predictably, improved continuously, and delivered reliably across contexts. It transforms product development from episodic bursts into sustained capability. The most resilient products emerge when execution systems support learning, quality, and adaptability—turning product creation into a disciplined, repeatable, and scalable process.

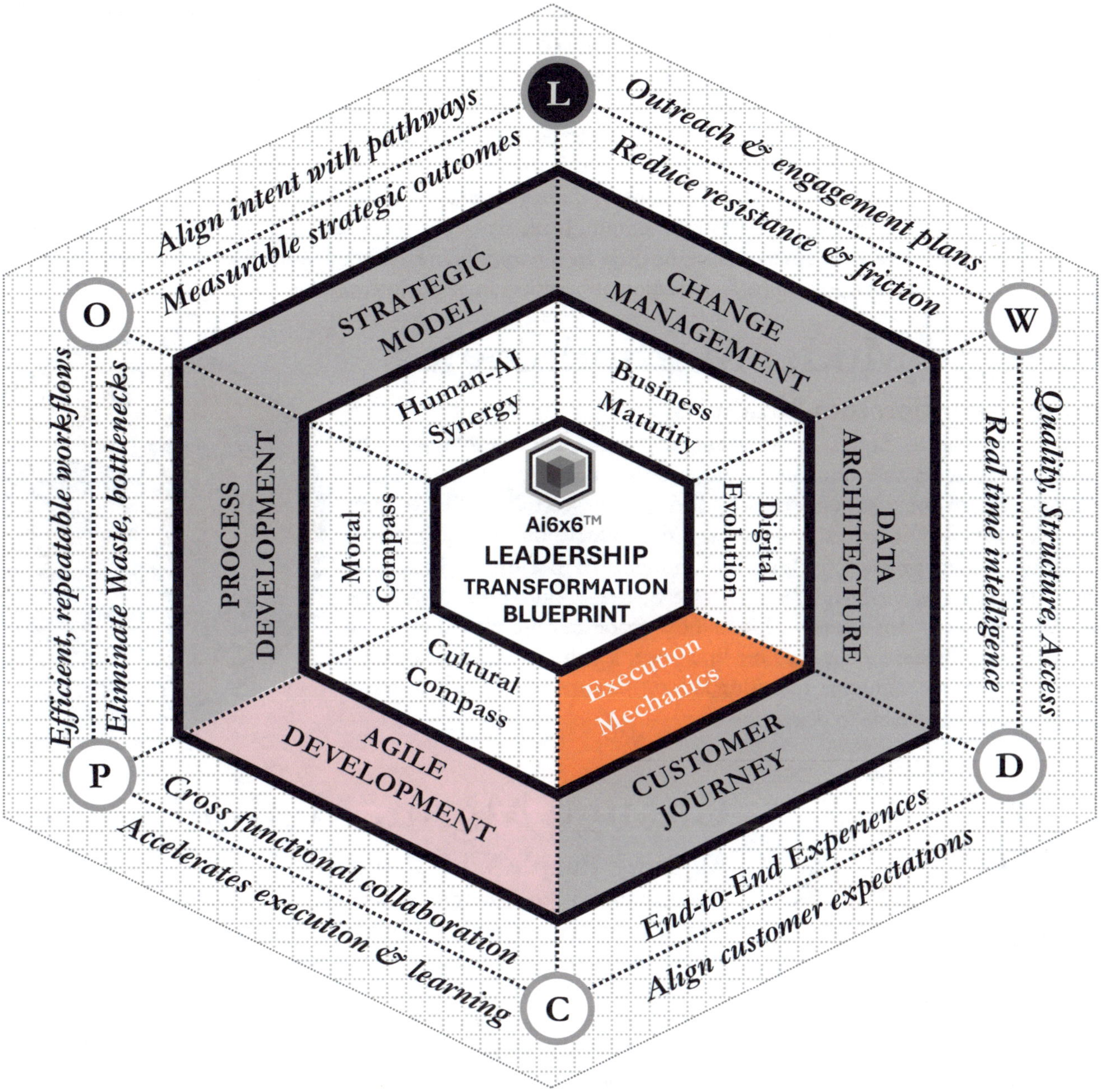

The **Ai6x6 Research Lab** identified this lens as the primary antidote to the "developmental drift" that often plagues complex engineering projects. We chose it because the human mind requires a disciplined, repeatable system to navigate the high-stakes demands of **Scrum** and **Kanban** without losing the rhythm of progress. This lens isn't merely about the mechanics of **CI/CD Pipelines**; it is about the psychological certainty and mental clarity that come from a proven methodology like the **Lean Startup** or the **SAFe Framework**. It is vital because it provides a mental scaffolding, turning the chaos of a build into a series of surgical, confident releases.

This lens allows the product team's heart to remain calm while the development cycle moves at terminal velocity. By relying on **Feature Flags** and **Testing**, the leader replaces "launch-day anxiety" with evidence-based conviction. We believe that execution mechanics are the ultimate form of creative mindfulness—the practice of being fully present in the "how" of the flow so that the "what" of reliability is achieved with excellence. This lens ensures that every sprint has a defined purpose and a coherent structure, providing the steady footing necessary to scale globally with the assurance of a master coordinator.

Execution has always been the crucible where strategy proves its worth. Leaders can articulate bold visions, design elegant frameworks, and outline ambitious roadmaps, but none of it matters if the organization cannot execute with precision, coherence, and resilience. In the age of AI and accelerated change, **Execution Mechanics** becomes the lens that determines whether transformation becomes real or remains theoretical. It forces leaders to confront the uncomfortable truth that most failures are not failures of strategy but failures of execution. The gap between intention and impact is rarely philosophical; it is operational. Execution Mechanics exposes that gap and provides the architecture to close it. This lens reframes execution as a system—an interconnected set of methodologies that convert ambition into momentum. It recognizes that execution is not the final step of strategy but the environment in which strategy lives or dies. In a world where AI accelerates decision cycles, compresses timelines, and amplifies both strengths and weaknesses, leaders must design execution as they design vision.

Strategic Model Methodologies

Turning Ambition Into Structure

Execution Mechanics begins with **strategic model methodologies**, the structured approaches that translate ambition into actionable pathways. These methodologies are not academic exercises; they are the scaffolding that ensures strategy is grounded. Leaders who neglect this dimension find themselves improvising under pressure, relying on intuition rather than disciplined frameworks. Strategic models define the logic of execution—how choices are made, how priorities are sequenced, and how resources are allocated. They ensure that every initiative has a defined purpose, a coherent structure, and a measurable outcome. Without this foundation, execution becomes reactive, fragmented, and vulnerable to disruption.

Change Mgmt. Methodologies

Aligning People With Progress

The next dimension, **change management methodologies**, recognizes that execution is ultimately a human endeavor. No transformation succeeds unless people understand, accept, and commit to the change. Leaders who underestimate this reality discover that resistance is not a barrier but a predictable response to uncertainty. Effective change management acknowledges the emotional, cultural, and psychological dimensions of transformation. It ensures that communication is transparent, expectations are clear, and support is available. It turns change from something imposed into something shared. In this dimension, execution becomes relational—rooted in trust, empathy, and the ability to guide people toward possibility.

Data Architecture Methodologies

Building the Backbone of Execution

Data architecture provide the backbone for modern execution. In a world where decisions must be informed by evidence, organizations cannot afford fragmented data, inconsistent definitions, or unreliable pipelines. Data architecture is not a technical concern; it is a strategic one. It determines whether insights are trustworthy, whether systems can scale, and whether AI can operate responsibly. Leaders who invest in strong data architecture create the conditions for intelligence to flow freely across the enterprise. In this dimension, execution becomes intelligent—powered by clarity, consistency, and evidence.

Customer Journey Methodologies

Anchoring Execution in Value

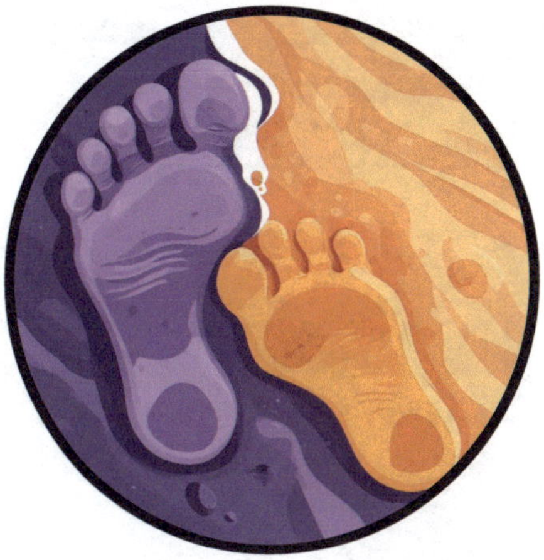

Customer journey ensure that execution remains anchored in value creation. Too many organizations design processes around internal convenience rather than customer experience. Customer journey thinking forces leaders to see the organization from the outside in, identifying friction, gaps, and opportunities that internal perspectives overlook. It ensures that execution is not only efficient but meaningful. It aligns operations with the expectations, needs, and emotions of the people the organization serves. In this dimension, execution becomes empathetic—shaped by the reality of customers rather than the assumptions of leaders.

Agile Development Methodologies

Enabling Adaptation at Speed

Agile development introduce the adaptability required in a world where conditions shift rapidly. Traditional execution models assume stability, predictability, and linear progress. Agile assumes the opposite. It embraces iteration, experimentation, and continuous learning. It allows teams to adjust quickly, respond to feedback, and deliver value incrementally. Leaders who adopt agile mindsets create organizations that move with speed without sacrificing quality. In this dimension, execution becomes dynamic—capable of evolving in real time as new information emerges.

Process Improvement Methodologies

Sustaining Excellence Over Time

The final dimension, **process improvement methodologies**, completes the system by ensuring that execution evolves rather than stagnates. They provide the discipline to identify inefficiencies, eliminate waste, and refine workflows. They ensure that execution is not a one-time effort but a continuous practice. Leaders who embrace process improvement build organizations that get better over time, not simply bigger or busier. In this dimension, execution becomes self-correcting—driven by curiosity, discipline, and a commitment to excellence.

Execution Mechanics is the lens that transforms strategy into reality by ensuring that ambition is supported by structure, that change is supported by people, and that progress is supported by systems. It reveals that execution is not the final step of leadership but the continuous arena in which leadership is tested. In a world where AI accelerates both opportunity and exposure, leaders must design execution as deliberately as they design vision, building the operational muscle that allows the organization to move with coherence under pressure. This lens makes clear that execution is not a matter of effort but of engineering—of constructing the models, architectures, and behaviors that convert intention into repeatable performance. Execution becomes a living capability rather than a periodic push, a discipline that strengthens with use rather than erodes under stress.

EXECUTION MECHANICS ANALOGY – STREAMLINING AIRPORT

STRATEGIC MODEL
Flight Schedule

Establishing a universal cadence ensures every department operates with synchronized precision. The master timeline that dictates when every resource must be in place to ensure a coherent, global operation.

DATA ARCHITECTURE
Air Traffic Control

Maintaining a high-fidelity signaling system eliminates the risk of operational collisions and blind spots. The invisible infrastructure that provides the real-time telemetry needed for confident, evidence-based decisions at high velocity.

AGILE DEVELOPMENT
Quick Turnaround

Compressing execution cycles minimizes the cost of delay and keeps strategic assets in constant motion. The ability to de-plane, refuel, and re-load (iterate) with extreme speed, ensuring the organization never "idles" on the tarmac.

CUSTOMER JOURNEY
The Terminal Flow

Eliminating friction at every touchpoint transforms the user experience into a seamless journey of value. Redesigning the "Gate" flow to remove bottlenecks and demonstrate respect for the customer's time and needs.

CHANGE MANAGEMENT
Ground Crew Synergy

Cultivating organizational elasticity allows the workforce to pivot instantly without losing mechanical momentum. Leaders create psychological safety for the "crew" to adapt during disruptions, synchronized under pressure.

PROCESS IMPROVEMENT
Logistics Optimization

Embedding a habit of marginal gains compounds small refinements into a structural advantage. Identifying every millisecond lost in baggage handling or fueling to maximize the total capacity and throughput of the hub.

COMMON FRAMEWORK - TOGAF (The Open Group Architecture Framework)

Book 1: Leadership Transformation—Strategic Direction : The frameworks in this pillar ensure that all efforts are strategically aligned and that leaders are focused on the right outcomes for value creation.

1. **McKinsey 7-S Framework:** Aligns the organization's strategy with its structure, systems, staff, and skills to ensure all components support the desired transformation.
2. **Business Model Canvas:** Creates a shared understanding of how value will be created, delivered, and captured, providing a clear map for all teams.
3. **McKinsey Three Horizons of Growth:** Structures innovation efforts across short-term improvements (Horizon 1), scaling existing successes (Horizon 2), and future-proofing the business (Horizon 3), ensuring a sustainable pipeline of value.
4. **Balanced Scorecard:** Translates the strategic vision into measurable performance objectives, connecting operational activities to high-level financial and non-financial goals.
5. **SWOT Analysis:** Provides a structured way to evaluate the strategic context, enabling leaders to leverage strengths, mitigate weaknesses, seize opportunities, and counter threats.
6. **Blue Ocean Strategy:** Drives leaders to look for uncontested market space, shifting the focus from competing for value to creating new value.

Book 2: Workforce Transformation—Adoption and Engagement : This pillar addresses the human element of change, ensuring that employees are prepared, engaged, and able to sustain new ways of working.

1. **Kotter's 8-Step Change Model:** Provides a step-by-step process for leading major change, from creating a sense of urgency to solidifying new approaches in the corporate culture.
2. **ADKAR Model:** Guides change managers in ensuring that individual employees have the Awareness, Desire, Knowledge, Ability, and Reinforcement needed for successful change.
3. **Lewin's Change Management Model:** Conceptualizes change as a three-stage process: "Unfreeze" existing habits, implement the "Change," and "Refreeze" the new processes to make them stick.
4. **Bridges' Transition Model:** Focuses on the psychological journey of individuals, helping them navigate the emotional stages of an ending, a neutral zone, and a new beginning.
5. **Prosci Change Management Methodology:** Provides a structured, research-based methodology for applying change management to any project or initiative.
6. **The Satir Change Model:** Helps leaders understand and anticipate the emotional and performance impacts of change on individuals and teams.

Book 3: Data Transformation—Measurement and Insights - This pillar provides the tools and architectures to manage data as a strategic asset, ensuring that value can be reliably measured and tracked.

1. **Zachman Framework:** Offers a comprehensive, logical structure for classifying and organizing enterprise architecture, ensuring data management is part of the larger business strategy.
2. **DAMA-DMBOK:** Provides a set of best practices for managing data throughout its lifecycle, ensuring the quality and integrity of data used for value measurement.
3. **Data Mesh:** Promotes a decentralized, domain-driven approach to data ownership, empowering teams to create and manage data as a product.
4. **Lambda Architecture (or Kappa Architecture):** Establishes a technical architecture for processing large volumes of data efficiently, supporting real-time value tracking and analytics.
5. **Master Data Management (MDM) Frameworks:** Ensure that an organization's critical business data is consistent, accurate, and provides a "single source of truth" for reliable reporting.
6. **Data Fabric:** Creates a flexible, interconnected data architecture that supports seamless access and integration of data, regardless of its location.

Book 4: Customer Transformation—External Value Delivery: This pillar focuses on defining and delivering value from the customer's perspective, ensuring that transformation efforts translate to measurable customer outcomes.

1. **Customer Journey Mapping:** Visualizes the customer's end-to-end experience, identifying pain points and opportunities for value creation.
2. **Service Blueprinting:** Expands on journey mapping by detailing the front-stage and backstage processes, people, and systems that support a customer experience.
3. **Jobs to be Done (JTBD) Framework:** Helps define "value" from the customer's perspective by focusing on the functional and emotional outcomes they seek.
4. **Design Thinking Methodology:** Provides a human-centered approach to innovation that focuses on empathizing with customers to design meaningful solutions.
5. **CX (Customer Experience) Maturity Models:** Provides a benchmark for tracking and improving customer experience over time, linking strategic initiatives to better customer outcomes.
6. **Persona Development Frameworks:** Creates detailed profiles of target users, ensuring that transformation efforts are grounded in a deep understanding of customer needs.

Book 5: Product Transformation—Internal Value Delivery : This pillar establishes an agile, iterative process for continuously building and improving the products and services that deliver customer value.

1. **Scrum:** Enables cross-functional teams to deliver value incrementally in short sprints, allowing for continuous feedback and adaptation.
2. **Kanban:** Helps manage and visualize workflow, limiting work-in-progress and improving the flow of value to the customer.
3. **Lean Startup:** Applies a "build-measure-learn" feedback loop to product development, ensuring teams are constantly validating their assumptions about value creation.
4. **Scaled Agile Framework (SAFe):** Provides a system for coordinating and aligning multiple agile teams across the enterprise to deliver large-scale solutions.
5. **Feature Flags / Toggles Architecture:** Enables teams to deploy new features and capabilities quickly and safely, allowing for real-time testing of value propositions.
6. **Continuous Integration / Continuous Delivery (CI/CD) Pipeline Architecture:** Automates the software delivery process, allowing teams to deliver new value to customers efficiently and reliably.

Book 6: Operations Transformation—Efficiency and Sustainability
This pillar focuses on optimizing the internal processes and systems that support the entire value chain, ensuring efficiency and sustainability.

1. **Lean Management:** Aims to eliminate waste and optimize the flow of value through the organization's processes, generating cost savings and efficiency gains.
2. **Six Sigma (DMAIC Methodology):** Uses a data-driven approach to improve processes by defining, measuring, analyzing, improving, and controlling them.
3. **BPM (Business Process Management) Lifecycle:** Provides a structured approach to continuously improving business processes, ensuring sustained value delivery over the long term.
4. **SCOR Model (Supply Chain Operations Reference):** Offers a standard framework for managing and improving supply chain efficiency, a key driver of operational value.
5. **Digital Twin Architecture:** Creates virtual models of physical assets or processes, enabling simulation and optimization before real-world changes are made.
6. **Robotic Process Automation (RPA) Frameworks/Methodologies:** Automates repetitive, rule-based tasks, freeing up human resources for higher-value activities.

TOGAF (The Open Group Architecture Framework) is a widely recognized and comprehensive enterprise architecture methodology that provides a structured approach for developing and managing enterprise architecture. It helps organizations design, plan, implement, and govern an enterprise information architecture. At its core is the Architecture Development Method (ADM), a detailed, iterative process for developing various architectural domains: Business, Data, Application, and Technology. For leadership transformation, particularly in achieving digital maturity, TOGAF is indispensable. Digital transformation is not just about implementing new technologies; it requires aligning business strategy with IT capabilities across the entire enterprise. Without a coherent architecture, digital initiatives can become fragmented, leading to integration challenges, redundant systems, security vulnerabilities, and a lack of scalability. TOGAF provides the necessary rigor to ensure that digital investments are strategically aligned, well-integrated, and support the organization's long-term digital vision. It helps leaders understand the current state of their digital landscape, define the desired future state (e.g., cloud-native infrastructure, API-driven integration, AI-enabled applications), and create a roadmap to get there. By fostering a common language and framework for architects and stakeholders, TOGAF enables effective communication and collaboration across complex digital initiatives. It ensures that the digital transformation is not just a series of isolated projects but a cohesive, well-governed evolution of the entire enterprise, a critical component of achieving and sustaining comprehensive digital maturity. It provides the blueprint for how an organization's digital capabilities (applications, data, infrastructure) support its strategic business goals, ensuring that every technological investment contributes to a unified, adaptable, and future-ready digital enterprise. This structured approach helps avoid costly misalignments and ensures that digital initiatives are built on a solid, scalable foundation, allowing the organization to truly harness the power of technology for competitive advantage.

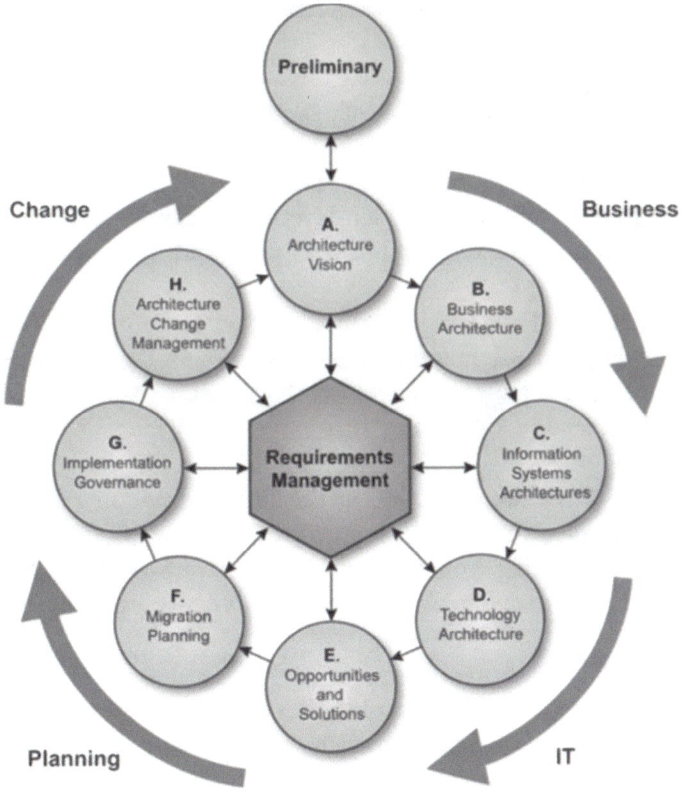

TOGAF Frameworks

1. **Preliminary Phase: Framework and Principles:** Establish the architecture capability for the enterprise. This involves defining the organizational context for architecture work, identifying key stakeholders, establishing the architecture principles (e.g., "cloud-first," "API-driven," "data-driven decision making"), and defining the scope and governance of the architecture effort.

2. **Phase A: Architecture Vision:** Develop a high-level vision for the target architecture that aligns with business strategy and digital transformation goals. This involves identifying key stakeholders, defining the business value of the architecture work, and securing approval for the architecture development cycle.

3. **Phase B: Business Architecture:** Develop a baseline (current state) and target (future state) Business Architecture that describes the organization's processes, functions, and organizational structure. This ensures that the digital transformation efforts are aligned with and support the core business capabilities and strategic objectives, identifying where digital enablement can optimize business value, streamline workflows, and enhance customer interactions through digital means.

4. **Phase C: Information Systems Architectures (Data & Application):** Develop baseline and target Data Architecture (logical and physical data models, data governance, data flows) and Application Architecture (applications, their interactions, and how they support business functions). This phase is critical for designing integrated digital systems, ensuring data flows seamlessly across digital platforms, and supporting new digital applications that drive business value and customer engagement.

5. **Phase D: Technology Architecture:** Develop baseline and target Technology Architecture that describes the logical software and hardware capabilities required to support the Business, Data, and Application components. This includes defining infrastructure (e.g., cloud platforms, edge computing), platforms (e.g., microservices, AI/ML platforms), and security standards necessary for a robust, scalable, and resilient digital enterprise.

6. **Phase E: Opportunities & Solutions:** Identify and evaluate various implementation options and opportunities to deliver the target architecture. This involves assessing gaps between baseline and target architectures, identifying potential solutions (e.g., Commercial Off-The-Shelf (COTS), custom development, cloud migration, SaaS adoption), and defining high-level implementation projects. For digital maturity, this prioritizes digital solutions that offer the greatest business value and strategic alignment.

7. **Phase F: Migration Planning:** Develop a detailed implementation and migration plan. This involves prioritizing projects, estimating costs, defining a timeline, and outlining the sequence of activities required to transition from the baseline to the target architecture. This ensures a structured and managed approach to digital transformation implementation, minimizing disruption and maximizing efficiency.

8. **Phase G: Implementation Governance:** Oversee the implementation of the architecture. This involves establishing governance processes to ensure that projects adhere to the defined architecture, managing architectural contracts, and providing architectural guidance during development and deployment. This continuous oversight is vital for maintaining architectural integrity and consistency throughout the digital transformation.

9. **Phase H: Architecture Change Management:** Establish processes for managing changes to the architecture. This ensures that the architecture remains relevant and responsive to evolving business needs, technological advancements (e.g., new AI capabilities, quantum computing), and market shifts.

10. **Requirements Management:** This phase runs concurrently throughout all ADM phases. It involves managing the architectural requirements, ensuring they are consistent, complete, and aligned with stakeholder needs. For digital maturity, this ensures that evolving digital requirements (e.g., real-time data access, scalability for new digital services) are captured and integrated into the architectural design and implementation.

Ai6x6™ Transformation Blueprint

Agile Development Frameworks

"The only way to win is to learn faster than anyone else."

Eric Ries. This captures the core Agile philosophy of continuous iteration, feedback, and rapid learning.

Agile Development Frameworks (such as Scrum or SAFe) are indispensable for Product and Operations Transformations, replacing rigid methodologies with a focus on speed, iteration, and customer-centricity. These frameworks provide the **structural rhythm and cadence**—through short cycles (sprints), constant feedback loops, and daily stand-ups—necessary to manage the uncertainty of digital change. They are critical for fostering cross-functional team collaboration and guaranteeing that development efforts are continuously aligned with shifting business priorities and user needs. By enforcing transparency and constant reassessment, Agile frameworks drastically **reduce the risk of building the wrong product** or incurring massive technical debt. Ultimately, a strong Agile structure institutionalizes **organizational learning**, ensuring the enterprise can rapidly develop, test, and release new features and services, accelerating time-to-market and sustaining an edge in the digital economy.

The **Agile Development Frameworks** pillar establishes an iterative, flexible process essential for continuously building and enhancing the products and services that directly deliver customer value. Core to this is Scrum, which empowers cross-functional teams to deliver increments of value in short, time-boxed cycles called sprints, facilitating rapid feedback and adaptation. Kanban focuses on workflow management and visualization, optimizing the flow of value by actively limiting the amount of work in progress. For product innovation, the Lean Startup methodology enforces a strict "build-measure-learn" feedback loop, which ensures that development efforts are continually validating assumptions about what truly creates value for the customer. To scale these practices across a large organization, frameworks like the Scaled Agile Framework (SAFe) provide a system for coordinating and aligning multiple agile teams to deliver complex, enterprise-level solutions. The technical capabilities that support this agility include Feature Flags / Toggles Architecture, which allows new features to be deployed quickly and safely for real-time testing, and the Continuous Integration / Continuous Delivery (CI/CD) Pipeline Architecture, which automates the entire software delivery process to efficiently and reliably push new value to customers.

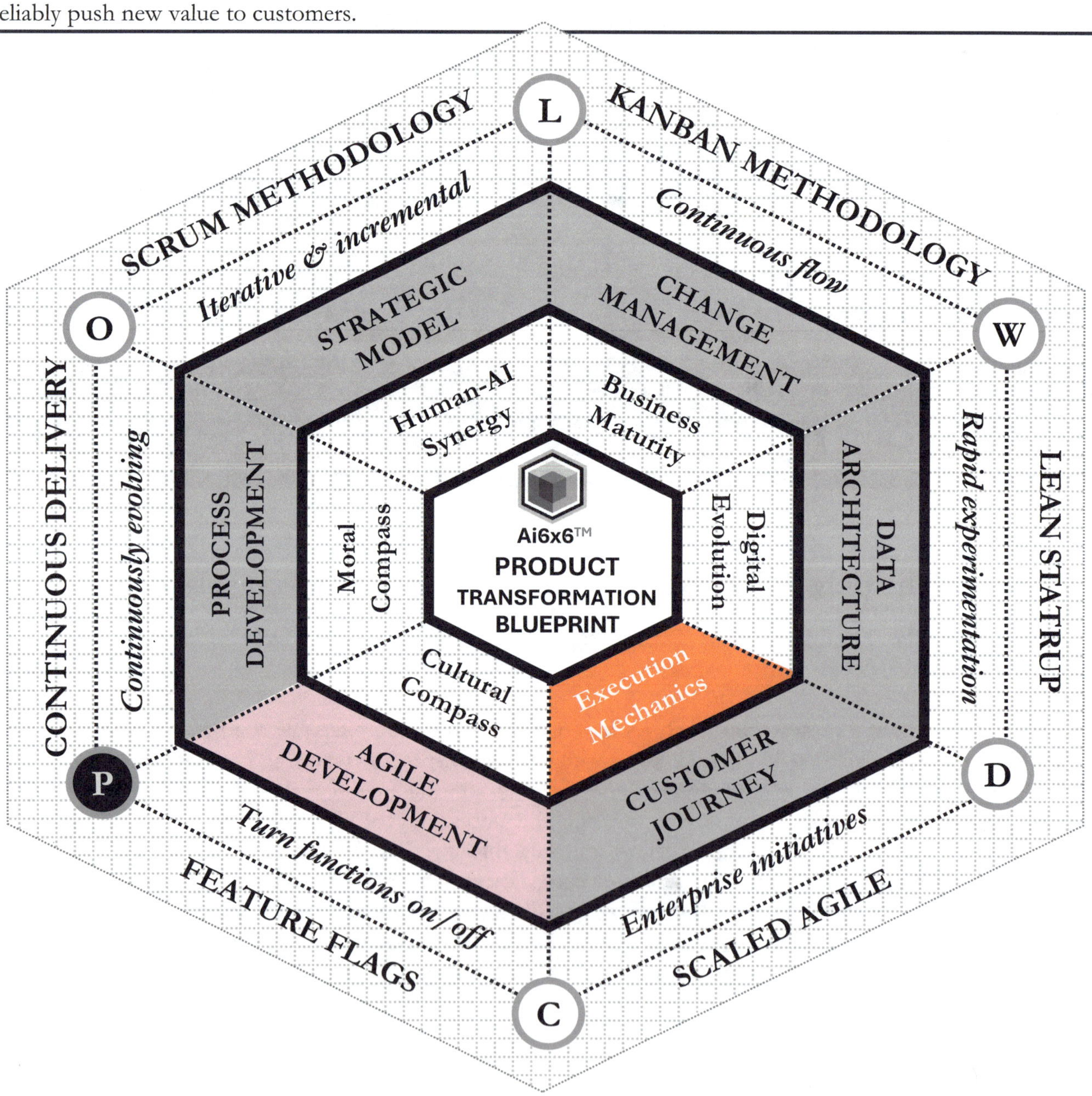

Ai6x6 AGILE DEVELOPMENT FRAMEWORK | PRODUCT TRANSFORMATION

Scrum is the most widely adopted agile framework for managing complex product development. It is an iterative, incremental approach designed to deliver value in short, fixed-length cycles called "Sprints" (typically 1-4 weeks). At its core, Scrum emphasizes collaboration, self-organizing teams, and continuous feedback.

For product transformation, particularly in the context of digital maturity, Scrum is invaluable. Traditional "waterfall" development often struggles to keep pace with the rapidly evolving digital landscape and changing customer expectations. Scrum's iterative nature allows digital product teams to quickly build, test, and release new features, gather real-world feedback, and adapt course rapidly. This minimizes the risk of building the "wrong" digital product and maximizes the chances of delivering what customers truly need. It fosters a culture of continuous learning and improvement, where teams regularly inspect their processes and adapt. By breaking down large digital product initiatives into smaller, manageable increments,

Scrum provides transparency, predictability, and the ability to respond to emerging digital opportunities or threats. This agility in product development, coupled with a relentless focus on delivering working software frequently, is a hallmark of digitally mature organizations that excel at continuous digital innovation and responsiveness.

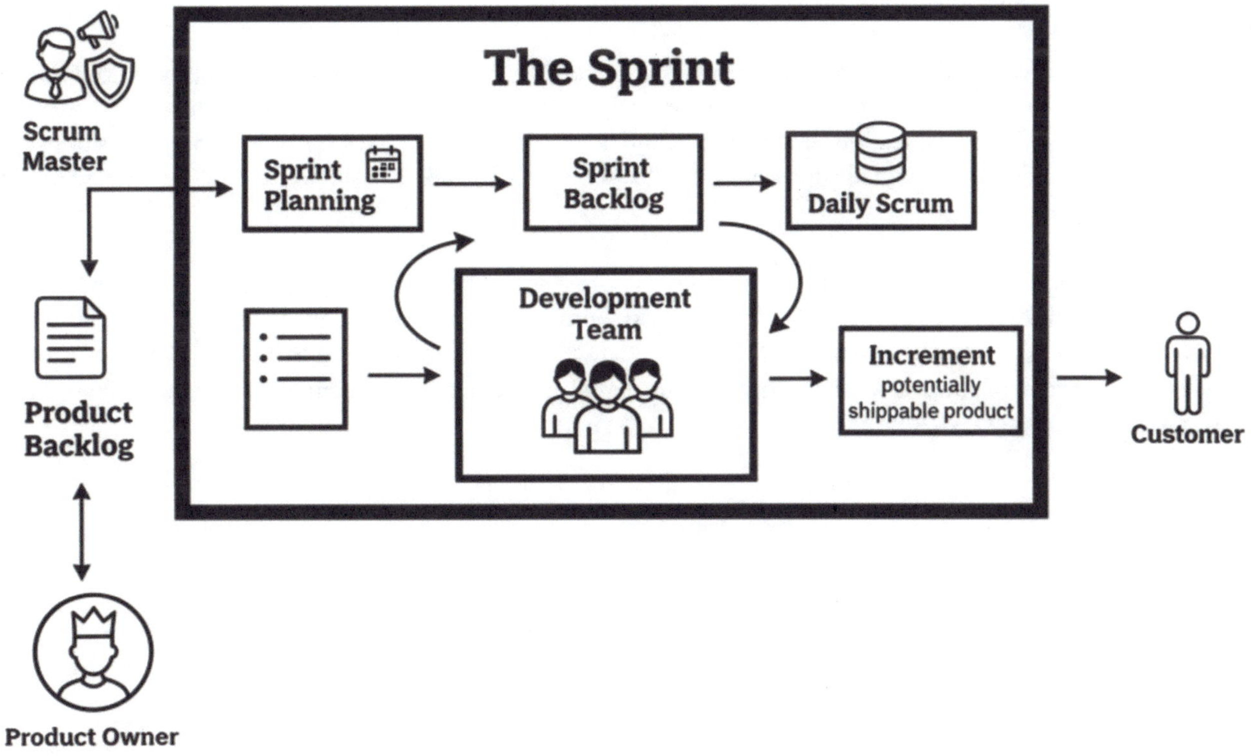

Scrum Methodology

1. **Product Backlog:** The Product Owner creates and continuously refines a single, prioritized, dynamic list of all desired features, requirements, and enhancements for the product. This backlog represents the "what" that needs to be built, ordered by value, risk, or necessity. In a digitally mature context, this backlog is often heavily influenced by data analytics, A/B testing results, and customer feedback from digital channels, ensuring features are truly valuable.

2. **Sprint Planning:** At the start of each Sprint (a fixed-length iteration, typically 1-4 weeks), the Scrum Team (Product Owner, Scrum Master, Development Team) collaboratively selects a subset of high-priority items from the Product Backlog to complete within that Sprint. They also define *how* they will achieve this work. This short planning cycle is crucial for adapting to rapidly changing digital market demands and integrating new technologies.

3. **Daily Scrum:** The Development Team holds a short (15-minute), daily meeting to synchronize activities, discuss progress towards the Sprint Goal, and identify any impediments that are blocking their work. This fosters self-organization and rapid problem-solving within the team, enabling quick adjustments to digital development tasks and ensuring continuous flow of work.

4. **Development Work:** The Development Team works collaboratively during the Sprint to build, test, and integrate the selected Product Backlog items. They focus on delivering a "Done," potentially shippable increment of the product by the end of the Sprint, adhering to quality standards. For digital products, this involves writing code, designing user interfaces, integrating APIs, and ensuring cybersecurity, all within the sprint.

5. **Sprint Review:** At the end of the Sprint, the Scrum Team demonstrates the completed, potentially shippable product increment to stakeholders (e.g., business owners, customers). This is an informal meeting to gather feedback, discuss progress, and adapt the Product Backlog based on new insights and market conditions. For digital products, this allows for rapid validation of new digital features with real users.

6. **Sprint Retrospective:** After the Sprint Review, the Scrum Team holds a retrospective to inspect its process and identify improvements for the next Sprint. They discuss what went well, what could be improved, and commit to specific actions to enhance their effectiveness and quality. This continuous self-improvement loop is vital for optimizing digital product development processes and fostering a culture of digital excellence.

> "Scrum is simple. It's just a framework for getting work done."
> — Ken Schwaber, Co-creator of Scrum

Kanban is an agile method that focuses on visualizing workflow, limiting work in progress (WIP), and maximizing efficiency through continuous flow. Originating from Toyota's production system, it's a pull-based system that helps teams manage work by making it transparent and optimizing the delivery process.

For product transformation, particularly in achieving digital maturity, Kanban is highly effective. In digital product development, work can be highly variable and unpredictable. Kanban helps teams manage this variability by providing a clear visual representation of all tasks, from ideation to deployment. By setting explicit WIP limits, it prevents teams from becoming overwhelmed, reduces context switching, and forces them to focus on completing tasks before starting new ones. This leads to faster cycle times and more predictable delivery of digital features.

Kanban's emphasis on continuous flow and incremental improvement allows digital product teams to quickly adapt to changing priorities, respond to emerging digital trends, and continuously deliver value to customers without the fixed iterations of Scrum. It fosters a culture of continuous delivery and operational excellence in digital product development, a key aspect of digital maturity where products are constantly evolving and improving based on real-time feedback and market needs.

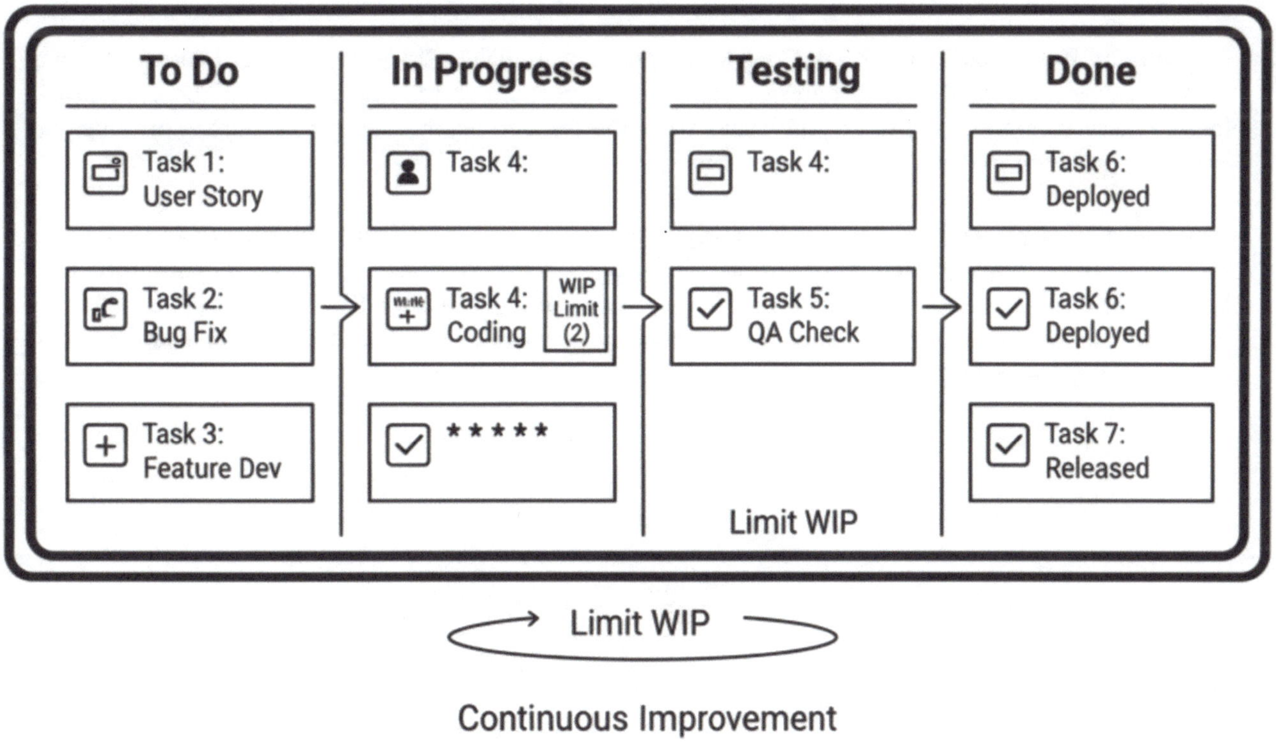

Kanban Framework

1. **Visualize Workflow:** Create a Kanban board (physical whiteboard or digital tool like Jira, Trello) with columns representing each distinct stage of the product development workflow (e.g., "Backlog," "Analysis," "Development," "Code Review," "Testing," "Deployment," "Done"). This visual representation provides transparency and a shared understanding of the work being done, making digital product development processes visible to all.

2. **Limit Work in Progress (WIP):** Set explicit limits on the number of tasks that can be in each "In Progress" column at any given time. This is a core principle that prevents multitasking, reduces context switching, identifies bottlenecks, and encourages teams to focus on completing work before starting new tasks. For digital product teams, this means ensuring that developers are not overwhelmed and can efficiently deliver digital features.

3. **Manage Flow:** Continuously track items as they move through the board. The focus is on optimizing the flow of value from start to finish. Identify and address any blockages, slowdowns, or queues that impede progress, striving for a smooth and predictable delivery cadence. For digital product development, this means ensuring new features move quickly from idea to deployment, reducing time-to-market.

4. **Make Policies Explicit:** Clearly define the rules for moving tasks from one column to the next. This includes criteria for when a task is "done" in one stage and ready for the next (e.g., "Definition of Done" for development, "Entry Criteria" for testing, "Deployment Checklist" for release). Explicit policies ensure consistency and clarity for the team, streamlining digital delivery processes.

5. **Implement Feedback Loops:** Hold regular meetings (e.g., daily stand-ups, service delivery reviews, operations reviews) to discuss progress, issues, and opportunities for improvement. These feedback loops ensure continuous communication and allow the team to adapt quickly to changing priorities or emerging problems in digital product development.

6. **Improve Collaboratively:** Continuously analyze the workflow and relevant metrics (e.g., lead time, cycle time, throughput) to identify opportunities for process improvement. The team works together to implement changes, experiment with new approaches, and evolve the Kanban system to enhance efficiency and effectiveness over time. This continuous optimization of the digital delivery pipeline is a hallmark of digital maturity.

> "Deciding what not to do is as important as deciding what to do."
> — Steve Jobs, Co-founder, Apple

The Lean Startup methodology, developed by Eric Ries, provides a scientific approach to creating and managing startups and new products. It emphasizes rapid experimentation, validated learning, and iterative product releases to minimize risk and accelerate learning in uncertain environments.

For product transformation, particularly in achieving digital maturity, Lean Startup is critical. In the digital age, market conditions, customer preferences, and technological capabilities evolve at an unprecedented pace. Building large, complex digital products based on lengthy upfront planning carries significant risk of failure. Lean Startup addresses this by advocating for the creation of a Minimum Viable Product (MVP)—the smallest possible version of a new digital feature or product that can be released to early customers to gather validated learning. This "Build-Measure-Learn" feedback loop allows digital product teams to quickly test hypotheses about customer needs and market demand, learn from real-world usage data, and rapidly pivot or persevere based on evidence. This approach minimizes wasted development effort on features customers don't want and accelerates the discovery of truly valuable digital solutions. By fostering a culture of continuous experimentation, data-driven decision-making, and rapid iteration, Lean Startup enables organizations to develop digital products that are highly responsive to market needs and continuously deliver value, a defining characteristic of digitally mature product development.

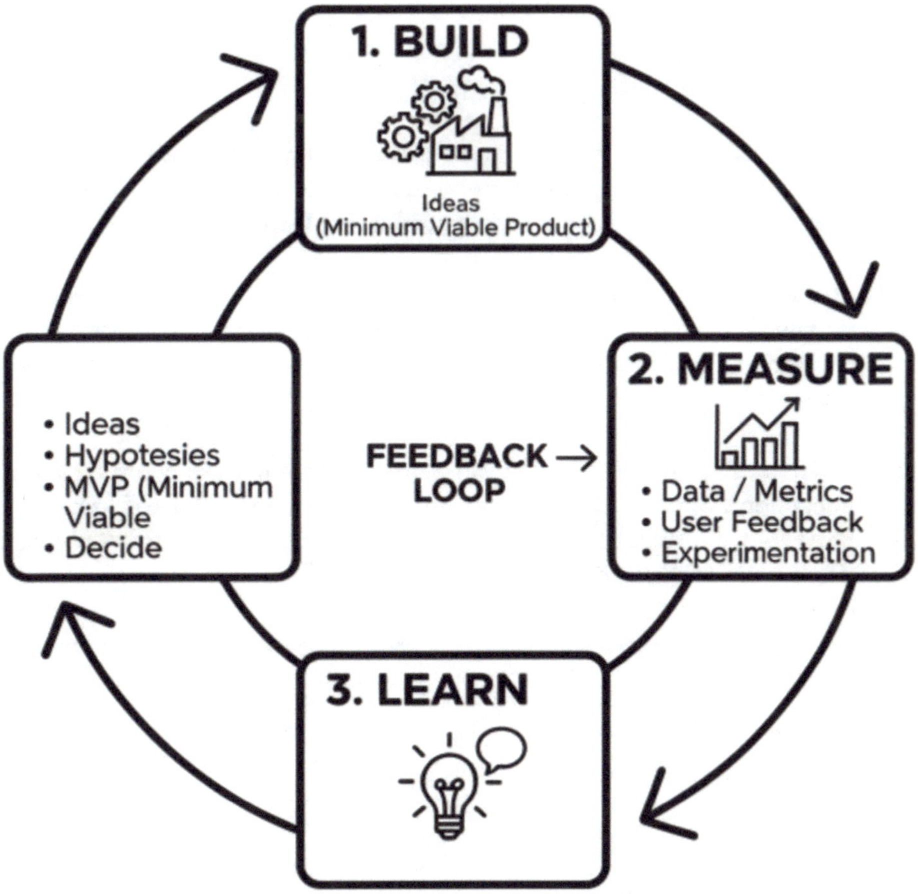

: Lean Startup Framework

1. **Build (MVP):** Instead of building a full-featured product, create a Minimum Viable Product (MVP). An MVP is the smallest possible version of a new product or feature that can be released to early customers to gather validated learning. For digital products, this means launching a core digital feature or a basic digital service that addresses a key customer problem, rather than waiting for a perfect, comprehensive solution.

2. **Measure**: Once the MVP is released, systematically collect quantitative and qualitative data on how customers interact with it. Focus on actionable metrics that directly prove or disprove the initial hypotheses about customer behavior, market demand, and product value. For digital products, this involves leveraging analytics tools to track user engagement, conversion rates, feature usage, and A/B test results, avoiding "vanity metrics" that don't offer true insights.

3. **Learn**: Analyze the collected data to gain insights into customer behavior and market demand. This involves critically evaluating whether the initial assumptions and hypotheses were validated or invalidated. The learning phase is about understanding *why* customers behaved the way they did and what that implies for the product's future, particularly for digital features. This rapid learning cycle is a cornerstone of digital maturity.

4. **Pivot or Persevere:** Based on the learning, make a strategic decision:

   - Pivot: Change a fundamental assumption about the product, strategy, or market if the initial hypotheses are invalidated. This could involve changing the target customer, value proposition, or even the underlying digital technology.

   - Persevere: Continue in the current direction with minor adjustments if the hypotheses are validated, focusing on optimizing and scaling the existing approach. This agility in decision-making is vital for navigating the fast-paced digital landscape.

5. **Iterate Continuously:** The Build-Measure-Learn cycle is not a one-time event but a continuous loop. The insights gained from one iteration feed directly into the next, allowing for rapid refinement, adaptation, and continuous improvement of the product based on real-world customer feedback. This relentless iteration is how digitally mature organizations stay ahead in innovation.

6. **Minimize Waste:** A core principle of Lean Startup is to minimize waste by avoiding building features that customers don't want or need. By focusing on MVPs and validated learning, organizations reduce the risk of investing significant time and resources into digital products that ultimately fail in the market, ensuring efficient use of digital development resources.

> "The only way to win is to learn faster than anyone else."
> — Eric Ries, Author of "The Lean Startup"

The Scaled Agile Framework (SAFe) is a comprehensive framework for applying Lean-Agile principles and practices at enterprise scale. It provides guidance for coordinating multiple Agile teams, managing large solutions, and aligning strategy with execution across an entire organization.

For product transformation, particularly in large enterprises pursuing digital maturity, SAFe is crucial. While Scrum and Kanban are excellent for individual teams, scaling Agile across hundreds or thousands of people working on complex digital products and services introduces new challenges: maintaining alignment, managing dependencies, and ensuring consistent delivery across diverse teams. SAFe addresses these by organizing teams into "Agile Release Trains" (ARTs) that deliver solutions collaboratively within defined "Value Streams." It provides structured events like Program Increment (PI) Planning, where all teams in an ART align on a shared vision and plan for upcoming increments. This ensures that individual digital product development efforts contribute to larger strategic digital goals. SAFė also emphasizes the role of Lean-Agile leadership and a Lean-Agile Center of Excellence (LACE) to champion the transformation. By providing a common operating model and a structured approach to scaling agility, SAFe helps large organizations overcome the complexities of digital transformation, accelerate the delivery of large-scale digital solutions, and achieve enterprise-wide digital maturity, where product development is highly responsive and strategically aligned.

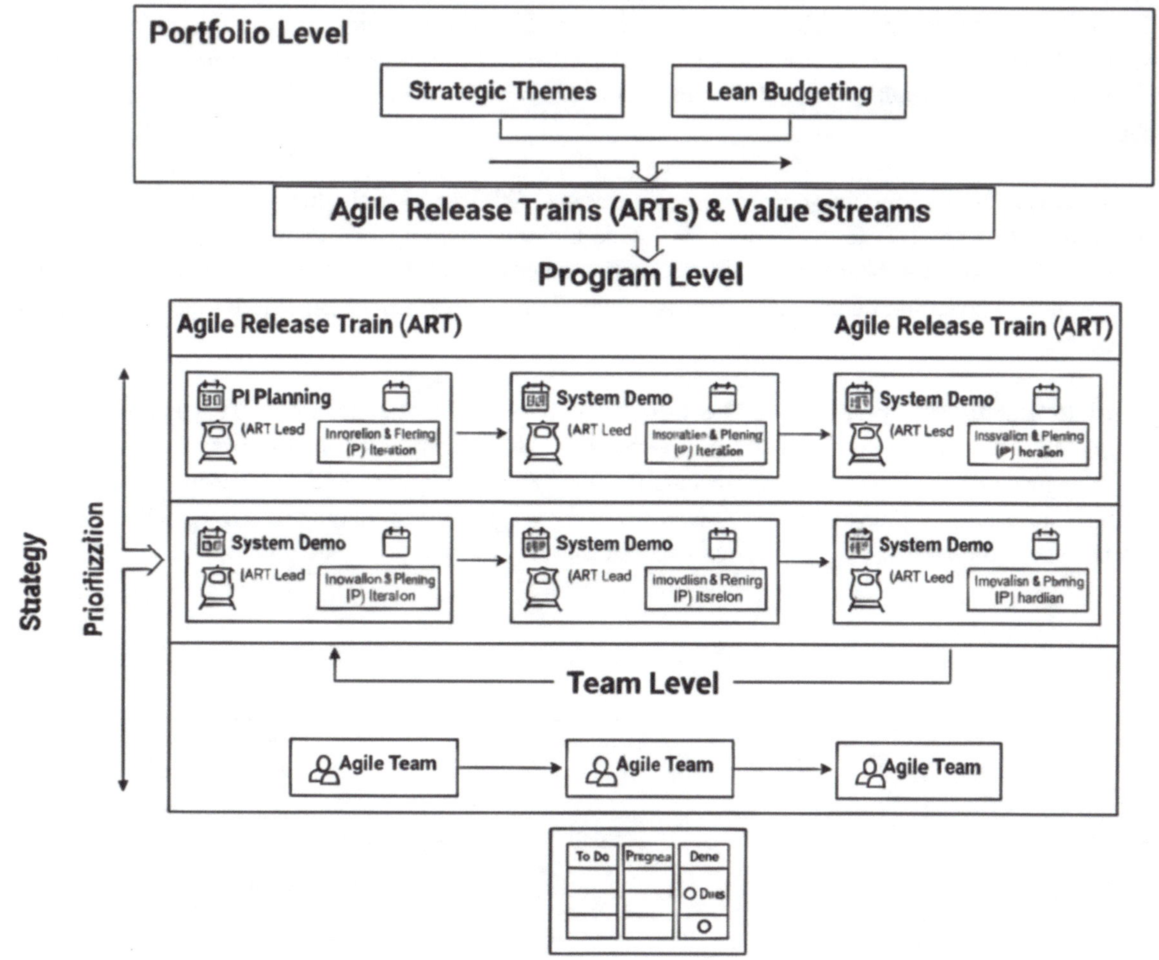

: Scaled Agile Framework (SAFE)

1. **Train Lean-Agile Leaders:** Begin by educating senior leadership and managers on Lean-Agile principles and practices. This foundational step is crucial to foster a supportive environment for transformation, ensuring that leaders understand their role in enabling agility and removing impediments at scale. For digital maturity, this means developing leaders who can champion digital initiatives and foster a culture of continuous digital delivery.

2. **Establish a Lean-Agile Center of Excellence (LACE):** Create a small, dedicated team (the LACE) to champion and guide the SAFe implementation across the organization. The LACE acts as a central hub for training, coaching, and facilitating the adoption of Lean-Agile practices, providing ongoing support and expertise. This center is vital for standardizing digital development practices and spreading digital fluency across the enterprise.

3. **Identify Value Streams and Agile Release Trains (ARTs):** Organize teams around the flow of value to the customer, rather than traditional functional silos. Form Agile Release Trains (ARTs), which are virtual teams of 5-12 Agile teams (50-125 people) that collaboratively deliver solutions within a defined value stream. For digital maturity, this means organizing around digital product lines or digital service offerings to optimize end-to-end digital value delivery.

4. **Launch ARTs:** Conduct an "ART Launch" event, typically a Program Increment (PI) Planning session. This involves all teams in an ART aligning on a shared vision, planning for the upcoming increment (usually 8-12 weeks), and committing to a set of objectives. This big-room planning fosters collaboration and alignment across multiple digital development teams, ensuring they work towards common digital goals.

5. **Coach ARTs:** Provide ongoing coaching and support to ARTs and individual Agile teams to help them adopt and mature their Lean-Agile practices. This includes training Scrum Masters, Product Owners, and development teams, facilitating continuous improvement, and addressing challenges as they arise. This continuous coaching helps teams improve their digital development capabilities and adhere to agile best practices.

6. **Extend to Portfolio:** Apply Lean-Agile principles to portfolio management. This involves aligning strategic investments with defined value streams and ARTs, ensuring that the organization's highest-level strategic goals are translated into executable work for Agile teams, and that resources are allocated effectively across the entire portfolio. This ensures that digital investments are strategically aligned and deliver maximum business value, a key indicator of digital maturity.

> "SAFe helps enterprises achieve business agility."
> — Dean Leffingwell, Creator of SAFe

Feature Flags (also known as feature toggles or feature switches) represent an architectural pattern that allows developers to turn specific functionalities on or off in a software application without deploying new code. This is achieved by wrapping new features within a "flag" that can be controlled dynamically.

For product transformation, particularly in achieving digital maturity, feature flags are a powerful enabler. In a world of continuous delivery and rapid digital innovation, organizations need the ability to release new features frequently with minimal risk. Feature flags decouple code deployment from feature release, meaning new code can be pushed to production but remain inactive until toggled on. This enables a range of advanced practices: A/B testing (releasing a feature to a small segment of users to compare performance), controlled rollouts (gradually exposing a feature to more users), kill switches (quickly disabling a problematic feature without a full rollback), and personalized experiences (showing different features to different user groups). By reducing the risk associated with new digital product launches and enabling real-time experimentation, feature flags accelerate the learning cycle, allow for data-driven product decisions, and ensure that digital products are continuously optimized based on live user feedback. This agility in feature management is a clear indicator of a highly mature digital product development capability.

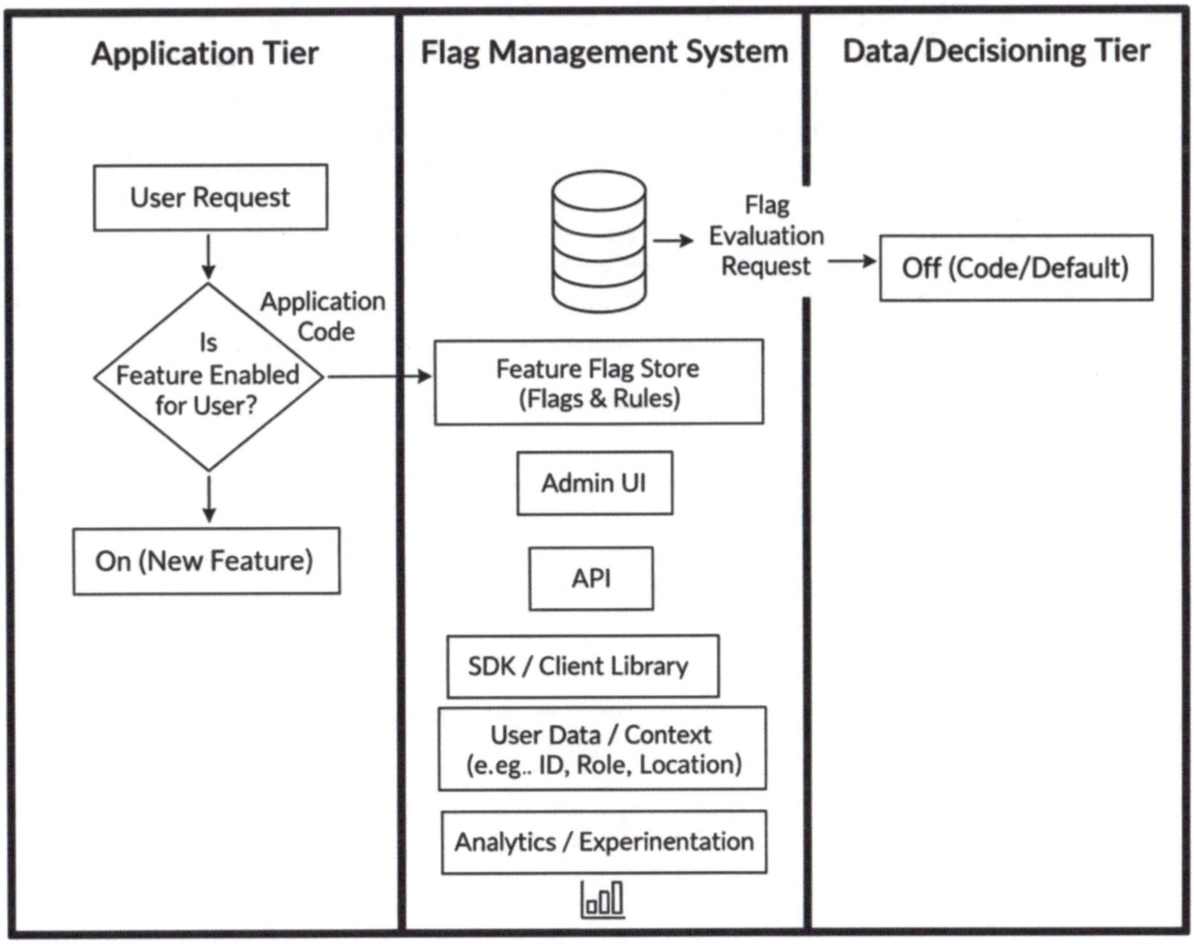

: Feature Flags / Toggles Architecture

1. **Integrate Flag Management System:** Choose and integrate a feature flag management tool or library into your product's codebase. This system provides the infrastructure to define, manage, and control the state of feature flags (on/off, for whom). For digital products, this system is critical for dynamic control over features in a live environment, supporting rapid iteration and experimentation.

2. **Wrap Features with Flags:** When developing new product features, encapsulate the new code within a feature flag. This means the new functionality is deployed to the production environment but remains inactive or "dark" by default, preventing it from being visible or accessible to all users immediately. This allows for continuous integration and deployment of digital features without immediate public exposure.

3. **Configure Flag Rules:** Define the conditions under which a feature flag will be "on" or "off." These rules can be based on various criteria, such as user ID, geographic location, subscription tier, A/B test group assignment, or internal testing environments. This allows for granular control over feature exposure, enabling targeted digital feature rollouts and personalized experiences.

4. **Deploy Code:** Release the new code (containing the hidden or inactive features wrapped in flags) to the production environment. This allows for continuous deployment without requiring a full product release every time a new feature is ready, decoupling deployment from release. This is a fundamental practice for achieving continuous delivery in digital product development.

5. **Toggle Flags:** Dynamically turn features on or off for specific segments of users based on business needs, testing requirements, or incident response. This can be done through the feature flag management system's user interface, enabling rapid activation or deactivation without code changes or redeployments. This real-time control is a powerful enabler for agile digital product management.

6. **Monitor & Analyze:** Continuously monitor the performance and user behavior of features that are under flags. Gather data (e.g., usage patterns, error rates, conversion metrics) to inform decisions on full rollout, iteration, or discontinuation. This data-driven approach ensures features are validated before broad release, minimizing risk in digital product launches and enabling data-informed digital maturity.

7. **Clean Up:** Once a feature is fully adopted, stable, and no longer requires dynamic control (e.g., it's a permanent part of the product), its corresponding feature flag should be removed from the codebase. This prevents "flag debt" and keeps the code clean and maintainable, ensuring the digital product's architecture remains streamlined and efficient.

> "Continuous Delivery is about making releases boring."
> — Martin Fowler, Chief Scientist, ThoughtWorks (Influential Software Engineer)

Continuous Integration (CI) and Continuous Delivery (CD) represent a set of practices that automate the process of building, testing, and deploying software code changes rapidly and reliably. CI involves frequently merging code changes into a central repository, where automated builds and tests are run. CD extends this by ensuring that the software can be released to production at any time, often automatically.

For product transformation, particularly in achieving digital maturity, CI/CD pipelines are fundamental. In the digital age, speed to market and the ability to respond quickly to customer feedback are paramount. Manual, infrequent releases are slow, error-prone, and limit innovation. CI/CD automates the entire software delivery lifecycle, from code commit to production deployment. This drastically reduces the time it takes to get new digital features into the hands of users, enabling rapid iteration and continuous improvement. It also improves code quality by catching defects early through automated testing, reducing technical debt. By fostering a culture of frequent, small releases, CI/CD minimizes risk, increases predictability, and allows digital product teams to experiment and learn faster. This highly automated, efficient, and reliable software delivery capability is a cornerstone of digitally mature organizations that excel at continuous digital innovation and maintaining a competitive edge.

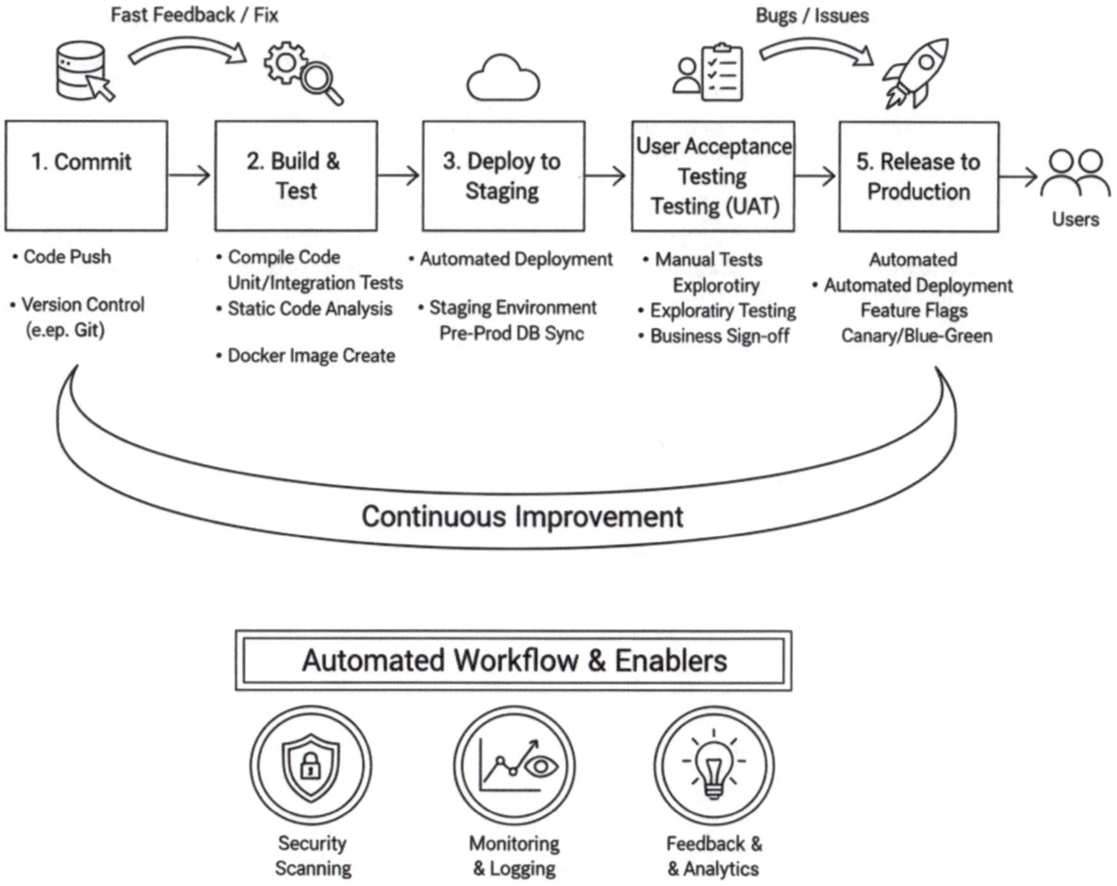

Continuous Integration / Continuous Delivery

1. **Version Control:** All product code is stored in a shared version control system (e.g., Git, SVN), serving as the single source of truth. This allows multiple developers to collaborate efficiently, track changes, and revert to previous versions if needed. For digital products, this ensures that all code for web applications, mobile apps, and backend services is managed centrally, enabling collaborative digital development.

2. **Automated Builds (Continuous Integration - CI):** Every time a developer commits code changes to the version control system, an automated build process is triggered. This compiles the code, checks for syntax errors, and runs basic static code analysis, ensuring that new code integrates seamlessly with existing code. For digital products, this means quickly identifying integration issues in complex digital systems, preventing larger problems down the line.

3. **Automated Tests (CI):** A comprehensive suite of automated tests (unit tests, integration tests, API tests, security scans, performance tests) runs on the newly built code. This crucial step catches defects and regressions early in the development cycle, providing rapid feedback to developers and maintaining code quality. For digital products, robust automated testing ensures the reliability and security of digital features before they reach users.

4. **Automated Deployment to Staging (CI/CD):** If all automated tests pass, the code is automatically deployed to a staging or pre-production environment. This environment mirrors the production environment, allowing for further automated or manual testing (e.g., user acceptance testing, penetration testing) in a realistic setting before deployment to live users. This ensures digital products are thoroughly validated in a near-production environment.

5. **Automated Release (Continuous Delivery - CD):** Once validated in the staging environment, the code can be automatically deployed to the production environment. In a pure CD model, this deployment happens without manual intervention. Some organizations may have a manual "go/no-go" gate here for critical releases. This capability enables rapid iteration and deployment of new digital features, allowing organizations to respond quickly to market demands.

6. **Monitoring & Feedback:** After deployment to production, continuous monitoring tools track the application's performance, stability, and user behavior in real-time. Any errors, performance degradation, or unexpected user interactions trigger alerts and feed insights back into the development cycle, enabling rapid response and continuous improvement. This real-time feedback loop is essential for maintaining the health and optimizing the performance of digital products, a key aspect of digital maturity.

> "The goal of DevOps is to shorten the system development life cycle and provide continuous delivery with high software quality."
> — Gene Kim, Author & DevOps Thought Leader

Ai6x6™ Transformation Blueprint

The Pivot Latency Delta (PLD)

"It is not the strongest of the species that survive, nor the most intelligent, but the ones most responsive to change." --- Leon C. Megginson, the scholar who summarized Darwin's theory of evolution identified that survival is a function of "Responsiveness," not just raw power. For Product transformation, the PLM is the diagnostic for "Materiality." Megginson's summary is the ultimate justification for reducing pivot latency; in the Net Zero era, the winner is the organization that can re-design its entire product architecture the moment planetary boundaries shift.

In the era of traditional manufacturing, "Product Strategy" was a heavy, linear commitment. You designed, you built, and you launched. Because the cost of retooling a factory was astronomical, the goal was to be "Right the First Time." This created a culture that rewarded persistence, long-term planning, and the "Finish What You Start" mentality. But in the AI era, the cost of retooling a digital product—a feature, an algorithm, or a service—has plummeted. The bottleneck is no longer the "Cost of Change"; it is the "Cost of Ego." This is the Identity Chasm of product development: the psychological friction between Commitment and Course-Correction. When a product team receives a high-fidelity data signal that a feature is failing, but they continue to fund it because they've already spent millions on it, they are suffering from a Sunk Cost Anchor. The primary metric for this evolution is the Pivot Latency Delta (PLD).

The product / legacy chasm is the psychological friction between **Sunk-Cost Commitment** and **Perpetual Beta**. It is the "Creator's Ego" that refuses to kill a failing feature because of the blood, sweat, and capital already invested in it. Naming this chasm allows product teams to decouple their self-worth from the code and attach it to the outcome. It is the only way to achieve a low "Pivot Latency Delta," ensuring that human creativity is always flowing toward the highest-value opportunities.

The greatest barrier to product transformation is the Creator's Ego. We often tie our professional identity to the "Success" of a specific feature. To admit a product is failing feels like admitting the creator has failed. This leads to "Persistence Bias," where teams tweak the edges of a dying product rather than making the radical pivot the data demands. They spend months in "Validation Hell," trying to prove the data is wrong so they can keep their project alive.

This hesitation creates "Resource Stagnation." If the machine identifies a product-market misfit in three days, but the organization takes ten months to reallocate the budget, the company is effectively burning capital on a corpse. To bridge this chasm, the team must undergo a shift: moving from being "Feature Builders" to being "Value Discoverers." You must realize that your worth is not in the code you wrote yesterday, but in the speed with which you move toward the opportunity of tomorrow.

THE RAISON D'ÊTRE:

"TO EXPRESS"

By mastering the PLD, the product team fulfills its true purpose: To Express. In a world where AI can generate the "Function" (the code, the wireframes, the testing), the human's role is to express the Vision. When the fear of "being wrong" is replaced by the speed of "being agile," creativity is liberated. We are not killing projects; we are Recapturing Potential. This allows the human to move from the "Maintenance" of the past to the "Expression" of the future.

DEFINING THE PLD: THE LOGIC OF RECAPTURE

The Pivot Latency Delta is a measurement of Strategic Fluidity. It asks one simple, uncomfortable question: How many days passed between the moment we knew it was failing and the moment we stopped funding it?

In a legacy firm, the PLD is measured in quarters or years. Projects die only when the budget runs out. In a high-velocity culture, the PLD is measured in weeks. The goal is to minimize this "Delta" so that talent and capital are never "Trapped." The PLD turns "Failure" from a permanent mark into a temporary data point. It ensures that the organization's best minds are always working on the highest-probability wins.

THE OBJECTIVE: TOWARD PERPETUAL BETA

The goal of the PLD intervention is to reach a state of Equilibrium we call Perpetual Beta. In this state, the product is not a "thing" that is finished; it is a "stream" that is constantly evolving.

Perpetual Beta allows for infinite experimentation. By shrinking your PLD, you aren't just "failing fast"—you are "learning at scale." You are allowing the "Logic" of the machine to handle the "Feedback," so the "Vision" of the product leader can handle the "Evolution." This is the only way to build products that remain relevant in a market that moves at the speed of an algorithm.

CASE STUDY	THE CHASM REALITY	ACTION & RESULT
1: A LEGACY SOFTWARE GIANT struggled to compete with AI-native startups. Their flagship product was bloated with 20 years of features, making it too slow and complex for modern users who wanted simple, automated outcomes.	Leadership was trapped in **Feature-Led Identity**. They believed "more is better" and feared that simplifying the product would alienate long-time users, even as those users were already leaving.	They enacted a **"Sunsetting Protocol,"** killing off the bottom 40% of unused features. They pivoted to a "Results-First" interface powered by AI, which recaptured 15% of lost market share within a year.
2: A GLOBAL AUTOMOTIVE BRAND spent billions developing an in-house dashboard OS. By the time it launched, it was already obsolete compared to mobile-integrated AI assistants that provided better voice and navigation.	This was a **Sunk-Cost Trap**. Executives refused to pivot or partner because they had "invested too much to stop," even though the data showed customers preferred third-party apps over the factory system.	The CEO mandated **Strategic Self-Cannibalization**, halting the internal project and pivoting to an open-source AI platform. This saved $200M in annual R&D and boosted "Tech Satisfaction" scores by 50%.
3: A PROMINENT NEWS PUBLISHER saw traffic drop as AI search engines summarized their articles. They doubled down on "Gated Content" and paywalls, hoping to force users back to their traditional long-form reading experience.	The publisher suffered from **Platform Rigidity**. They were trying to protect a "Reading Habit" that had already evolved into a "Discovery Habit," making their product feel like an obstacle rather than a service.	They shifted to **Atomic Content Strategy**, using AI to break long stories into "Search-Ready" snippets and audio summaries. This "Product-as-a-Service" model increased mobile engagement by 80%.
4: A SMART HOME DEVICE MAKER faced a PR crisis when their AI-powered cameras were flagged for privacy leaks. The engineering team had prioritized "Cool Features" over "Safety Guardrails" during the sprint to launch.	This was **Performance over Purpose**. The team was so focused on the "What" (facial recognition) that they ignored the "Why" (security), leading to a product that failed its primary customer promise.	They implemented **Ethical Product Guardrails**, making privacy the lead feature. They gave users "Physical Kill-Switches" for the AI, which restored consumer trust and led to a 30% increase in holiday sales.
5: A CONSUMER ELECTRONICS FIRM launched a new wearable that flopped because it required too much "Manual Input" from the user. People wanted the device to tell them what to do, not ask them for data all day.	The firm was in **Manual Override Mode**. They built a tool that required the user to be the "Architect," failing to realize that in the AI era, the product itself must be the "Agent" that delivers the outcome.	They pivoted to **Autonomous Feedback Loops**, where the device used sensors to "predict" user needs without being asked. This "No-UI" approach led to the product becoming the top-rated wearable in its category.
6: A RETAIL BANKING APP noticed younger users were leaving for "Neobanks." Their app was just a digital version of a physical branch—clunky, slow, and full of "Digital Paperwork" that frustrated mobile users.	This was **Analog-Digital Parity**. The bank had "digitized" its old processes instead of "reimagining" them for AI, resulting in a product that felt like an ancient relic in a high-speed world.	They launched a **"Zenith" Opportunity Team** to build a mobile-first AI assistant that handled all transfers and budgeting automatically. This "Invisible Bank" reduced churn among Gen Z users by 65%.

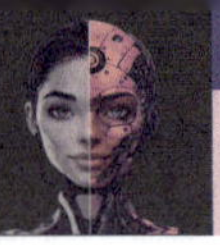

ASSIGNMENT : THE PRODUCT CULTURE DIAGNOSTIC AUDIT

The "Identity Chasm" in product development is most visible when a team is "Polishing a Sinking Ship." You see it in features that no one uses, but everyone is too afraid to cancel. To bridge this gap, we must diagnose Resource Stagnation—the state where your best people are held hostage by your worst ideas. This audit is designed to expose where "Pride" is standing in the way of "Pivot." On the left are the Symptoms of a product culture anchored by the past. On the right are the Strategic Interventions required to recalibrate for Equilibrium.

The Assignment: Identify the "Sunk Cost Anchors" in your roadmap and match them to the agility-centric remedy.

THE SYMPTOMS (THE ANCHORS)	MATCH	THE INTERVENTIONS (THE REMEDY)
1. The Frankenstein Feature: Adding more "bells and whistles" to a failing product instead of admitting the core premise is wrong.		**A. The Algorithmic Sunset:** Establishing automated "Kill-Switches" where funding stops if engagement misses an AI-predicted target.
2. Pride-Based Persistence: Continuing a project solely because a senior executive publicly announced it at a conference.		**B. Resource Recapture Protocol:** Publicly rewarding teams who "kill" their own projects early to move to higher-value work.
3. The Validation Loop: Requesting "more data" or "one more pilot" to delay the inevitable cancellation of a feature.		**C. Shadow Backlog Readiness:** Keeping a "Ready-to-Go" list of high-value pivots that can be funded the moment a failure is flagged.
4. Sunk Cost Storytelling: Reframing a failure as a "learning opportunity" while continuing to spend millions on it.		**D. Radical Capital Fluidity:** Moving budgets from "Project-Based" to "Signal-Based" where money follows the traction, not the plan.
5. The "Maintenance" Trap: Spending 80% of your talent's time fixing bugs on a legacy product that has no growth potential.		**E. Pivot-First Incentives:** Measuring Product Managers on their "Unlearning Rate" and "Pivot Latency" rather than "Feature Launches."
6. The Launch Illusion: Celebrating the "Go-Live" date as the finish line, rather than the "Signal" date as the starting line.		**F. Perpetual Beta Roadmaps:** Replacing rigid 12-month plans with 30-day "Experimental Sprints" that require data-proof to continue.
7. Consensus Drag: Requiring seven departments to agree before a project can be officially "Sunset."		**G. The Kill-Switch Commission:** An independent board empowered to pull funding based on data, bypassing internal politics.
8. Fear of the "Pivot": A culture where changing direction is viewed as "admitting defeat" or "lack of vision."		**H. Vision-over-Feature Training:** Shifting the team's identity from "Product Owners" to "Problem Solvers."
9. The Feedback Lag: Relying on quarterly customer surveys instead of real-time AI usage-telemetry to judge success.		**I. Real-Time Signal Integration:** Connecting the product roadmap directly to AI sentiment and usage feeds for instant visibility.
10. Specialized Silos: Talent is so specialized in "Product A" that they resist moving to "Product B," even if A is dying.		**J. Agile Skill-Mapping:** Training teams to be "Generalist-Augmented" so they can shift across the portfolio as signals dictate.

ANSWER MATCH: 1-A, 2-B, 3-C, 4-G, 5-D, 6-F, 7-I, 8-H, 9-E, 10-J

SECTION F

BLUEPRINT SERIES

Cultural Compass

Ai6x6™ Transformation Blueprint

Story 1 — The Post-War Japanese Industrial Ethos (1945–1960): Japan emerged from the war with limited resources but a collective commitment to national rebuilding. Companies embraced a culture of discipline, apprenticeship, and shared responsibility. Workers identified deeply with their organizations, reinforcing loyalty and craftsmanship. Leaders cultivated humility and long-term thinking. This cultural foundation enabled rapid industrial recovery. It demonstrated how shared values can accelerate capability even before systems mature.

Story 2 — NASA's Mission Culture During Apollo (1961–1972): NASA built a culture defined by precision, accountability, and collective purpose. Engineers and astronauts operated with a shared belief that failure was unacceptable, yet learning was essential. Communication was open, direct, and cross-disciplinary. Rituals, reviews, and norms reinforced psychological safety and technical rigor. This culture enabled teams to navigate uncertainty and complexity. It became a global example of mission-driven cohesion.

Story 3 — The Civil Rights Movement's Organizing Ethos (1954–1968): The movement relied on a culture of discipline, nonviolence, and moral clarity. Communities trained in de-escalation, collective action, and strategic communication. Leaders modeled courage and restraint, shaping norms that guided behavior under extreme pressure. Shared values created unity across diverse groups. This cultural coherence amplified impact far beyond organizational structure. It showed how culture can mobilize change at societal scale.

Story 4 — The Nordic Social Trust Model (1970s–1990s): Nordic countries cultivated cultures rooted in transparency, equality, and civic trust. Institutions operated with high accountability and low hierarchy. Citizens expected fairness and contributed to collective welfare. This cultural fabric enabled strong public services, stable governance, and economic resilience. Trust reduced friction in decision-making and collaboration. It demonstrated how cultural norms can become strategic national assets.

Story 5 — The Rise of Silicon Valley's Innovation Culture (1980s–2000s): Silicon Valley developed a culture that celebrated experimentation, risk-taking, and rapid iteration. Informal networks, open information flow, and meritocratic ideals shaped behavior. Failure was reframed as learning, attracting talent willing to push boundaries. Venture capital reinforced speed and ambition. This culture accelerated technological breakthroughs and entrepreneurial growth. It showed how cultural norms can amplify innovation beyond structural advantages.

Story 6 — The Globalization of Corporate Diversity & Inclusion (2000s–2020s): Organizations worldwide recognized that diverse teams outperform homogeneous ones when supported by inclusive cultures. Companies invested in psychological safety, equitable processes, and representation. Employees gained platforms to voice perspectives and challenge norms. Inclusion became a driver of creativity, retention, and ethical integrity. This cultural shift reshaped leadership expectations and organizational identity. It demonstrated how culture evolves to meet societal and competitive demands.

LESSONS FROM A PRODUCT TRANSFORMATION PERSPECTIVE

Product maturity is shaped by the cultural norms that influence craftsmanship, experimentation, and accountability. Organizations learned that products reflect the values of the teams that build them—precision cultures produce reliability, learning cultures produce innovation, and trust-based cultures produce integrity. Mature cultures encourage cross-functional collaboration, constructive critique, and pride in quality. They balance creativity with discipline, ensuring that products evolve without losing coherence. Culture becomes the engine that drives iteration, ethical choices, and long-term stewardship. Maturity is achieved when products embody the organization's identity, translating cultural principles into tangible outcomes that customers can experience and trust.

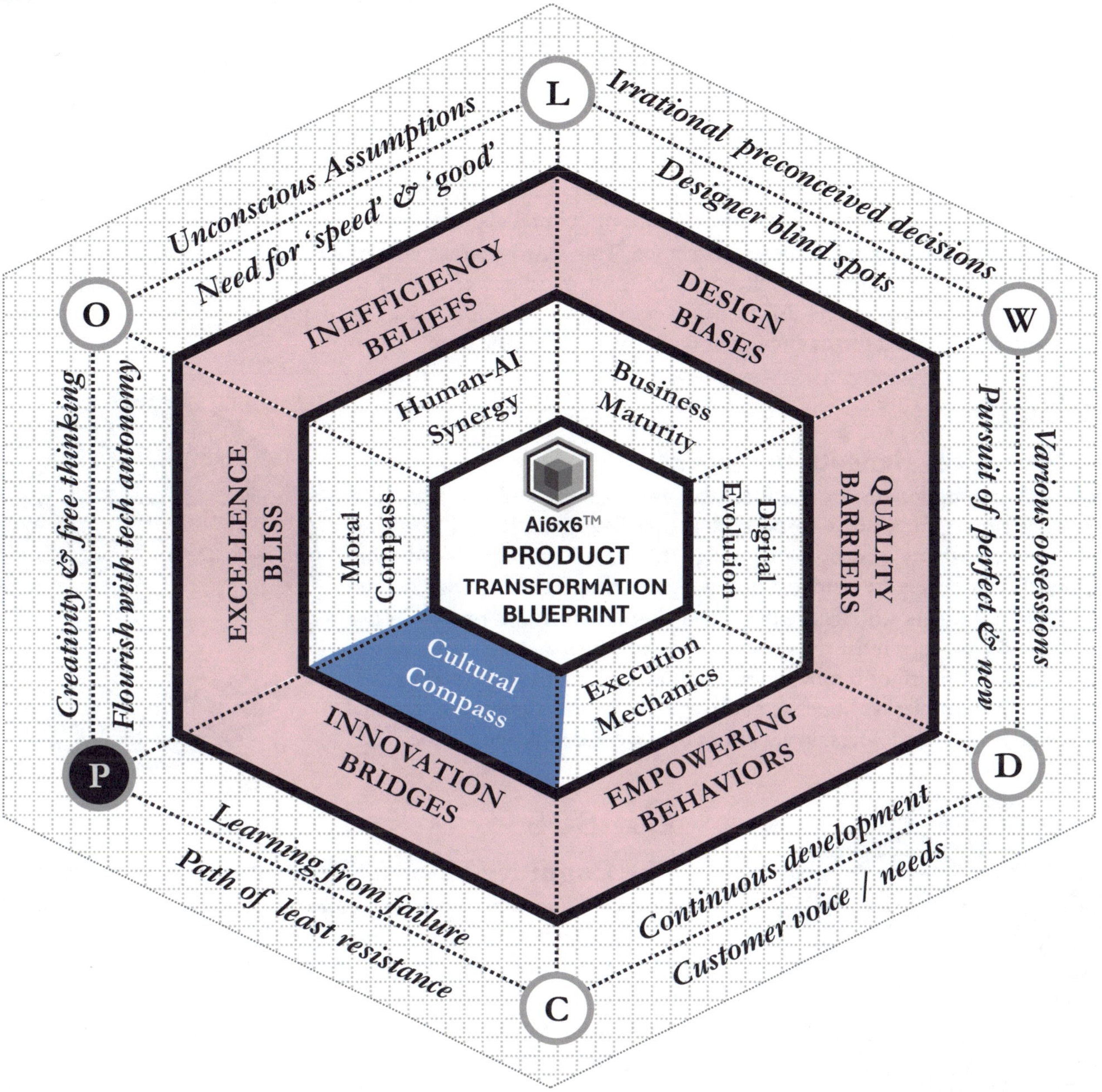

The **Ai6x6 Research Lab** chose this lens because the ultimate vitality of a product is found in the collective "Bliss" of the team that builds it and the market that adopts it. We reframed this dimension as the stewardship of the **Product Vision**, recognizing that culture is the invisible operating system that dictates whether a product is built with obsession or with purpose. It is vital because it addresses the invisible shadows—the **Product Development Biases** and **Obsessions**—that live in the minds of the creators. This lens is the bridge between the leader's strategic roadmap and the team's creative reality.

We selected it to ensure leaders build a visionary community rather than just a feature-heavy roadmap. By dismantling **Barriers** to transformation and building **Product Experience Bridges**, the leader creates a "greenhouse" for innovation where **Transformation Behaviors** become the standard. The Lab believes that in an age of automated production, the human connection to the vision is the only self-sustaining source of energy. This lens forces a look at how beliefs about the product impact its collective trajectory. When culture is treated as a discipline, the organization achieves a state of "Flow Bliss," where the entire team feels fundamentally connected to the product's impact.

Culture has always been the invisible architecture of an organization, shaping how people think, behave, and make decisions long before any formal strategy takes hold. In the age of AI, this architecture becomes even more consequential. Technology may accelerate capability, but culture determines whether that capability becomes meaningful progress or wasted potential. The **Cultural Compass** lens forces leaders to confront the truth that transformation is not sustained by systems alone; it is sustained by the beliefs people hold, the biases they carry, the barriers they navigate, the behaviors they exhibit, the bridges they build, and the moments of profound satisfaction that anchor their commitment. Without cultural alignment, even the most sophisticated AI strategy collapses under the weight of human resistance. This lens reframes culture as a dynamic system rather than a static identity. It recognizes that culture is not a set of values on a wall but a living pattern of interpretation, interaction, and intention. As AI reshapes workflows, decision cycles, and power dynamics, leaders must understand how cultural forces amplify or undermine transformation.

Beliefs

The Narratives Beneath the Surface

Beliefs and assumptions form the first dimension of the Cultural Compass. They are the quiet forces that shape how people interpret change, how they respond to uncertainty, and how they judge the intentions behind leadership decisions. When beliefs are outdated or misaligned, they distort perception and create friction. Leaders who ignore this dimension discover that people do not resist technology—they resist the meaning they attach to it. Addressing beliefs requires leaders to surface the narratives that operate beneath the surface and reshape them with clarity, honesty, and purpose. In this dimension, cultural truth-telling becomes a leadership act rather than a communication tactic.

Biases

The Cognitive Patterns That Shape Judgment

Biases and mental shortcuts represent the second dimension. These cognitive patterns influence how people evaluate information, assign trust, and make decisions. In an AI-enabled environment, biases can distort both human judgment and the interpretation of machine-generated insights. Addressing them requires intentional design, transparent dialogue, and environments where people feel safe challenging their own assumptions. Leaders must create conditions where cognitive unlearning becomes possible. In this dimension, cultural evolution is driven by awareness, humility, and the willingness to question long-held mental models.

Barriers

The Structural Forces That Restrict Change

Barriers and systemic impediments form the third dimension. These are the structural forces—policies, hierarchies, incentives, and legacy processes—that constrain behavior even when people are willing to change. Leaders often misinterpret resistance as emotional when it is structural. Removing barriers is not about demanding more effort; it is about redesigning the environment so that the desired behaviors become the natural ones. In this dimension, organizational de-friction becomes a strategic priority rather than an operational afterthought, enabling people to move with clarity rather than struggle against constraints.

Behaviors

The Visible Expression of Culture

Behaviors and observable actions represent the fourth dimension. Culture is not what people say; it is what they do consistently. In the AI era, behaviors must evolve to reflect curiosity, adaptability, collaboration, and responsible experimentation. Leaders must model these behaviors visibly and repeatedly. They must demonstrate that learning is valued, that questioning is encouraged, and that accountability is shared. In this dimension, behavioral signaling becomes a powerful tool for shaping norms, reinforcing the idea that culture is enacted through daily choices rather than declared through slogans.

Bridges

The Pathways That Connect People and Purpose

Bridges and cultural pathways form the fifth dimension. These are the connections that allow ideas, trust, and collaboration to flow across the organization. Bridges are built through shared language, shared rituals, and shared purpose. They transform isolated teams into aligned communities. Leaders who invest in cultural pathways create environments where people feel connected to something larger than their role. In this dimension, collective identity-building becomes a catalyst for cohesion, enabling the organization to move as a unified whole rather than a collection of disconnected parts.

Bliss

The Emotional Anchor of Commitment

Bliss and profound satisfaction complete the compass. Bliss is not about comfort; it is about meaning. It is the sense of fulfillment that emerges when people feel their work matters, their contributions are recognized, and their growth is supported. Bliss is the emotional anchor that sustains commitment through uncertainty. Leaders who cultivate it create cultures where people do not merely comply—they care. In this dimension, purposeful fulfillment becomes a strategic asset, strengthening resilience and deepening engagement.

The Cultural Compass lens reminds leaders that transformation is not sustained by technology or process alone. It is sustained by the human experience of change. Culture determines whether people interpret transformation as threat or opportunity, whether they respond with resistance or curiosity, and whether they move in alignment or fragmentation. In a world where AI accelerates both progress and pressure, leaders must treat culture as the primary operating system of the enterprise. When leaders align beliefs, challenge biases, remove barriers, shape behaviors, build bridges, and cultivate meaning, they create cultures capable of evolving continuously. They build environments where people feel psychologically safe, intellectually stimulated, and emotionally invested. Culture becomes the force that amplifies strategy rather than undermining it. It becomes the stabilizing anchor in moments of disruption and the accelerating engine in moments of opportunity.

CULTURAL COMPASS ANALOGY – CULTIVATING JAPANESE GARDEN

The False Climate — BELIEFS

Dismantling "Weather Myths" ensures that unrealistic, universal expectations don't stifle individual growth. Exposing the false belief that all "plants" require the same sun, water, or timing—allowing for specialized care.

The Invasive Weeds — BIASES

Identifying and removing hidden mental shortcuts prevents old habits from choking out innovation. Surgically exposing the leadership judgment and systemic biases that create "shadow" cultures and block progress.

The Hard Clay — BARRIERS

Breaking through structural resistance allows the organizational roots to expand and stabilize. Removing the deeply embedded obstacles like data, automation and AI anxiety that act as physical walls against digital growth.

Calibrated Care — BEHAVIORS

Executing specialized nurturing routines ensures each distinct "species" receives what it needs to thrive. Moving away from "one-size-fits-all" mode to model specific watering, feeding, and light-balancing behaviors required for a diverse ecosystem.

The Trellis System — BRIDGES

Constructing pathways for upward growth provides the support needed for teams to reach their full potential. Building the specific leadership and development paths that allow the workforce to climb through the maturity stages.

The Full Harvest — BLISS

Securing a state of permanent organizational harmony results in a self-sustaining ecosystem of excellence. Reaching the ultimate destination where the "Cultural Soul" operates in a high-fidelity hum, delivering constant value.

Ai6x6™ Transformation Blueprint

Product Inefficiency Beliefs

"Be stubborn on vision but flexible on details." - Jeff Bezos

Jeff Bezos (b. 1964) is an American entrepreneur and the founder of Amazon. He is known for his "customer obsession" philosophy, which prioritizes a seamless and satisfying customer experience, and is a key figure in the e-commerce revolution.

Product inefficiency beliefs are negative assumptions about product development that constrain innovation and lead to mediocre products. A product team might hold the belief that they know what customers want without needing to ask them, leading to products that are brilliant in design but useless in practice. Another limiting belief is that technology is an end in itself, rather than a tool to solve a customer's problem. These beliefs foster an inward-focused, technology-obsessed mindset. To truly transform, an organization must cultivate a new mindset that values deep customer understanding, thoughtful design, and a commitment to creating tangible value above all else. This requires a shift in perspective, as Jeff Bezos's quote suggests, from being rigid on the minor details to being "stubborn on vision," ensuring that the core purpose of the product remains aligned with customer needs.

Dedicated to delivering the tangible results of the Mission Core, this chapter delves into the Product Inefficiency Beliefs that fuel Innovation and drive successful Value Realization. It shifts focus from merely listing features to instilling the deep conviction required to build market-leading products. This begins with Empathy Beliefs—the foundational principle that understanding the user's struggle, context, and unmet need is paramount to responsible product conception. Simplicity and Utility Beliefs then guide the design process, stressing that elegant functionality and ease-of-use are superior to feature-rich complexity. To sustain momentum, the book emphasizes Iterative Beliefs, codifying the conviction that products should be continuously tested, improved, and evolved in rapid cycles of feedback and delivery. Finally, it addresses market differentiation through Aesthetic Beliefs (recognizing beautiful design as a key value differentiator) and reinforces the ethical and market necessity of Accessibility Beliefs, ensuring products are usable by everyone and thus maximizing potential reach and impact. These convictions are the essential engine for realizing value in the hands of the end-user.

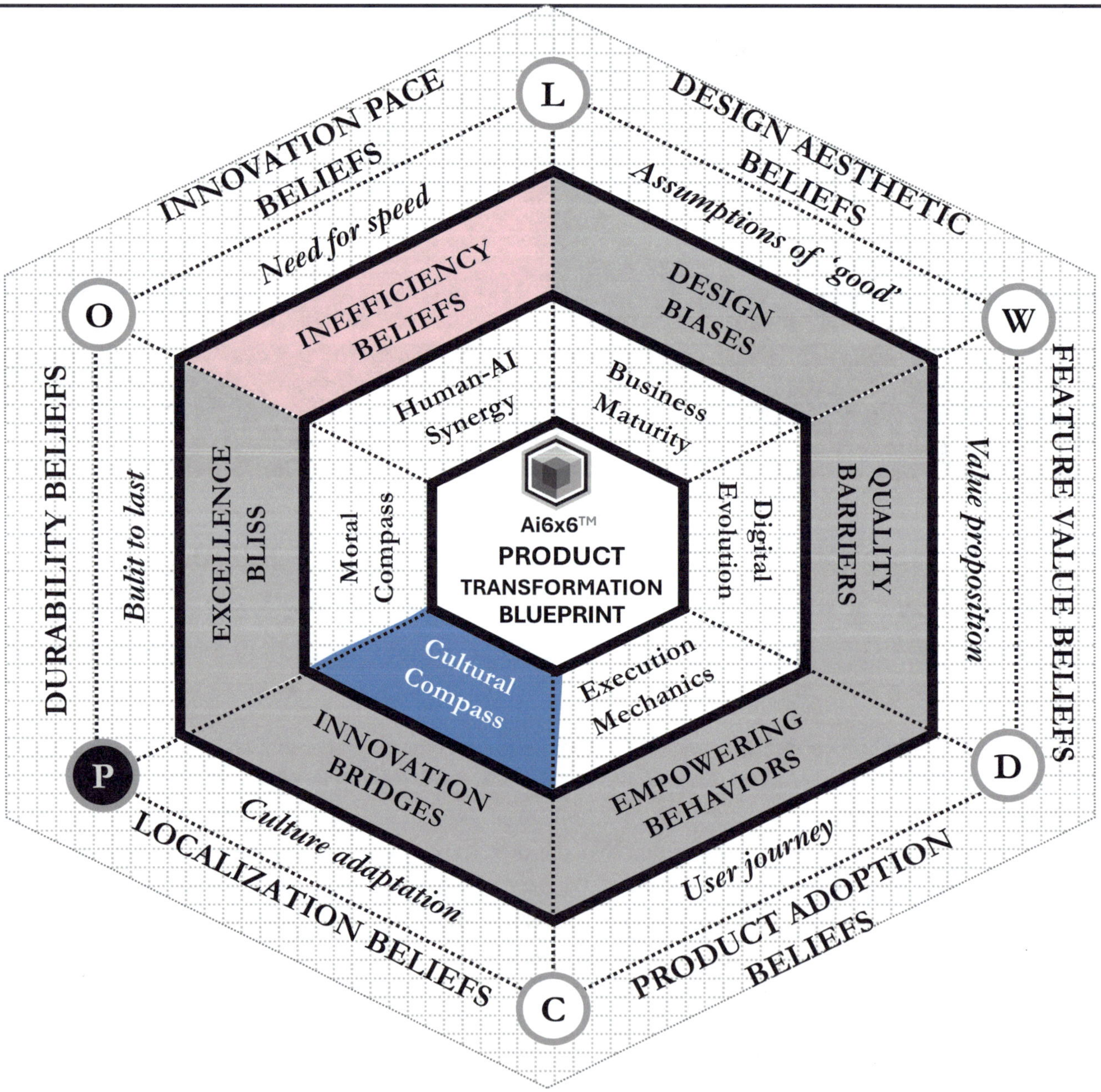

Ai6x6 CULTURAL COMPASS FOR BELIEFS | PRODUCT TRANSFORMATION

If you believe that "more releases equals more innovation," you are confusing activity with progress. In the AI era, the pace of release must be balanced by the pace of absorption. If you release features faster than your users—or your own AI models—can learn from them, you create a "feature soup" that dilutes your core value proposition. Daily, this manifests as a team that is constantly rushing to meet arbitrary deadlines, leading to "shallow" features that lack the depth of data integration required for AI to be effective. Transformation requires believing that the "right" pace is the one that allows for meaningful feedback loops. You must shift from a "sprint-to-finish" mindset to a "sprint-to-learn" mindset that prioritizes quality over sheer volume.

DO'S FOR THE LEADER	DON'TS FOR THE LEADER
Balance speed with quality.	Rush to release features that break.
Use iterative development cycles.	Release everything at once.
Prioritize a few key features.	Try to build too many things at once.
Set realistic timelines and goals.	Set aggressive, unachievable deadlines.
Release a minimum viable product.	Wait for the "perfect" product.
Use data to inform your speed.	Let competitors dictate your release speed.

SITUATION	SUGGESTED ACTION
Teams rush releases because they believe customers expect constant updates regardless of stability impact	Set release cadences, validate readiness, and reinforce discipline by prioritizing reliability over speed when customer workflows depend on predictable performance
Leaders delay launches because they assume customers will reject anything imperfect even during early testing	Use controlled pilots, gather feedback, and reinforce learning by releasing incremental improvements aligned with customer tolerance
Teams overbuild features because they believe rapid innovation signals market leadership regardless of adoption	Track usage patterns, measure impact, and reinforce focus by aligning innovation with demonstrated customer needs
Product managers accelerate roadmaps because they assume competitors' announcements require immediate responses	Validate urgency, assess relevance, and reinforce strategy by anchoring decisions in customer value rather than competitive noise
Engineers push experimental technologies because they believe novelty alone drives customer excitement	Evaluate use cases, test prototypes, and reinforce practicality by proving real customer benefit before committing resources
Teams slow innovation because they assume customers prefer stability over meaningful improvements	Communicate value, manage transitions, and reinforce confidence by guiding customers through change with clear support

The belief that good design is purely about "visual minimalism" can be a trap in the AI era. While a clean UI is important, the core of AI-driven design is "anticipatory." If you believe that design ends at the surface level, you will fail to build the underlying data pathways that make a product truly intuitive. Daily, this leads to products that look beautiful but feel "dumb" because they don't anticipate the user's next move. Transformation requires believing that design is about "cognitive load reduction." You must value "invisible design"—where the AI handles the complexity in the background, so the user experience feels effortless. This shift ensures your product is not just a pretty interface, but a truly intelligent and helpful partner.

DO'S FOR THE LEADER	DON'TS FOR THE LEADER
Create beautiful, functional designs.	Make products that are ugly or clunky.
Use a consistent design language.	Use a jumble of different styles.
Prioritize user-friendly design.	Focus on looks over usability.
Get feedback on the design.	Assume your design is perfect.
Design for accessibility.	Forget to design for disabilities.
Follow a clear brand identity.	Create a design that doesn't fit the brand.

SITUATION	SUGGESTED ACTION
Designers prioritize minimalism because they believe fewer elements always improve usability regardless of context	Test variations, gather feedback, and reinforce balance by matching visual simplicity with functional clarity
Teams favor dark themes because they assume modern aesthetics universally appeal across industries	Offer theme options, study preferences, and reinforce inclusivity by accommodating diverse visual expectations
Leaders push brand-heavy visuals because they believe strong identity outweighs functional clarity	Validate comprehension, simplify layouts, and reinforce usability by reducing visual noise
Designers avoid bold colors because they assume customers perceive them as unprofessional in enterprise tools	Test palettes, analyze reactions, and reinforce confidence by using color intentionally for guidance
Teams overuse animations because they believe motion enhances delight regardless of task complexity	Measure task efficiency, adjust timing, and reinforce focus by minimizing unnecessary movement
Designers resist accessibility changes because they assume adjustments compromise aesthetic quality	Apply inclusive principles, test with diverse users, and reinforce excellence by proving accessibility enhances overall design

If you believe that "the most complex features provide the most value," you are likely over-engineering your product. In the AI era, the most valuable features are often the ones that are the most "seamlessly helpful." Daily, this belief leads to product roadmaps cluttered with high-effort, low-impact technical experiments that confuse the user. Transformation requires believing that value is measured by the "utility-to-effort" ratio for the customer. When you believe that simplicity is the ultimate sophistication, you use AI to automate the mundane, allowing your core features to shine. This ensures your development resources are focused on the "critical few" capabilities that actually drive retention and solve the user's primary pain points with precision.

DO'S FOR THE LEADER	DON'TS FOR THE LEADER
Prioritize features customers need.	Build features you think are cool.
Validate new features with users.	Make assumptions about feature value.
Use data to rank feature ideas.	Let a single person decide on features.
Focus on solving a key problem.	Build features that solve nothing.
Build features that add new value.	Just copy your competitors' features.
Listen to your sales team's needs.	Ignore your sales team's feedback.

SITUATION	SUGGESTED ACTION
Teams prioritize advanced features because they believe power users represent the entire customer base	Segment needs, analyze behavior, and reinforce alignment by balancing advanced capabilities with essential workflows customers rely on daily across varied contexts effectively
Leaders push AI features because they assume automation always increases value regardless of context	Validate use cases, measure outcomes, and reinforce relevance by focusing on customer-defined success metrics rather than novelty alone in real environments consistently
Product managers add dashboards because they believe more data automatically improves decision-making quality	Simplify metrics, highlight insights, and reinforce clarity by surfacing only actionable information tied directly to decisions customers must make in their workflows
Teams build integrations because they assume customers require broad ecosystem coverage without confirming demand	Prioritize integration candidates using evidence, validate partner demand, and reinforce focus by selecting connections with measurable, recurring customer impact over purely symbolic
Engineers enhance customization because they believe flexibility always increases satisfaction for every customer	Limit configuration complexity, guide choices with opinionated defaults, and reinforce usability by protecting customers from overwhelming, low-value options and decision fatigue risks
Teams expand feature sets because they assume breadth signals maturity even when adoption remains low	Track adoption rigorously, retire underused capabilities, and reinforce discipline by reallocating investment toward features demonstrably improving outcomes for specific segments over time

4. PRODUCT ADOPTION BELIEFS | The User Journey

Assuming that "build it and they will come" is a belief that leads to massive market failure. If you believe adoption is purely a matter of marketing, you will ignore the "onboarding friction" that kills AI products. Daily, this manifests as high "drop-off" rates after the first session because the user didn't see the value of the AI immediately. Transformation requires believing that adoption is an ongoing "education journey." You must design your product to "demonstrate value" within the first sixty seconds of use. This belief shift ensures you build "hooks" into the product that show the user exactly how the AI will make their life better, turning skeptical first-time users into loyal, long-term advocates who understand the product's true power.

DO'S FOR THE LEADER	DON'TS FOR THE LEADER
Create a smooth onboarding process.	Make it hard for new users to start.
Provide clear instructions and help.	Let users figure it out on their own.
Offer a product demo or tutorial.	Expect people to adopt it instantly.
Launch with a strong marketing plan.	Just release a product and hope for the best.
Get feedback from early adopters.	Ignore what your first users say.
Make it easy to see product value.	Make users search for the value.

SITUATION	SUGGESTED ACTION
Teams expect rapid adoption because early testers respond enthusiastically during controlled environments only	Broaden testing cohorts, analyze segment differences, and reinforce realism by validating adoption patterns across diverse, production-like conditions before committing aggressive forecasts publicly
Leaders assume enterprise customers adopt quickly because contracts mandate usage regardless of readiness	Provide structured onboarding, track activation milestones, and reinforce success by addressing organizational barriers early with targeted training and change-management support for stakeholders
Teams underestimate adoption time because they assume intuitive design eliminates learning curves entirely	Offer layered guidance, monitor behavior analytics, and reinforce learning by supporting gradual skill development through tutorials, prompts, and contextual help within everyday workflows
Product managers expect viral growth because they assume network effects activate immediately after launch	Design deliberate sharing mechanics, measure referral behavior, and reinforce patience by iterating on activation levers instead of assuming automatic exponential expansion in practice
Teams assume customers will migrate instantly because new versions appear objectively better on paper	Communicate concrete benefits, provide side-by-side comparisons, and reinforce confidence by offering phased migration options with rollback safety nets for critical workloads and teams
Leaders expect uniform adoption because they assume all segments share similar motivations and constraints	Segment customers by context, tailor enablement journeys, and reinforce alignment by addressing distinct incentives, risks, and decision dynamics within each group over time

The belief that localization is just "translating text" is a major barrier to global AI success. If you believe your product is culturally neutral, you will fail to account for the different ways people in different regions interact with technology and trust algorithms. Daily, this leads to products that feel "foreign" or "irrelevant" in international markets, even if the language is correct. Transformation requires believing that localization is an act of "cultural empathy." You must adapt the AI's logic, the UI's flow, and even the "tone of voice" to match local expectations. This shift ensures your product feels "native" everywhere, allowing you to capture global markets by respecting the nuances of how diverse cultures perceive value and interact with machines.

DO'S FOR THE LEADER	DON'TS FOR THE LEADER
Adapt your product for local needs.	Use a single product for all cultures.
Research cultural norms and laws.	Use an insensitive or offensive design.
Translate your content accurately.	Use poor or literal translations.
Offer local payment options.	Force everyone to use the same payment.
Test your product with local users.	Assume it will work everywhere.
Respect local holidays and customs.	Send promotions on local holidays.

SITUATION	SUGGESTED ACTION
Teams assume simple translation is enough because they underestimate cultural differences affecting comprehension	Conduct cultural research, test localized flows, and reinforce accuracy by adapting content, visuals, and workflows to regional expectations effectively
Leaders believe global customers prefer identical experiences regardless of local norms or regulatory requirements	Map regional constraints, tailor features, and reinforce compliance by aligning product behavior with local expectations and required standards
Teams delay localization because they assume early adopters speak dominant languages comfortably across markets	Prioritize key regions, release localized versions, and reinforce inclusivity by supporting diverse language needs early in product expansion
Designers assume color choices carry universal meaning across cultures without validating interpretations beforehand	Test palettes regionally, gather reactions, and reinforce cultural sensitivity by adjusting visual cues to match local expectations
Product managers assume pricing models work globally without considering local purchasing power differences	Analyze affordability, adjust tiers, and reinforce fairness by aligning pricing structures with regional economic realities consistently
Teams assume onboarding flows translate directly without adapting examples or metaphors to local contexts	Localize scenarios, rewrite references, and reinforce clarity by using culturally relevant examples throughout onboarding

In the fast-moving AI era, the belief that a product should "last forever" without change is a recipe for obsolescence. Conversely, believing that products are "disposable" leads to technical debt. Daily, this manifests as either a refusal to pivot or a series of "quick fixes" that eventually break the system. Transformation requires believing in "evolutionary durability." You must build products that are designed to be "modular" and "updatable" at their core. When you believe that the core of the product is its "learning capacity," you build a foundation that can survive multiple waves of technological change. This ensures your product remains relevant and reliable for years, not by staying the same, but by being the absolute best at adapting to new data.

DO'S FOR THE LEADER	DON'TS FOR THE LEADER
Build for quality and long-term use.	Build products that break quickly.
Offer a good warranty or guarantee.	Offer no support for broken products.
Use high-quality materials.	Use cheap parts to save money.
Design for easy repair.	Make products impossible to repair.
Test for long-term wear and tear.	Only test for short-term functionality.
Release timely software updates.	Abandon products after the sale.

SITUATION	SUGGESTED ACTION
Teams assume customers replace hardware frequently because technology cycles move quickly in competitive markets	Study replacement patterns, design longevity, and reinforce trust by supporting extended product lifespans through reliable components
Leaders believe software stability matters less because updates can always fix issues after release	Strengthen testing, prioritize reliability, and reinforce quality by preventing failures before deployment through disciplined engineering
Teams underestimate durability expectations because they assume customers tolerate minor degradation over time	Monitor performance, address wear patterns, and reinforce dependability by maintaining consistent experience across product lifespan
Product managers assume cloud services require minimal maintenance because infrastructure scales automatically	Track resource usage, optimize workloads, and reinforce resilience by planning capacity proactively across environments
Engineers assume AI models remain accurate indefinitely without retraining or monitoring drift	Implement monitoring, schedule retraining, and reinforce reliability by updating models regularly to maintain accuracy
Teams assume customers accept shorter lifecycles because subscription models imply constant upgrades	Communicate support timelines, maintain compatibility, and reinforce value by honoring long-term commitments transparently

Ai6x6™ Transformation Blueprint

Product Design Biases

"Don't find customers for your products, find products for your customers."
- Seth Godin

Seth Godin (b. 1960) is an American author and public speaker. He is one of the most influential thinkers in modern marketing and is known for his work on "permission marketing," which emphasizes building trust with customers.

Product design biases are unconscious cognitive shortcuts that can derail product development and lead to solutions that fail to resonate with users. A common and dangerous pitfall is Usability Bias, where a designer or engineer assumes a product is easy to use because they find it so, failing to recognize that the average user lacks their specialized knowledge. Another critical bias is Survivorship Bias, where a product team focuses only on the products that have succeeded, ignoring the valuable lessons that could be learned from the products that have failed. Overcoming these biases requires a commitment to a rigorous, evidence-based approach to product design, a willingness to learn from all outcomes, and a disciplined focus on solving a core problem for the customer. As Seth Godin suggests, the most successful approach is to "find products for your customers," not the other way around.

This chapter delves into **Product Design Biases**, revealing how cognitive shortcuts lead to flawed conception, execution, and user interaction, thereby compromising Innovation and hindering Value Realization. It shifts focus from technical skill to the psychological traps inherent in the design process. Sections address the insidious Usability Bias (Designer's Curse), where product creators overestimate the ease-of-use of their own designs for new users. The critical flaw of Survivorship Bias is explored, emphasizing the danger of only studying successful products while ignoring the lessons embedded in failures. The chapter addresses internal friction caused by Procedural Injustice Bias (perceptions of unfairness in the design process) and the financial and temporal drain of the Scope Creep Effect (uncontrolled requirement expansion). Furthermore, it warns against the superficial attraction of Gamification Bias (over-relying on non-essential game elements) and addresses the fundamental usability barrier of Hick's Law Bias, ensuring designers understand that maximizing choices increases decision time and decreases user engagement, which is essential for successful value capture.

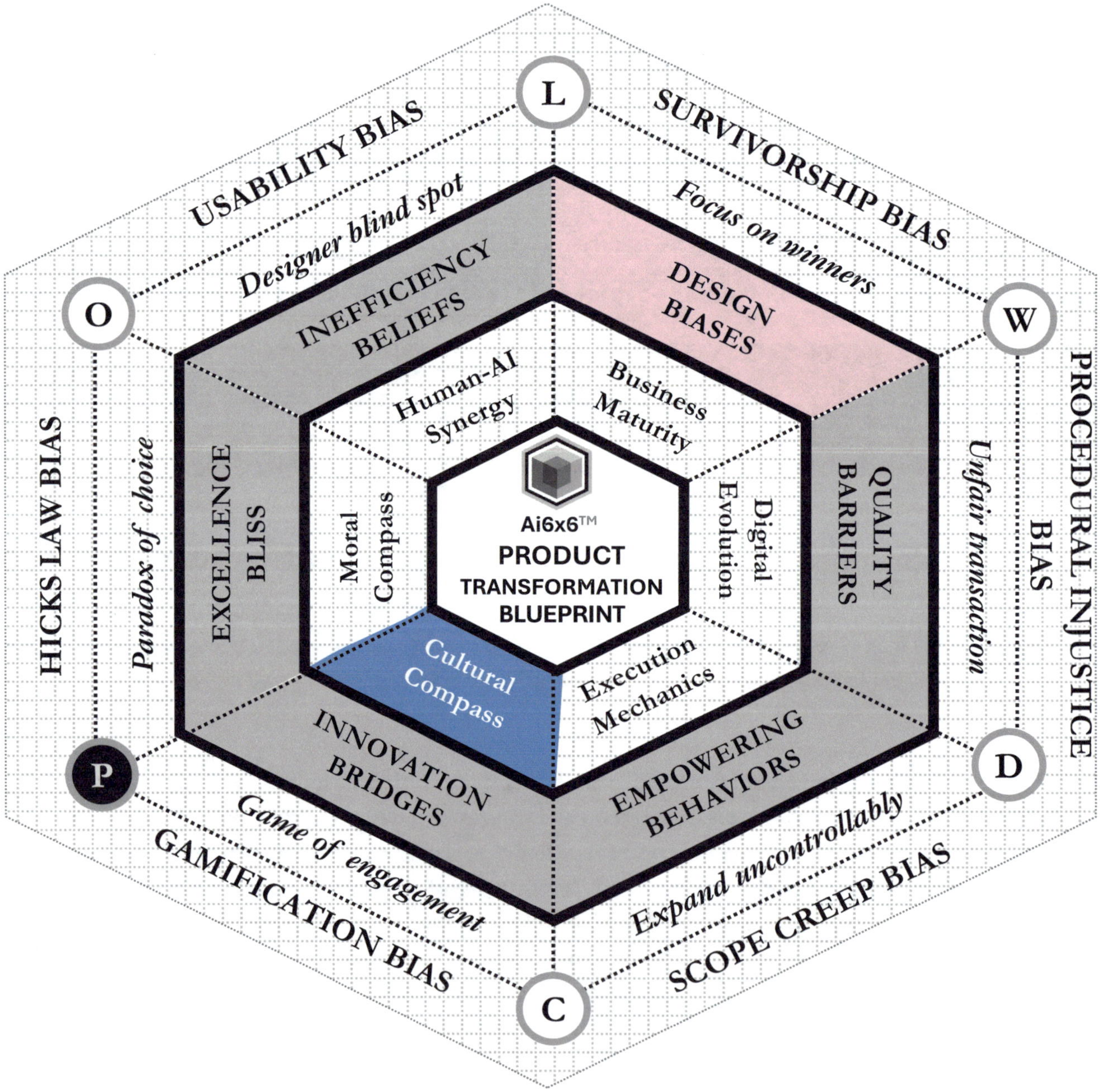

Ai6x6 CULTURAL COMPASS FOR BIASES | PRODUCT TRANSFORMATION

1. USABILITY BIAS | The Designer's Blind Spot

This bias occurs when designers and developers assume a product is easy to use simply because they are "too close" to the project. Because they know where every button is, they cannot see the friction that a new user experiences. Daily, this leads to "feature blindness," where obvious UX flaws are ignored until they become major customer complaints. Transformation requires a commitment to "radical user testing." You must intentionally seek out the most "non-technical" users to break your product. By recognizing usability bias, you move from "designing for yourself" to "designing for the human." This ensures your AI-driven features are accessible to everyone, not just the experts who built them, making the technology truly inclusive and useful.

DO'S FOR THE LEADER	DON'TS FOR THE LEADER
Test your product with real users.	Assume the product is easy to use.
Get feedback from a diverse group.	Only get feedback from your team.
Use simple and clear design.	Use a complex and confusing design.
Observe how users interact with the product.	Believe the way you use it is the only way.
Focus on the user's journey.	Focus on how you want them to use it.
Watch for friction and fix it.	Ignore signs of user frustration.

SITUATION	SUGGESTED ACTION
Designers assume workflows are intuitive because they understand internal logic deeply from development involvement	Conduct usability tests, observe behavior, and reinforce clarity by simplifying steps customers struggle with repeatedly
Teams believe onboarding is simple because they skip steps unconsciously due to familiarity	Map first-time journeys, remove friction, and reinforce empathy by designing for beginners rather than experts
Product managers assume navigation is obvious because menu structures reflect internal mental models	Validate structure, reorganize content, and reinforce usability by aligning navigation with customer expectations
Engineers assume configuration screens are clear because they understand technical terminology customers rarely recognize	Replace jargon, add explanations, and reinforce comprehension by using customer-friendly language consistently
Teams assume error messages are helpful because they know underlying causes customers cannot see	Rewrite messages, provide guidance, and reinforce support by explaining next steps clearly and simply
Designers assume gestures feel natural because they use them daily across multiple devices	Test interactions, gather feedback, and reinforce accessibility by offering alternative controls for diverse users

2. SURVIVORSHIP BIAS | The Hall of Fame

This error involves focusing only on the "success stories" of past products while ignoring the many failures that could provide even more valuable lessons. In an executive context, this leads to "copy-catting" the features of market leaders without understanding the context of their success. Daily, this manifests as a strategy that is purely "reactive" rather than "innovative." Overcoming survivorship bias requires a "post-mortem" culture where failures are studied with the same intensity as wins. By learning from what didn't work, you avoid the "hidden traps" of the AI era. This ensures your product strategy is built on a comprehensive understanding of the market landscape, rather than a narrow and misleading view of a few high-profile outliers.

DO'S FOR THE LEADER	DON'TS FOR THE LEADER
Learn from both successes and failures.	Only study products that succeeded.
Ask, "what product didn't make it?"	Ignore products that failed.
Study the entire market, not just leaders.	Only study market leaders.
Get feedback from churned customers.	Only talk to your most loyal customers.
Use a wide range of data for decisions.	Only use the data you already have.
Acknowledge luck in success.	Believe success is purely skill.

SITUATION	SUGGESTED ACTION
Teams prioritize successful features because they ignore abandoned experiments revealing important customer constraints	Review failures, extract insights, and reinforce learning by analyzing unsuccessful attempts systematically
Leaders assume market demand is strong because they focus on top-performing segments exclusively	Study underserved groups, analyze gaps, and reinforce strategy by addressing overlooked needs thoughtfully
Product managers assume onboarding works because successful users complete it despite hidden friction	Track drop-offs, interview non-completers, and reinforce improvements by fixing early barriers thoroughly
Teams assume pricing is effective because high-value customers renew consistently	Analyze churned segments, test alternatives, and reinforce fairness by addressing affordability issues directly
Designers assume interface patterns work because expert users navigate easily without difficulty	Observe novices, simplify flows, and reinforce inclusivity by designing for broader audiences
Engineers assume performance is adequate because heavy users rarely complain about speed	Monitor diverse usage, test edge conditions, and reinforce reliability by optimizing for all segments

3. PROCEDURAL INJUSTICE BIAS | The Unfair Transaction

This bias occurs when a customer develops a negative perception of a product not because of the product itself, but because of an "unfair or opaque process" related to it. Daily, this leads to "untraceable churn," where customers leave even though the product "worked" perfectly. Transformation requires "procedural transparency." You must ensure that every process—from billing to algorithmic output—is fair and easy to understand. By recognizing this bias, you protect your brand's integrity. You move from a "black box" operation to an "open-book" one, ensuring that the customer's entire experience with your product ecosystem feels just, respectful, and worth their continued investment. This builds a moral bridge between your technology and your users.

DO'S FOR THE LEADER	DON'TS FOR THE LEADER
Be transparent about your processes.	Use a complex, black box system.
Clearly explain how decisions are made.	Make decisions without explanation.
Treat all customers fairly.	Show favoritism to certain clients.
Use clear, fair rules for everyone.	Have different rules for different people.
Get feedback on your processes.	Assume your processes are fair.
Correct unfair processes quickly.	Let unfair processes continue.

SITUATION	SUGGESTED ACTION
Customers feel mistreated when refund workflows require unnecessary steps unrelated to actual eligibility	Simplify criteria, remove redundant checks, and reinforce fairness by aligning processes with transparent customer-centric standards consistently across journeys
Users abandon verification because identity checks feel excessive compared to the actual risk involved	Calibrate verification, match rigor to risk, and reinforce trust by explaining why specific steps are required clearly
Customers complain when support escalations require repeating information multiple times across different teams	Unify case histories, streamline handoffs, and reinforce respect by ensuring customers never repeat previously provided details unnecessarily
Users perceive bias when automated decisions deny access without clear explanations or appeal paths	Add transparency, provide reasoning, and reinforce accountability by offering human review options for disputed automated decisions
Customers feel punished when subscription cancellations require contacting support instead of simple self-service	Enable self-service cancellation, clarify terms, and reinforce fairness by removing unnecessary friction from exit processes
Users distrust pricing when discounts apply inconsistently across similar customer profiles	Audit rules, standardize eligibility, and reinforce equity by ensuring consistent application of promotional logic across segments

4. SCOPE CREEP BIAS | The Expanding Horizon

Scope creep is the tendency for a project's requirements to expand uncontrollably, often driven by the desire to "just add one more thing."2 In the AI era, this is particularly dangerous because adding a small feature can have massive implications for your underlying data models. Daily, this leads to missed deadlines and a "bloated" product that tries to do everything but does nothing well. Transformation requires "disciplined scoping." You must be willing to say "no" to good ideas to protect the "great" ones. By recognizing scope creep as a bias toward "more," you regain control over your roadmap. This ensures your product remains focused, fast, and capable of delivering its core promise with high-quality execution and zero technical waste.

DO'S FOR THE LEADER	DON'TS FOR THE LEADER
Create a clear, focused roadmap.	Let the project's scope expand wildly.
Say "no" to features that don't fit.	Say "yes" to every single request.
Prioritize based on core value.	Get distracted by the latest trends.
Use a disciplined change process.	Allow new features to be added on a whim.
Focus on the primary problem.	Try to solve every problem at once.
Have a clear product vision.	Have no clear product vision.

SITUATION	SUGGESTED ACTION
Teams add features mid-cycle because stakeholders request enhancements unrelated to original goals	Define guardrails, enforce change reviews, and reinforce discipline by approving additions only when aligned with core objectives
Engineers expand technical ambitions because new tools appear exciting during development cycles	Evaluate necessity, assess impact, and reinforce focus by limiting experimentation to validated use cases with measurable value
Designers introduce new patterns because inspiration arises late in the design process unexpectedly	Freeze patterns, document rationale, and reinforce consistency by updating design systems only after structured reviews
Product managers widen scope because competitor announcements create perceived urgency to respond	Reassess priorities, validate customer need, and reinforce strategy by resisting reactive expansions lacking evidence
Teams add integrations because partners request visibility without confirming actual customer demand	Analyze usage, validate value, and reinforce alignment by prioritizing integrations with measurable customer benefit
Leaders expand timelines because they assume more features guarantee stronger market reception	Set launch criteria, measure readiness, and reinforce impact by focusing on essential value delivery rather than breadth

5. GAMIFICATION BIAS | The Game of Life

Assuming that making a product "a game" with points and badges will automatically make it engaging is a common bias that often leads to "shallow engagement." If the core utility of the product is missing, gamification feels like a manipulative gimmick. Daily, this manifests as users who "play" for a week and then disappear forever. Transformation requires "meaningful engagement." You must focus on the "intrinsic value" of the product first. When you use AI to help users achieve their real-world goals more effectively, engagement happens naturally. By recognizing gamification bias, you move from "tricking" the user to "empowering" them, ensuring that any game-like elements are used to enhance the user's true journey, rather than distract from a lack of value.

DO'S FOR THE LEADER	DON'TS FOR THE LEADER
Make products truly useful and engaging.	Assume badges and points equal engagement.
Use gamification to add real value.	Add game mechanics for their own sake.
Get feedback on how people use it.	Assume your game mechanics are fun.
Focus on solving a key problem.	Use gamification to hide a bad product.
Use gamification to empower users.	Use gamification to manipulate users.
Design for long-term engagement.	Design for a short-term hit of dopamine.

SITUATION	SUGGESTED ACTION
Teams add points because they assume scoring systems motivate all users equally across contexts	Study motivations, test mechanics, and reinforce relevance by designing incentives aligned with real user goals
Designers add badges because they believe visual rewards increase engagement universally across segments	Validate reactions, measure behavior, and reinforce effectiveness by using rewards only where they drive meaningful outcomes
Product managers add streaks because they assume daily usage is always desirable for every workflow	Assess task frequency, respect natural rhythms, and reinforce balance by avoiding pressure-based engagement patterns
Teams add leaderboards because they assume competition motivates all customer segments consistently	Segment audiences, offer alternatives, and reinforce inclusivity by supporting collaborative engagement modes
Engineers add challenges because they assume difficulty increases satisfaction across all workflows	Match challenge to context, test difficulty, and reinforce usability by avoiding unnecessary cognitive load
Leaders push gamification because they believe it modernizes products regardless of customer purpose	Clarify intent, validate fit, and reinforce integrity by using gamification only when it enhances actual value

6. HICK'S LAW BIAS | The Paradox of Choice

Failing to simplify a product's interface leads to a bias toward "over-choice," which significantly slows down user decision-making. In an AI era, where we can generate infinite options, this is a massive risk. Daily, this manifests as "choice paralysis," where the user is so overwhelmed by the AI's suggestions that they do nothing at all. Transformation requires a "pathway to simplicity." You must use AI to "curate" options, not just generate them. By recognizing this bias, you prioritize a "minimalist decision path." You ensure that your product guides the user toward the best possible outcome with the least amount of cognitive effort, turning a complex tool into a blissful and efficient experience that feels effortless to navigate.

DO'S FOR THE LEADER	DON'TS FOR THE LEADER
Simplify choices for users.	Create a confusing interface with too many options.
Remove all unnecessary features.	Add features that serve no purpose.
Use a clear, logical interface.	Use a messy, jumbled interface.
Get feedback on user flow.	Assume your product is easy to navigate.
Guide users through the experience.	Let users get lost in a sea of options.
Create a clear path to the goal.	Have many different paths to a goal.

SITUATION	SUGGESTED ACTION
Users hesitate because screens present too many choices without clear prioritization cues	Reduce options, group actions, and reinforce clarity by highlighting primary tasks prominently across interfaces
Customers abandon flows because menus contain excessive nested categories difficult to navigate quickly	Flatten hierarchy, simplify labels, and reinforce usability by reducing unnecessary depth in navigation structures
Users misclick frequently because button density overwhelms available visual space on critical screens	Increase spacing, remove redundancies, and reinforce precision by presenting fewer actions per screen
Customers struggle during onboarding because too many setup decisions appear simultaneously without guidance	Sequence steps, offer defaults, and reinforce comfort by guiding users gradually through early configuration
Users delay actions because advanced settings appear alongside essential controls without separation	Separate complexity, hide advanced options, and reinforce simplicity by surfacing only necessary controls initially
Customers feel overwhelmed when dashboards display excessive metrics without clear hierarchy or grouping	Prioritize indicators, remove noise, and reinforce insight by surfacing only actionable information clearly

Ai6x6™ Transformation Blueprint

Product Quality Barriers

"The chains of habit are too weak to be felt until they are too strong to be broken." - Samuel Johnson

Samuel Johnson (1709–1784) was an English writer and a towering figure of the 18th century. His most monumental achievement was the compilation of A Dictionary of the English Language in 1755, a work of immense influence on the English language.

Product quality barriers are cultural and procedural impediments that slow development and compromise the quality of the final product. A Perfection Obsession Barrier is a culture of seeking perfection, which endlessly delays product launches and prevents the organization from getting valuable customer feedback. A Technology Obsession Barrier is an over-eagerness to adopt new technology for its own sake, rather than with a clear customer use case in mind, leading to products that are technologically advanced but solve no real problems. A Competitor Obsession Barrier is a culture of focusing on copying competitors instead of creating unique value. Overcoming these barriers requires a focused effort to redesign the product development process, instill a culture of learning and rapid iteration, and always prioritize the customer's problem over internal or competitive pressures.

This framework delves into Product Quality Barriers, revealing how self-imposed obstacles prevent effective conception, execution, and user interaction, thereby compromising Innovation and hindering Value Realization. A common failure point is Feature Bloat, the tendency to endlessly add features, confusing users and draining development resources without increasing core value. This often leads directly to Poor User Experience, where frustrating interfaces or confusing workflows sabotage customer adoption regardless of the product's underlying utility. This section stresses the long-term detriment of Technical Debt, where quick fixes and poor architectural choices slow future development to a crawl and destroy agility. Critically, Lack of User Feedback is examined, emphasizing the organizational failure to systematically listen to and integrate external insights, leading to the creation of products nobody needs. Furthermore, the framework addresses Inconsistent Design, where a lack of a cohesive look, feel, or navigational logic across product lines erodes brand trust, and highlights the risk of Ineffective Testing, which allows critical usability or functional flaws to escape into the production environment.

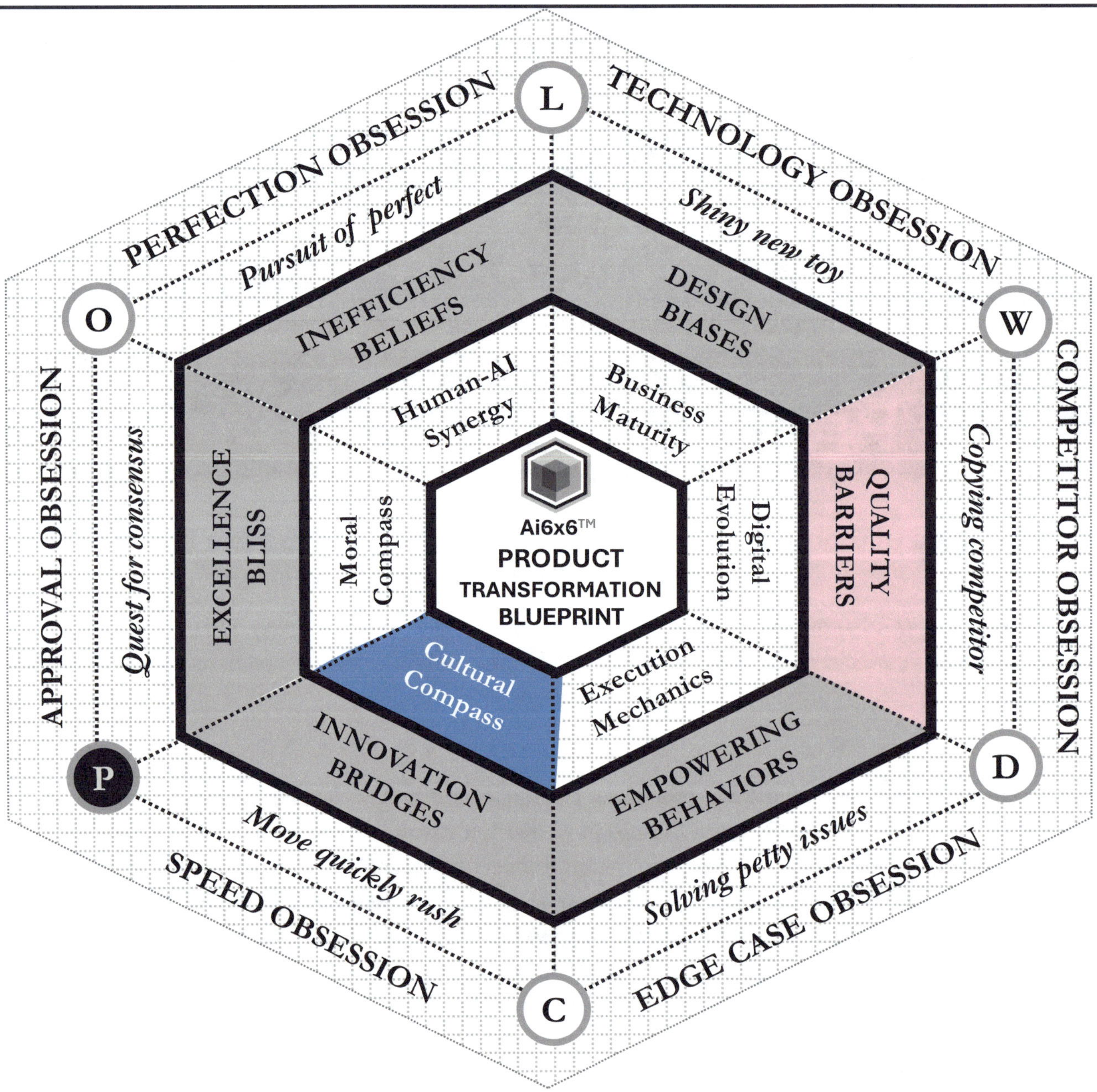

Ai6x6 CULTURAL COMPASS FOR BARRIERS | PRODUCT TRANSFORMATION

1. PERFECTION OBSESSION | The Tyranny of Perfection

The barrier of seeking "perfection" before launch is a major killer of AI innovation.[3] Since AI models learn from real-world data, holding back a product until it is "perfect" means you are losing out on the data needed to actually make it better. Daily, this manifests as endless "beta" cycles and a fear of "shipping." Transformation requires embracing the "Minimum Viable Model." You must be comfortable launching with "good enough" and using live feedback to iterate toward excellence. By breaking this barrier, you shift from a "static launch" mindset to a "continuous learning" mindset. This ensures your product gets to market faster and evolves more rapidly than competitors who are still polishing their initial version in a vacuum.

DO'S FOR THE LEADER	DON'TS FOR THE LEADER
Use an iterative, fast approach.	Wait for a "perfect" product to launch.
Launch an MVP and get feedback.	Never launch because it's not perfect.
Balance quality with speed.	Sacrifice speed for perfect quality.
Be willing to release a good product.	Only release a flawless product.
Use feedback to make it better.	Assume your product is already perfect.
Focus on getting value to customers.	Get stuck on tiny details.

SITUATION	SUGGESTED ACTION
Teams delay releases because they believe every feature must match idealized internal standards	Define launch thresholds, release iteratively, and reinforce momentum by improving based on real customer feedback gathered after launch
Designers postpone handoff because they want every pixel refined before engineering begins	Set collaboration checkpoints, share drafts early, and reinforce flow by enabling parallel progress across teams
Engineers extend timelines because they aim for flawless performance across all theoretical scenarios	Prioritize critical paths, test real conditions, and reinforce pragmatism by focusing on customer-visible reliability
Leaders block launches because they fear reputational damage from minor imperfections customers may never notice	Validate risk, assess impact, and reinforce confidence by aligning decisions with customer expectations
Teams rewrite content repeatedly because they want messaging perfectly polished before release	Freeze copy, test comprehension, and reinforce clarity by prioritizing usefulness over stylistic perfection
Product managers delay experiments because they want complete data before making directional decisions	Run small tests, gather signals, and reinforce learning by acting on early evidence

2. TECHNOLOGY OBSESSION | The Shiny New Toy

Adopting new technology just because it is "cool" or "new"—without a clear customer use case—is a massive waste of resources. This "tech for tech's sake" barrier leads to products that are technically impressive but solve no real problems. Daily, this looks like an engineering team that is constantly chasing the latest AI framework while the core user experience remains buggy or confusing. Transformation requires "problem-first" thinking. You must demand that every new technology solves a specific, documented customer pain point. By breaking this barrier, you ensure that your AI investments are "strategic" rather than "speculative." This keeps your product grounded in reality and focused on delivering the "utility" that customers are actually willing and eager to pay for.

DO'S FOR THE LEADER	DON'TS FOR THE LEADER
Use tech to solve customer problems.	Adopt new tech for its own sake.
Prioritize customer needs first.	Prioritize technology first.
Validate technology with user needs.	Assume new tech is always better.
Focus on the user, not the tool.	Focus on the tool, not the user.
Ask, "what problem does this solve?"	Ask, "what can we build with this?"
Use a technology that fits the problem.	Force a problem to fit the technology.

SITUATION	SUGGESTED ACTION
Teams adopt new frameworks because they believe modern stacks automatically improve product quality	Evaluate tradeoffs, test prototypes, and reinforce discipline by selecting technology based on customer impact
Leaders push AI features because they assume intelligence alone differentiates products regardless of usefulness	Validate use cases, measure outcomes, and reinforce relevance by solving real customer problems
Engineers integrate emerging tools because they want to experiment with cutting-edge capabilities	Assess necessity, limit experimentation, and reinforce focus by aligning tools with strategic goals
Teams migrate platforms because they assume newer infrastructure guarantees better performance	Benchmark performance, validate gains, and reinforce stability by migrating only when benefits outweigh disruption
Designers adopt novel interactions because they assume innovation requires abandoning familiar patterns	Test usability, gather feedback, and reinforce accessibility by retaining proven interaction models
Product managers prioritize technical novelty because they believe it signals market leadership	Anchor decisions in customer value, validate demand, and reinforce strategy by focusing on meaningful differentiation

Focusing too much on "copying competitors" leads to a "feature parity race" where no one wins and everyone looks the same. This barrier prevents you from creating unique, "blue ocean" value. Daily, this manifests as a product roadmap that is just a checklist of what the "other guy" did last month. Transformation requires "customer obsession" over "competitor obsession." You must look at the "unmet needs" of your users rather than the "latest moves" of your rivals. By breaking this barrier, you find the space to innovate in ways that others haven't even considered. This ensures your product has a "unique value proposition" that makes it the obvious choice for customers, rather than just another "me-too" option in the market.

DO'S FOR THE LEADER	DON'TS FOR THE LEADER
Provide unique value to customers.	Obsessively copy your competitors.
Focus on your own strengths.	Worry only about what others are doing.
Ask customers what they really want.	Ask what your competitors are doing.
Create your own product roadmap.	Let competitors dictate your strategy.
Focus on your own vision.	Lose your vision to competitive pressure.
Be a leader, not a follower.	Always follow the latest trend.

SITUATION	SUGGESTED ACTION
Teams replicate competitor features because they assume parity is required for market credibility	Analyze customer needs, prioritize differentiation, and reinforce strategy by building unique strengths
Leaders shift roadmaps because competitor announcements create pressure to respond immediately	Validate urgency, assess relevance, and reinforce focus by anchoring decisions in customer value
Designers mimic competitor interfaces because they assume familiarity guarantees adoption	Test alternatives, explore options, and reinforce creativity by designing for customer workflows
Product managers copy pricing models because they assume market norms dictate acceptable structures	Study willingness to pay, test tiers, and reinforce fairness by aligning pricing with customer outcomes
Engineers match performance benchmarks because they assume speed alone determines product success	Measure real usage, optimize bottlenecks, and reinforce impact by improving meaningful metrics
Teams chase competitor integrations because they assume ecosystem breadth signals maturity	Validate demand, prioritize value, and reinforce alignment by selecting integrations with measurable benefit

4. EDGE CASE OBSESSION | The Trivial Pursuit

Getting bogged down in solving minor, uncommon problems for a tiny fraction of users can derail your entire development cycle. This "80/20" barrier leads to a product that is overly complex for the vast majority of people because you tried to accommodate every possible scenario. Daily, this looks like "feature bloat" and an interface that is cluttered with "advanced settings" that no one ever uses. Transformation requires a focus on the "critical path." You must prioritize the features that benefit the 80% of your users. By breaking this barrier, you keep your product "clean" and "efficient." This ensures that your AI models are optimized for the most common and impactful use cases, providing the best experience for the most people.

DO'S FOR THE LEADER	DON'TS FOR THE LEADER
Focus on solving the primary problem.	Get bogged down in solving tiny problems.
Focus on the 80% of users.	Build features for a few edge cases.
Use data to prioritize features.	Let a few people dictate your features.
Build for the most common use case.	Build for a weird, one-off use case.
Fix the big problems first.	Fix all the small problems first.
Prioritize features for a large impact.	Focus on features with a small impact.

SITUATION	SUGGESTED ACTION
Teams delay launches because they want to handle rare scenarios affecting very few customers	Quantify impact, prioritize core flows, and reinforce momentum by addressing uncommon edge cases after validating real customer demand
Engineers over-engineer systems because they want to support unlikely usage patterns rarely encountered	Simplify architecture, focus on common paths, and reinforce efficiency by designing for typical behavior rather than hypothetical extremes
Designers complicate interfaces because they want to accommodate every possible exception within layouts	Streamline layouts, hide complexity, and reinforce clarity by supporting exceptions through secondary paths instead of primary screens
Product managers expand requirements because they want to satisfy hypothetical future needs without evidence	Validate assumptions, narrow scope, and reinforce discipline by focusing on current customer problems supported by measurable signals
Teams add configuration options because they want to support unusual workflows rarely used	Offer defaults, limit choices, and reinforce usability by reducing unnecessary flexibility that increases cognitive load
Leaders request additional features because they want to avoid criticism from niche segments with minimal influence	Assess segment size, prioritize impact, and reinforce strategy by focusing on majority needs rather than isolated requests

5. SPEED OBSESSION | The Finish Line Fallacy

A singular focus on "moving fast" at the expense of "moving correctly" is a barrier that leads to poor quality and high "rework" costs. While agility is important, "blind speed" in the AI era can lead to biased models and unstable code. Daily, this manifests as a "break things" culture that actually breaks the customer's trust. Transformation requires "sustainable speed." You must build "automated quality gates" into your process so that speed doesn't come at the cost of integrity. By breaking this barrier, you move from "hurrying" to "flowing." You ensure that your product launches are both rapid and reliable, allowing you to maintain a high pace of innovation without sacrificing the "stability" that professional users require.

DO'S FOR THE LEADER	DON'TS FOR THE LEADER
Balance quality with speed.	Rush to release features that break.
Prioritize a sustainable work pace.	Burn out your team with constant sprints.
Set realistic goals and deadlines.	Set impossible, aggressive deadlines.
Celebrate new innovations.	Stick to the old way of doing things.
Listen to new customer needs.	Assume old needs are the only needs.
Focus on long-term sustainability.	Only focus on short-term wins.

SITUATION	SUGGESTED ACTION
Teams ship features rapidly because they believe speed alone signals innovation to customers	Define quality baselines, enforce testing, and reinforce discipline by balancing rapid delivery with reliability expectations across releases
Engineers cut validation steps because they assume customers prefer fast releases over stable performance	Establish guardrails, automate checks, and reinforce trust by ensuring stability remains consistent across updates and environments
Designers skip research because they assume visual intuition is enough to move quickly	Conduct lightweight tests, gather signals, and reinforce accuracy by validating assumptions early before committing significant design effort
Product managers shorten discovery because they assume customer needs are already well understood	Interview segments, analyze patterns, and reinforce clarity by grounding decisions in current evidence rather than assumptions
Teams reduce documentation because they assume internal knowledge transfers happen naturally under pressure	Standardize templates, capture essentials, and reinforce continuity by maintaining shared understanding across teams
Leaders push aggressive timelines because they assume market pressure requires constant acceleration	Reassess priorities, protect quality, and reinforce sustainability by pacing delivery responsibly across product cycles

6. APPROVAL OBSESSION | The Quest for Consensus

The fear of upsetting internal stakeholders or "consensus-seeking" can lead to "design by committee," which results in a watered-down, uninspired product.6 This barrier prevents you from making the "bold decisions" that true transformation requires. Daily, this looks like endless meetings where every feature is debated until it loses its edge. Transformation requires "decisive leadership." You must empower product owners to make the final call based on data and vision, not just consensus. By breaking this barrier, you regain your "strategic edge." You ensure your product has a "clear soul" and a "strong point of view," which is essential for standing out in a crowded market and building a brand that people actually care about and trust.

DO'S FOR THE LEADER	DON'TS FOR THE LEADER
Make bold decisions that add value.	Avoid risk to please all stakeholders.
Empower teams to make their own choices.	Require approval for every small decision.
Use data to support your decisions.	Rely on the opinions of a few.
Focus on customer needs first.	Prioritize the needs of internal teams.
Be a leader, not a follower.	Avoid making any tough choices.
Communicate your vision clearly.	Let decisions be made in secret.

SITUATION	SUGGESTED ACTION
Product managers delay decisions because they want unanimous agreement across all stakeholder groups	Define decision rights, clarify ownership, and reinforce momentum by empowering accountable leaders to act decisively
Designers avoid simplification because they fear removing elements valued by specific internal teams	Validate usage, test alternatives, and reinforce clarity by prioritizing customer needs over internal preferences consistently
Engineers postpone refactoring because they worry stakeholders will resist temporary slowdowns required	Communicate benefits, plan increments, and reinforce alignment by linking improvements to long-term stability and performance
Teams avoid deprecating features because they fear backlash from small but vocal user segments	Analyze adoption, communicate rationale, and reinforce focus by retiring low-value capabilities responsibly and transparently
Leaders hesitate to pivot because they fear appearing inconsistent to executive peers and partners	Present evidence, outline impact, and reinforce confidence by framing pivots as strategic adjustments based on new information
Teams over-consult on minor decisions because they fear criticism for acting independently	Set thresholds, streamline reviews, and reinforce autonomy by enabling faster, informed choices within defined boundaries

Ai6x6™ Transformation Blueprint

Product Empowering Behaviors

"If you're not prepared to be wrong, you'll never come up with anything original." - Ken Robinson

Sir Ken Robinson (1950–2020) was a British author and international advisor on education. He is best known for his passionate advocacy for creativity and his famous TED Talk, "Do Schools Kill Creativity?".

Product empowering behaviors are positive and proactive actions that reorient the entire organization around the customer, leading to innovative and valuable products. Iterative Behavior is the practice of building and refining products through continuous cycles, getting real-world feedback from customers as early as possible. This is a direct antidote to the perfection obsession barrier. Customer-Focused Behavior is the commitment to using technology to solve customer problems, not for its own sake, which stands in stark contrast to technology obsession. Unique Value Behavior is the practice of consistently providing a unique value proposition to customers, which breaks the cycle of competitor obsession. As Ken Robinson's quote suggests, to come up with anything original, you must be prepared to be wrong. These behaviors, when adopted by the entire product team, create a virtuous cycle where feedback and iteration lead to more valuable products, which in turn leads to more customer loyalty.

Dedicated to delivering the tangible results of the Mission domain, this framework delves into Product Empowering Behaviors, revealing the actions that drive effective conception, execution, and user interaction, thereby fueling Innovation and accelerating Value Realization. These behaviors center on User-Centric Behavior, ensuring that design decisions are made with the user's documented needs in mind at every single step of the development process. Success is built on Iterative Behavior, continuously building, refining, and testing products in rapid, focused cycles based on actionable feedback from real-world usage. The model emphasizes Minimalist Behavior, focusing rigorously on core functionality to avoid complexity and prevent unnecessary feature bloat. This is reinforced by Testing Behavior, constantly employing A/B testing, usability studies, and functional tests to identify flaws and systematically improve performance before any major release. Critical to design integrity is Usability Behavior, guaranteeing the product is intuitive, easy to learn, and enjoyable to use, thus lowering the adoption barrier. Ultimately, teams must embrace Creative Behavior, exploring new and innovative solutions that disrupt the status quo and deliver breakthrough, unique value to the customer.

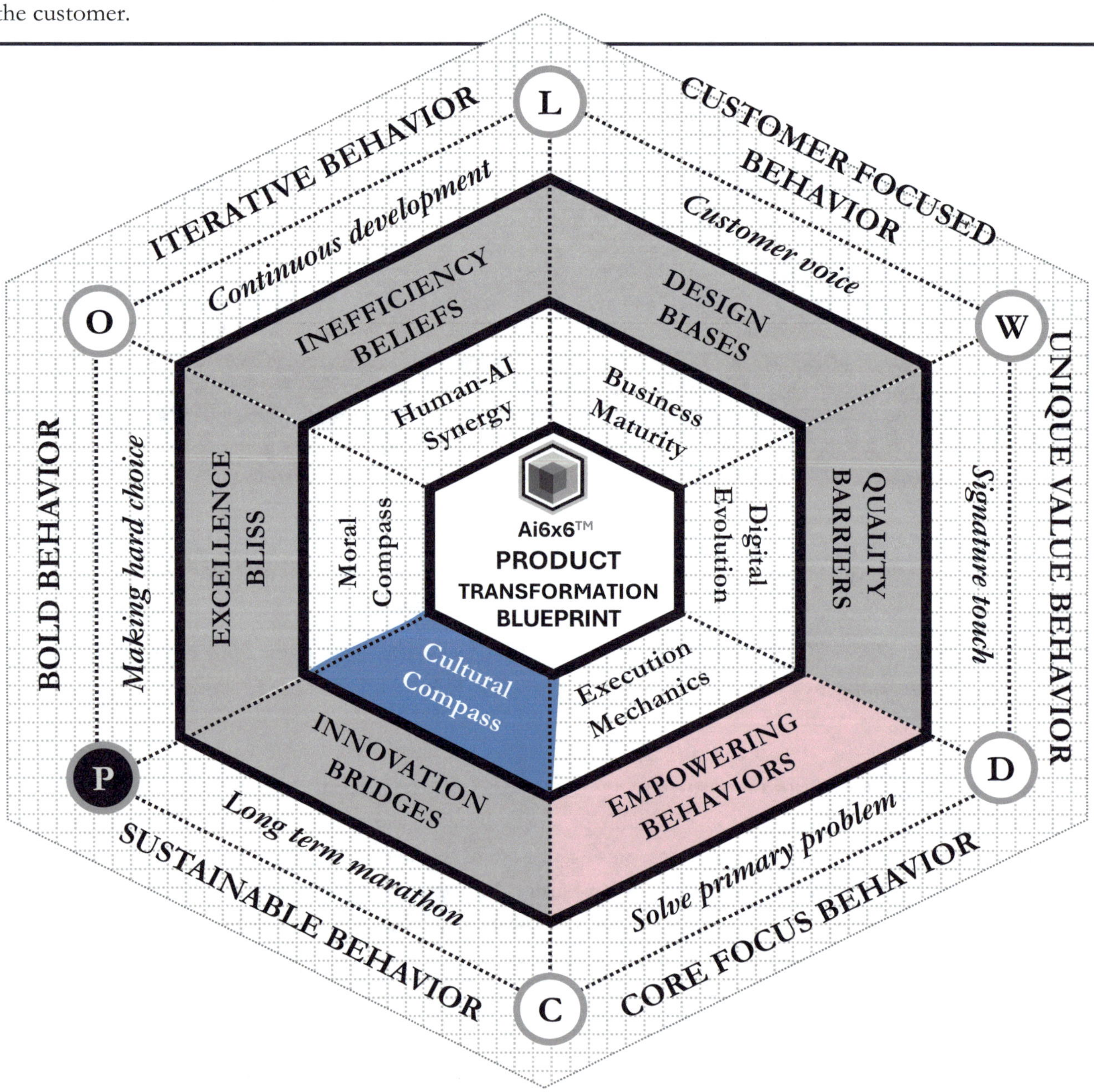

Ai6x6 CULTURAL COMPASS FOR BEHAVIORS | PRODUCT TRANSFORMATION

1. ITERATIVE BEHAVIOR | The Rhythm of Renewal

In a transformed product culture, building and refining through "continuous cycles" is a fundamental daily habit. It means you never consider a feature "done." This behavior involves constantly monitoring AI performance and user engagement to make small, rapid adjustments.[7] Daily, it looks like "micro-releases" that improve the product incrementally every day based on real-world usage. By modeling iterative behavior, you reduce the risk of large-scale failures. You turn the product into a "living organism" that is always getting smarter. This behavior ensures that your product is always "market-fresh" and that your team is constantly learning from real-world data, making the entire organization more agile and responsive to the "shifting sands" of the competitive AI era.

DO'S FOR THE LEADER	DON'TS FOR THE LEADER
Use continuous cycles of feedback.	Do a long, waterfall development.
Launch an MVP and get feedback.	Wait for the "perfect" product to launch.
Be willing to release a good product.	Only release a flawless product.
Use feedback to make it better.	Assume your product is already perfect.
Focus on getting value to customers.	Get stuck on tiny details.
Build and test in small increments.	Build everything at once.

SITUATION	SUGGESTED ACTION
Teams release small improvements regularly to validate assumptions before committing significant resources	Run experiments, measure outcomes, and reinforce learning by adjusting direction based on real customer behavior and insights
Designers test early concepts with customers before finalizing layouts or interaction patterns	Share prototypes, gather insights, and reinforce clarity by refining designs through repeated feedback cycles across sessions
Engineers deliver incremental updates to reduce risk and improve stability across environments	Break work into stages, monitor performance, and reinforce reliability by validating each change independently before rollout
Product managers adjust priorities frequently based on new evidence from usage patterns and interviews	Review data, reassess goals, and reinforce alignment by updating roadmaps responsively as conditions evolve
Teams refine onboarding flows continuously as they observe friction emerging in real customer journeys	Track drop-offs, test alternatives, and reinforce usability by improving steps iteratively based on observed behavior
Leaders encourage experimentation by rewarding learning outcomes rather than only successful results	Celebrate insights, document findings, and reinforce growth by normalizing iterative exploration across teams

Actively using technology to solve specific "customer problems"—rather than for its own sake—is the behavior that ensures long-term relevance. Daily, this looks like product managers spending half their time talking to users and the other half translating those needs into AI requirements. This behavior ensures that your technical "power" is always directed toward "user value." For you, it means rejecting a "cool" AI feature if it doesn't solve a documented pain point. Customer-focused behavior builds a product that people "love" because it actually works for them. By being the "voice of the customer" in every technical meeting, you ensure your AI transformation is a "human-success story" rather than just a "technical feat" of engineering.

DO'S FOR THE LEADER	DON'TS FOR THE LEADER
Use tech to solve customer problems.	Adopt new tech for its own sake.
Prioritize customer needs first.	Prioritize technology first.
Validate technology with user needs.	Assume new tech is always better.
Focus on the user, not the tool.	Focus on the tool, not the user.
Ask, "what problem does this solve?"	Ask, "what can we build with this?"
Use tech that fits the problem.	Force a problem to fit the tech.

SITUATION	SUGGESTED ACTION
Teams study customer workflows carefully before deciding which features deserve investment and prioritization	Interview users, analyze patterns, and align product decisions with real customer needs supported by measurable evidence and observed behavior
Designers validate assumptions early by testing prototypes directly with representative customer groups	Share concepts, gather reactions, and refine designs based on observed behavior rather than internal opinions or assumptions
Product managers prioritize problems customers experience frequently instead of chasing hypothetical opportunities without evidence	Review data, assess impact, and select initiatives that improve meaningful customer outcomes across essential workflows
Engineers shape technical decisions around customer value rather than internal preferences or convenience alone	Evaluate tradeoffs, measure benefits, and choose solutions that improve reliability and usability across customer journeys
Teams adjust roadmaps when customer feedback reveals new pain points requiring immediate attention and resolution	Reassess priorities, shift resources, and address emerging needs quickly to maintain alignment with customer expectations
Leaders encourage teams to challenge assumptions by asking how each decision benefits customers directly	Establish review rituals, document rationale, and link choices consistently to customer impact and measurable outcomes

3. UNIQUE VALUE BEHAVIOR | The Signature Touch

The practice of consistently providing "unique value" that cannot be easily replicated is a behavior that builds a powerful "product moat." Daily, this looks like looking for the "data advantage" or the "proprietary insight" that your AI can provide. It's about asking: "What can we do that no one else can?" This behavior encourages "creative differentiation." For you, it means investing in the features that set you apart, even if they are harder to build. Unique value behavior ensures that your product is not a "commodity." By focusing on your "unfair advantages," you create a product that is indispensable to your customers, making it much harder for competitors to displace you, regardless of how much they spend on marketing.

DO'S FOR THE LEADER	DON'TS FOR THE LEADER
Provide unique value to customers.	Obsessively copy your competitors.
Focus on your own strengths.	Worry only about what others are doing.
Ask customers what they really want.	Ask what your competitors are doing.
Create your own product roadmap.	Let competitors dictate your strategy.
Focus on your own vision.	Lose your vision to competitive pressure.
Be a leader, not a follower.	Always follow the latest trend.

SITUATION	SUGGESTED ACTION
Teams identify differentiators by studying unmet needs competitors consistently overlook across multiple markets	Analyze gaps, validate opportunities, and build capabilities that address underserved customer segments with meaningful advantages and measurable outcomes
Designers craft experiences emphasizing strengths customers cannot find in alternative products or competing services	Highlight advantages, refine interactions, and showcase unique benefits clearly across customer journeys and touchpoints
Product managers prioritize features that deliver measurable improvements customers immediately recognize and appreciate	Track outcomes, test messaging, and demonstrate value through real examples grounded in customer behavior and usage patterns
Engineers optimize performance in areas where customers experience the greatest friction or operational delay	Measure bottlenecks, improve responsiveness, and strengthen reliability by focusing on high-impact improvements across workflows
Teams refine positioning to emphasize the product's most compelling advantages across customer journeys and channels	Align messaging, update materials, and communicate strengths consistently across all customer-facing experiences
Leaders invest in capabilities that create long-term competitive advantage rather than short-term parity with competitors	Evaluate trends, assess durability, and choose initiatives with lasting value and strategic relevance across markets

4. CORE FOCUS BEHAVIOR | The North Star

A commitment to solving the "primary problem" for your customers is a behavior that prevents "product sprawl." Daily, this looks like having a "North Star" metric that every feature must support. It's the habit of "ruthless prioritization," where you prune away anything that distracts from the core mission. This behavior ensures that your product remains "intuitive" and "powerful." For you, it means your AI models are "specialized" rather than "generalized." Core focus behavior builds a brand that is the "go-to" for a specific need. By being the absolute best at one thing, you earn the right to expand later, ensuring that your initial transformation is built on a foundation of "unmatched excellence" in your chosen domain.

DO'S FOR THE LEADER	DON'TS FOR THE LEADER
Focus on solving one core problem.	Try to build every feature imaginable.
Prioritize based on customer value.	Build features that no one needs.
Say "no" to features that don't fit.	Say "yes" to every single feature request.
Focus on the product's core value.	Get distracted by the latest trends.
Use data to inform features.	Build features based on a whim.
Create a clear product vision.	Have no clear product vision.

SITUATION	SUGGESTED ACTION
Teams concentrate on essential workflows customers rely on daily instead of expanding into unrelated areas	Identify core tasks, measure usage, and prioritize improvements that strengthen foundational experiences across customer journeys
Designers simplify interfaces to highlight the most important actions customers must complete successfully	Remove distractions, refine layouts, and elevate primary tasks to improve clarity and usability across screens
Product managers decline feature requests that distract from solving the central customer problem effectively	Evaluate alignment, communicate rationale, and protect roadmap integrity by focusing on core value consistently
Engineers optimize core functionality before investing in secondary enhancements or experimental ideas	Address reliability, improve performance, and strengthen foundational systems supporting essential workflows and tasks
Teams revisit customer journeys regularly to ensure the product continues solving the primary problem effectively	Analyze behavior, identify gaps, and adjust priorities based on evolving customer needs and expectations
Leaders emphasize mission clarity by reminding teams which customer problem the product exists to solve	Communicate purpose, align decisions, and anchor strategy in core value consistently across initiatives

5. SUSTAINABLE BEHAVIOR | The Marathon, Not the Sprint

Building products at a "healthy pace" for both the team and the business is a behavior that ensures long-term viability. This looks like rejecting "crunch culture" in favor of "predictable delivery." Daily, it involves balancing "innovation work" with "maintenance work" to prevent technical debt.8 This behavior ensures that your team doesn't "burn out" and your product doesn't "break down." For you, it means setting "realistic expectations" with stakeholders. Sustainable behavior builds a "high-performance" team that can stay in the game for the long haul. By valuing the "well-being" of your developers and the "integrity" of your code, you ensure that your product transformation is a "marathon" of success rather than a "sprint" to failure.

DO'S FOR THE LEADER	DON'TS FOR THE LEADER
Prioritize a healthy work pace.	Burn out your team with constant sprints.
Use a sustainable development cycle.	Create a culture of burnout.
Set realistic deadlines and goals.	Set impossible, aggressive deadlines.
Protect your team from overwork.	Expect constant work and availability.
Communicate a long-term plan.	Have a short-term, sprint-based plan.
Focus on long-term maintainability.	Only focus on the present moment.

SITUATION	SUGGESTED ACTION
Teams plan realistic timelines that balance ambition with capacity across product cycles and releases	Estimate effort, manage scope, and pace delivery responsibly to maintain long-term sustainability across teams
Engineers adopt practices that reduce long-term maintenance burdens and technical debt accumulation	Standardize patterns, automate checks, and prevent unnecessary complexity through disciplined engineering practices across systems
Designers create systems that scale without requiring constant reinvention or redesign across teams	Build reusable components, document patterns, and enable consistent experiences across products and platforms
Product managers prioritize initiatives that deliver value without overwhelming teams or creating burnout	Assess workload, sequence projects, and distribute effort thoughtfully across cycles to maintain balance
Teams monitor operational metrics to ensure systems remain stable under increasing customer demand	Track performance, plan capacity, and address issues proactively before they impact customers or operations
Leaders encourage sustainable work habits by rewarding consistent progress rather than unsustainable bursts	Model balance, set expectations, and value steady improvement across teams and product lines

The courage to make "tough choices" that prioritize "core value" over "short-term noise" is the behavior of a visionary product leader. Daily, this looks like "killing" a feature that is popular but doesn't fit the long-term vision, or choosing to rebuild a legacy system that is holding the AI back. This behavior requires a "high tolerance for risk" and a "strong belief" in the mission. Bold behavior ensures that your product doesn't become a "compromised mess." By making the hard calls today, you clear the path for the "breakthroughs" of tomorrow. This behavior turns your product into a "market shaper," ensuring that you are the one defining the "future of the category" rather than just reacting to it.

DO'S FOR THE LEADER	DON'TS FOR THE LEADER
Make bold decisions that add value.	Avoid risk to please all stakeholders.
Empower teams to make their own choices.	Require approval for every small decision.
Use data to support your decisions.	Rely on the opinions of a few.
Focus on customer needs first.	Prioritize the needs of internal teams.
Be a leader, not a follower.	Avoid making any tough choices.
Communicate your vision clearly.	Let decisions be made in secret.

SITUATION	SUGGESTED ACTION
Teams pursue ambitious ideas even when outcomes remain uncertain but strategically aligned with goals	Define hypotheses, run experiments, and validate direction using measurable customer signals that support confident decision-making
Designers propose unconventional solutions when standard patterns fail to address customer needs effectively	Prototype concepts, test reactions, and refine bold ideas through structured feedback loops across representative segments
Product managers champion difficult tradeoffs when they improve long-term customer value significantly	Present evidence, clarify impact, and guide stakeholders toward decisions that support durable product outcomes
Engineers advocate architectural changes when existing systems limit scalability or reliability across environments	Assess constraints, model improvements, and implement changes incrementally to reduce risk while improving performance
Teams challenge outdated assumptions when customer behavior shifts meaningfully across markets and segments	Review data, update frameworks, and adjust strategy to reflect emerging patterns and validated insights
Leaders support decisive action when teams present clear rationale grounded in customer value and evidence	Reinforce ownership, remove obstacles, and maintain momentum by backing informed decisions consistently

Ai6x6™ Transformation Blueprint

Product Innovation Bridges

"The utmost thing is the user experience, to have the most useful experience." - Marissa Mayer

Marissa Mayer (b. 1975) is an American businesswoman and investor. She was an early and influential executive at Google and later served as the president and CEO of Yahoo!, where she focused on product development and user experience.

Product innovation bridges are practical frameworks and processes that connect a constrained product vision to a future of meaningful value. They are the strategic tools that turn good intentions into a seamless and satisfying product experience. A User-Centered Bridge is a formal process that ensures a product is designed with the user's needs at the very center, which is the most effective way to overcome usability bias. The Comprehensive Learning Bridge is a system for learning from both successful and unsuccessful products, ensuring that the organization does not fall prey to survivorship bias. The Disciplined Scoping Bridge is a framework that keeps a project's scope focused and manageable. As Marissa Mayer's quote suggests, "the utmost thing is the user experience, to have the most useful experience." These bridges are not just about collecting data but about creating a systematic and empathetic approach to every stage of product development, from ideation to launch.

This framework delves into Product Innovation Bridges, focusing on the mechanisms that connect ideas to realization, thereby fueling Innovation and accelerating Value Realization. User Research Bridges are the primary link, connecting design teams directly with customer needs and pain points through continuous qualitative and quantitative studies to ensure product relevance. Prototyping Bridges emphasize the tools and rapid cycles needed to quickly translate conceptual ideas into testable artifacts for validation, speeding up the build-measure-learn loop. The A/B Testing Bridges provide the necessary infrastructure to compare different product versions against specific performance metrics, basing design decisions on empirical evidence rather than intuition. Usability Testing Bridges involve observing real users interacting with the product to diagnose friction points and confirm intuitive flow. This framework addresses Feature Prioritization Bridges, detailing the disciplined mechanisms (like scoring models or RICE) used to decide which features yield the highest value-to-cost ratio. Lastly, Product Innovation Bridges are critical for scale, creating and enforcing a centralized library of reusable components and standards that ensure consistency, speed, and quality across all product lines.

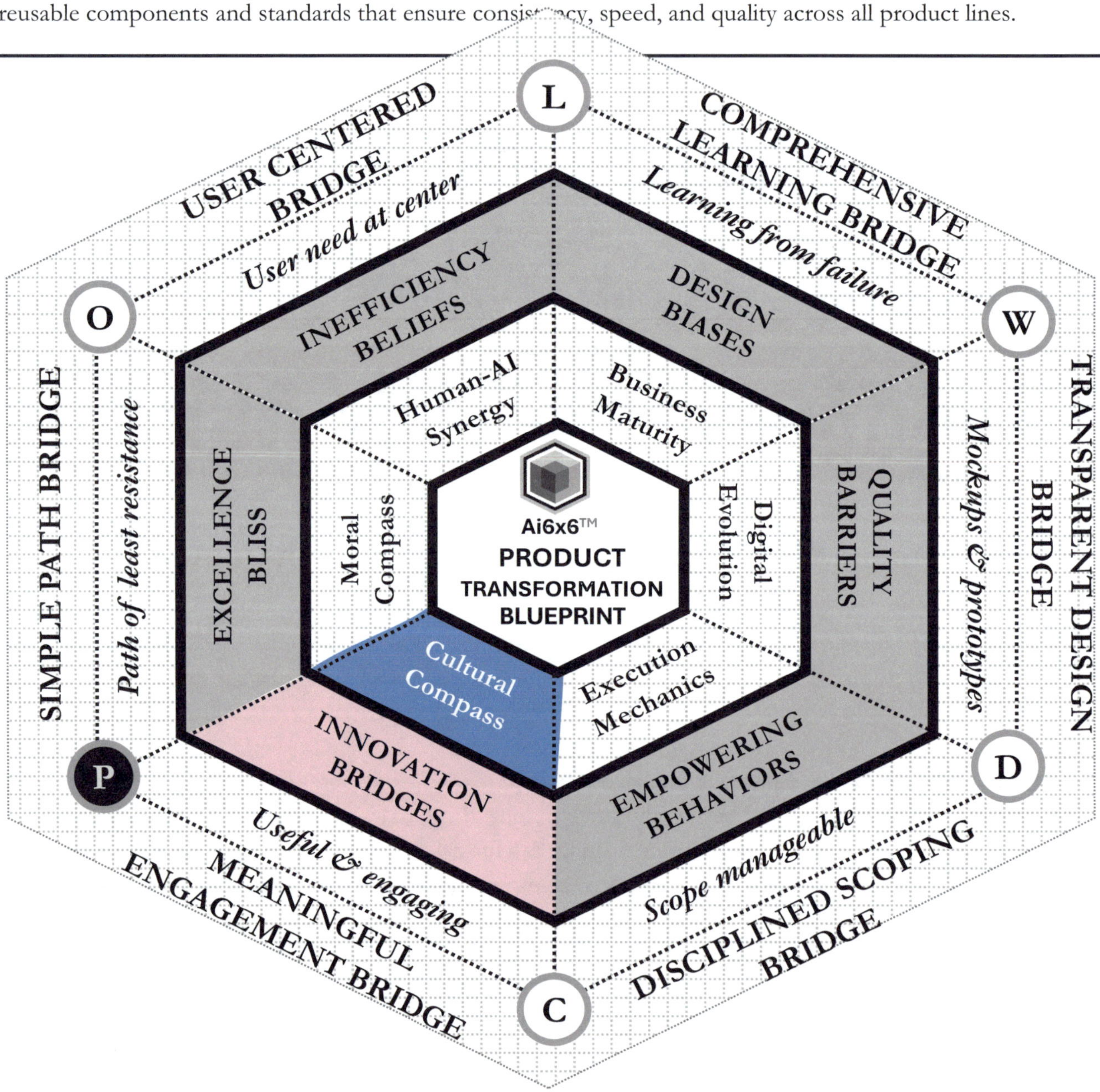

Ai6x6 CULTURAL COMPASS FOR BRIDGES | PRODUCT TRANSFORMATION

1. USER-CENTERED BRIDGE | The Empathy Engine

This bridge is a formal process—like "Design Sprints" or "User Co-creation"—that ensures every feature is designed with the "user's needs at the center." It's the bridge between "technical possibility" and "human utility." Daily, this bridge looks like "prototypes" being tested by real people before a single line of code is written. By building this bridge, you cross from the island of "assuming" to the mainland of "knowing." It ensures your AI is a "helper" rather than a "hindrance." For the executive, this bridge is a "resource optimizer"; it prevents you from building things that no one wants. This bridge provides the "human-centricity" that allows your product to build deep emotional resonance with your audience.

DO'S FOR THE LEADER	DON'TS FOR THE LEADER
Design with the user at the center.	Design with the developer at the center.
Test your product with real users.	Assume the product is easy to use.
Get feedback from a diverse group.	Only get feedback from your team.
Focus on the user's journey.	Focus on how you want them to use it.
Watch for friction and fix it.	Ignore signs of user frustration.
Make a product that is intuitive.	Make a product that requires a manual.

SITUATION	SUGGESTED ACTION
Teams observe real user behavior to understand motivations behind recurring workflow patterns and decisions	Conduct field studies, analyze recordings, and translate insights into actionable product improvements across journeys
Designers incorporate user feedback early to ensure concepts align with actual expectations and constraints	Share prototypes, gather reactions, and refine interactions based on observed behavior across representative groups
Product managers map user journeys to identify friction points affecting essential tasks across segments	Document steps, quantify impact, and prioritize improvements addressing high-value opportunities across workflows
Engineers consider accessibility requirements when implementing features supporting diverse user groups and needs	Follow guidelines, test variations, and ensure inclusive experiences across devices and environments
Teams validate assumptions regularly to prevent misalignment between product direction and user needs	Review evidence, adjust priorities, and maintain alignment through continuous discovery and structured learning
Leaders emphasize user outcomes when evaluating proposals competing for limited resources and attention	Compare benefits, assess reach, and allocate investment toward initiatives improving customer experience meaningfully

2. COMPREHENSIVE LEARNING BRIDGE | The Wisdom of Failure

This bridge is a system for systematically learning from "both successful and unsuccessful" products. It's the "data-driven feedback loop" that fuels evolution. Daily, this bridge looks like "A/B testing" every new model and conducting "blameless post-mortems" on every failed feature. By building this bridge, you turn "failure" into "intelligence." It allows you to move beyond "guessing" to "scientific product management." For the executive, this bridge is the "evolutionary engine." It ensures that your product is "self-correcting" and always getting better. This bridge provides the "intellectual honesty" that is required to lead a truly innovative AI product team, ensuring that every mistake is a steppingstone to a future breakthrough.

DO'S FOR THE LEADER	DON'TS FOR THE LEADER
Learn from both successes and failures.	Only study products that succeeded.
Ask, "what product didn't make it?"	Ignore products that failed.
Study the entire market, not just leaders.	Only study market leaders.
Get feedback from churned customers.	Only talk to your most loyal customers.
Use a wide range of data for decisions.	Only use the data you already have.
Acknowledge luck in success.	Believe success is purely skill.

SITUATION	SUGGESTED ACTION
Teams analyze usage data to understand how customers interact with core product workflows	Identify patterns, investigate anomalies, and translate findings into targeted improvements across customer journeys
Designers review experiment results to refine interface decisions affecting user comprehension and flow	Compare variants, interpret metrics, and adjust designs based on validated insights from structured testing
Product managers synthesize qualitative and quantitative inputs to guide roadmap decisions effectively	Combine interviews, analytics, and experiments to prioritize initiatives supported by strong evidence and measurable outcomes
Engineers monitor performance metrics to detect issues affecting reliability across environments and devices	Track trends, diagnose causes, and implement fixes improving stability and responsiveness across systems
Teams document learnings from failed experiments to strengthen future decision-making processes	Capture insights, share summaries, and integrate lessons into planning rituals across teams and functions
Leaders encourage evidence-based thinking by rewarding teams that test assumptions rigorously and transparently	Promote experimentation, highlight successes, and reinforce learning culture across groups and product areas

This bridge is the practice of making the "design process and decisions" clear to users, especially when it comes to AI. It's the bridge between "magic" and "trust." Daily, this looks like "tooltips" that explain why an AI made a suggestion, or "clear privacy toggles" that put the user in control. By building this bridge, you eliminate the "creepy factor" of AI. It ensures that your users feel like "partners" rather than "subjects." For the executive, this bridge is the "brand integrity" builder. It ensures that your product is seen as "ethical" and "reliable." This bridge provides the "clarity of purpose" that allows for a deep and lasting relationship between your technology and your customers.

DO'S FOR THE LEADER	DON'TS FOR THE LEADER
Share the design process with users.	Make design a closed-door process.
Use a consistent design language.	Use a jumble of different styles.
Explain design decisions.	Keep design decisions a secret.
Get feedback on the design.	Assume your design is perfect.
Design for accessibility.	Forget to design for disabilities.
Follow a clear brand identity.	Create a design that doesn't fit the brand.

SITUATION	SUGGESTED ACTION
Teams share design rationale openly to help stakeholders understand tradeoffs behind decisions	Present context, outline constraints, and explain reasoning clearly to build alignment across groups and teams
Designers document patterns thoroughly to ensure consistent implementation across teams and platforms	Provide examples, clarify rules, and maintain accessible references supporting shared understanding across contributors
Product managers communicate roadmap changes clearly to avoid confusion about shifting priorities	Explain drivers, outline impacts, and maintain trust through transparent updates across audiences and channels
Engineers expose system behavior through clear logs and diagnostics supporting troubleshooting efforts	Standardize formats, improve visibility, and enable faster resolution of issues across environments and deployments
Teams publish experiment results to help others understand what worked and what failed	Summarize findings, share implications, and encourage evidence-based iteration across teams and product groups
Leaders model transparency by explaining strategic decisions affecting product direction and investment	Share rationale, address concerns, and reinforce trust through open communication across the organization

This bridge is a framework—like "Objective and Key Results" (OKRs)—that keeps a project's scope "focused and manageable." It's the bridge between "ambition" and "delivery." Daily, this bridge looks like a "product backlog" that is ruthlessly pruned to maintain focus. By building this bridge, you ensure that your team is always working on the "most impactful" tasks. It prevents the "complexity trap" that slows down AI development. For the executive, this bridge is the "predictability engine." It ensures that you can hit your launch dates with high-quality features. This bridge provides the "operational discipline" that allows your product team to ship with "impact" and "speed" consistently.

DO'S FOR THE LEADER	DON'TS FOR THE LEADER
Create a clear, focused roadmap.	Let the project's scope expand wildly.
Say "no" to features that don't fit.	Say "yes" to every single request.
Prioritize based on core value.	Get distracted by the latest trends.
Use a disciplined change process.	Allow new features to be added on a whim.
Focus on the primary problem.	Try to solve every problem at once.
Have a clear product vision.	Have no clear product vision.

SITUATION	SUGGESTED ACTION
Teams expand scope when new ideas appear without evaluating alignment with original objectives	Define boundaries, assess relevance, and approve additions only when they strengthen core outcomes and support validated customer needs
Designers introduce extra variations when exploring alternatives without considering delivery constraints or timelines	Limit options, clarify requirements, and focus exploration on patterns supporting essential workflows while maintaining delivery feasibility
Product managers add features reacting to stakeholder pressure rather than validated customer needs	Review evidence, prioritize impact, and maintain discipline by protecting roadmap integrity and focusing on meaningful customer outcomes
Engineers propose technical enhancements unrelated to solving the primary customer problem effectively	Evaluate necessity, measure value, and defer improvements lacking meaningful customer benefit or alignment with core objectives
Teams widen timelines because they assume more features guarantee stronger market reception overall	Set launch criteria, emphasize essentials, and deliver incrementally to validate direction and reduce unnecessary complexity
Leaders request additions late because they fear missing opportunities competitors might pursue	Reassess urgency, validate demand, and maintain focus on initiatives with proven value and strategic relevance

This bridge is a method for designing products that are "truly useful," moving beyond "clicks" to "outcomes." It's the bridge between "attention" and "value." Daily, this bridge looks like "outcome-based metrics" rather than "vanity metrics." By building this bridge, you ensure that your AI is helping users achieve their "real-world goals." It prevents the "addiction-based" design that leads to user burnout. For the executive, this bridge is the "long-term loyalty" builder. It ensures that your product is a "valuable tool" in the user's life. This bridge provides the "purposeful engagement" that is the hallmark of a world-class, ethical, and high-value AI-era product.

DO'S FOR THE LEADER	DON'TS FOR THE LEADER
Make products truly useful and engaging.	Use game mechanics to manipulate users.
Use gamification to add real value.	Add game mechanics for their own sake.
Get feedback on how people use it.	Assume your game mechanics are fun.
Focus on solving a key problem.	Use gamification to hide a bad product.
Use gamification to empower users.	Use gamification to manipulate users.
Design for long-term engagement.	Design for a short-term hit of dopamine.

SITUATION	SUGGESTED ACTION
Teams add engagement features without understanding what motivates customers to return consistently	Study behavior, identify drivers, and design interactions that support genuine customer goals while reinforcing long-term value
Designers introduce visual rewards assuming they automatically increase satisfaction across different customer segments	Test reactions, measure outcomes, and apply incentives only when they reinforce meaningful progress and authentic engagement
Product managers prioritize stickiness metrics without considering whether increased usage benefits customers meaningfully	Evaluate value, refine goals, and align engagement strategies with outcomes that improve customer success and satisfaction
Engineers build complex interaction loops without validating whether customers find them helpful or intuitive	Prototype flows, gather feedback, and simplify mechanics to support natural usage patterns and reduce unnecessary cognitive load
Teams encourage daily activity even when workflows do not require frequent engagement or repeated actions	Respect rhythms, adjust prompts, and design usage patterns aligned with real customer needs and expectations
Leaders push gamification believing it modernizes products regardless of context or customer purpose	Clarify intent, validate fit, and apply engagement techniques only when they enhance usefulness and strengthen customer outcomes

6. SIMPLE PATH BRIDGE | The Path of Least Resistance

This bridge is a process that prioritizes a "simple, clear user experience" over "technical complexity." It's the bridge between "infinite AI options" and "one right choice." Daily, this bridge looks like "progressive disclosure"—only showing the user what they need at that specific moment. By building this bridge, you reduce the "cognitive effort" required to use your product. It ensures that your AI feels like a "natural extension" of the user's mind. For the executive, this bridge is the "market expander." It makes your product accessible to the "non-technical majority." This bridge provides the "blissful simplicity" that allows your product to become a "habit" for your customers, driving long-term adoption.

DO'S FOR THE LEADER	DON'TS FOR THE LEADER
Simplify choices for users.	Create a confusing interface with too many options.
Remove all unnecessary features.	Add features that serve no purpose.
Use a clear, logical interface.	Use a messy, jumbled interface.
Get feedback on user flow.	Assume your product is easy to navigate.
Guide users through the experience.	Let users get lost in a sea of options.
Create a clear path to the goal.	Have many different paths to a goal.

SITUATION	SUGGESTED ACTION
Users struggle when workflows contain unnecessary steps that complicate essential tasks across journeys	Remove friction, streamline actions, and design flows emphasizing clarity and efficiency while supporting predictable user behavior
Designers add options believing flexibility improves usability across diverse scenarios and customer needs	Limit choices, highlight essentials, and support complexity through secondary paths that avoid overwhelming primary experiences
Product managers introduce advanced settings alongside primary controls without appropriate separation or hierarchy	Group features, hide complexity, and surface only necessary elements initially to maintain clarity and reduce confusion
Engineers expose technical details customers do not need to complete tasks effectively or confidently	Abstract complexity, simplify interfaces, and present information aligned with user goals and expected mental models
Teams overload dashboards with metrics lacking clear hierarchy or actionable meaning for customers	Prioritize indicators, reduce noise, and present insights supporting confident decisions across essential workflows
Leaders request additional steps to capture data without considering user burden or workflow impact	Evaluate necessity, minimize input, and collect information only when essential for improving customer outcomes

Ai6x6™ Transformation Blueprint

Product Excellence Bliss

"The value is in what gets used, not in what gets built." - Kris Gale

Kris Gale is an American software engineer and technology executive. He was an original member of the engineering team at Yammer and is known for his contributions to scaling software products and his focus on product-market fit.

Product excellence bliss is the ultimate harmonious state where product development is a source of profound joy and success. It is the culmination of a successful product transformation journey. Blissful Ideation is the pleasure of a product team that can rapidly generate new ideas because they are not constrained by fear or bureaucracy. Blissful Design is the satisfaction of creating a product that is not only beautiful but also functional and intuitive for its users. Blissful Features is the joy of building features that make a real difference for customers, providing unique and meaningful value. As Kris Gale's quote suggests, "the value is in what gets used, not in what gets built." This state is not about a single successful product, but about a culture where the process of creating great products is both efficient and deeply satisfying.

Dedicated to delivering the tangible results of the Mission domain, this framework explores Product Excellence Bliss, defining the innovation outcomes that accelerate Value Realization. Product success begins with Intuitive Design, resulting in a product that is effortless to use, requiring minimal training or cognitive load from the user, drastically lowering support dependency. The goal is to achieve User Delight, where the product not only meets needs but also brings genuine satisfaction and positive emotion to its users, actively turning users into natural brand advocates. This leads directly to High Adoption Rates, meaning users widely embrace the product across the target market, securing rapid scaling. Product Bliss requires achieving Market Fit, ensuring the product perfectly meets a significant, unaddressed need in the market, eliminating costly development on superfluous features. Fundamental to trust is Unwavering Reliability, meaning the product consistently works as expected, eliminating performance anxiety and failure points, which is the foundation for all user trust and retention. The sustained outcome is High Retention, measured by users who continue to use the product over long periods, indicating integrated satisfaction and irreplaceable utility that maximizes Customer Lifetime Value (CLV).

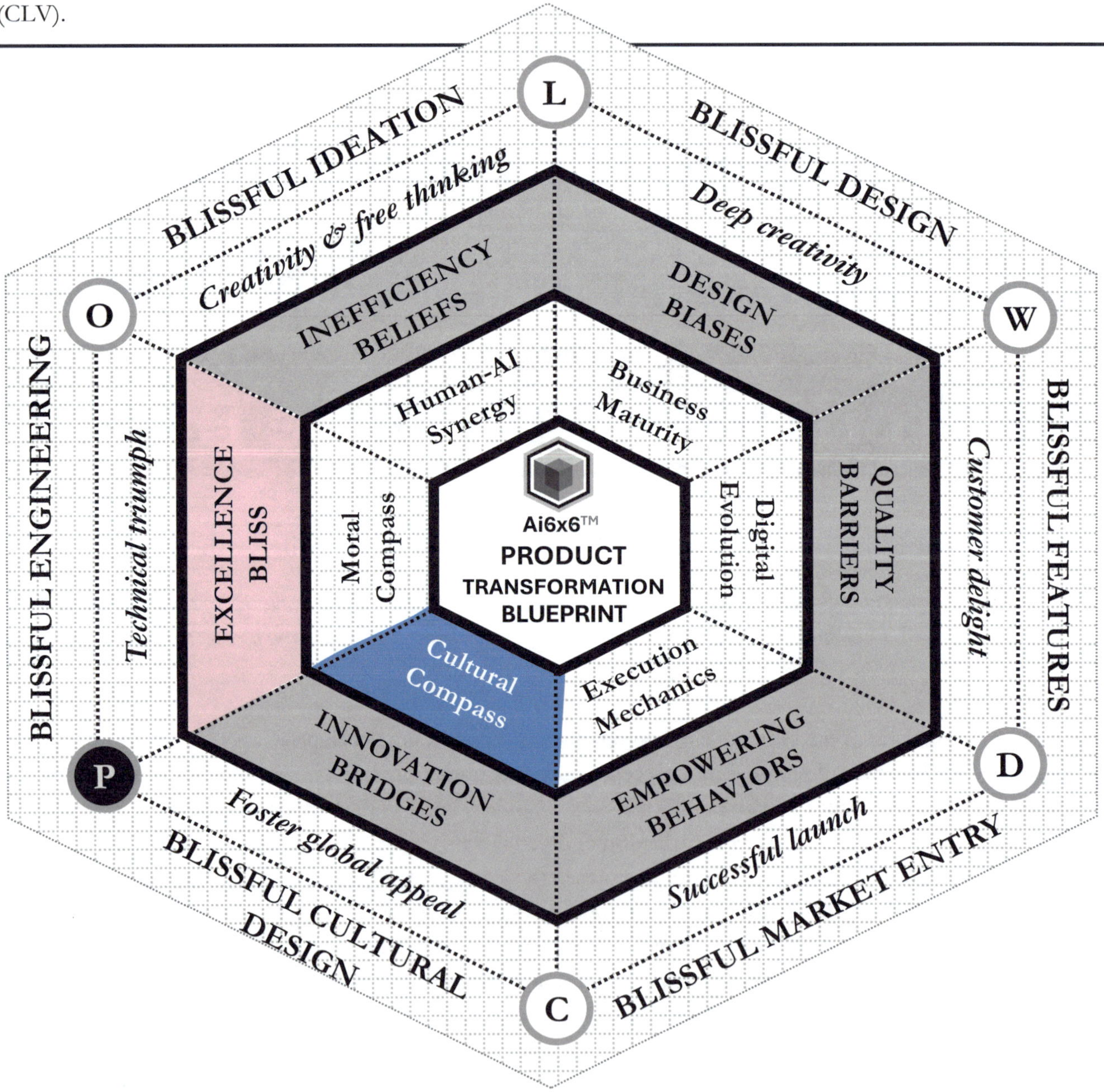

Ai6x6 CULTURAL COMPASS FOR BLISS | PRODUCT TRANSFORMATION

1. BLISSFUL IDEATION | The Creative Flow State

There is a profound sense of excitement that comes from a product team that can "rapidly generate and validate" new ideas. This bliss is the "creative flow" of innovation. In your daily life, it means you never feel "stuck." Your team is a constant source of "fresh possibilities." This bliss turns product development into a "joyful exploration." It provides a sense of "intellectual vitality" across the organization. This feeling of "anything is possible" is the ultimate motivator. This bliss is the reward for a culture that values "curiosity and experimentation"; it transforms the team from "order-takers" into "future-makers" who are genuinely excited by the challenges of the AI era.

DO'S FOR THE LEADER	DON'TS FOR THE LEADER
Create a culture of rapid ideation.	Have a slow, bureaucratic idea process.
Use diverse teams to generate ideas.	Let a single person generate all ideas.
Provide resources for new ideas.	Keep a tight control on all resources.
Foster a safe space for mistakes.	Punish people who try new things.
Use data to test new theories.	Rely on intuition over data.
Encourage bold innovation.	Reward conformity and old ideas.

SITUATION	SUGGESTED ACTION
Teams generate ideas quickly when psychological safety encourages open exploration without judgment	Facilitate sessions, welcome contributions, and refine concepts collaboratively to strengthen creativity and accelerate discovery
Designers propose bold concepts when experimentation is encouraged early in the product development process	Share sketches, test reactions, and iterate rapidly to identify promising directions grounded in customer needs
Product managers explore opportunities freely when constraints are clarified early and consistently across teams	Define boundaries, frame problems, and guide ideation toward impactful solutions aligned with strategic goals
Engineers contribute creative approaches when technical exploration is supported without premature restrictions or assumptions	Prototype options, evaluate feasibility, and integrate insights into broader ideation cycles supporting innovation
Teams build momentum when ideas are captured, organized, and revisited systematically across sessions	Document concepts, cluster themes, and prioritize opportunities using structured methods that maintain clarity
Leaders celebrate experimentation when teams pursue unconventional thinking responsibly and transparently	Recognize creativity, highlight learnings, and reinforce a culture supporting innovation across the organization

The satisfaction of creating a product that is both "beautiful and highly functional" is the bliss of "aesthetic utility." It's the feeling when the "form" and the "function" perfectly merge. This bliss is the "death of frustration." Daily, it means your users are not just "using" your product, they are "loving" it. It provides a sense of "craftsmanship and pride." This bliss is the realization that "great design" is a gift to the user. When your AI-driven interface feels like a "work of art" that just happens to solve complex problems, you have achieved a rare level of excellence. This bliss is the pinnacle of the design journey, where every pixel and every algorithm serves a clear and beautiful purpose.

DO'S FOR THE LEADER	DON'TS FOR THE LEADER
Create beautiful, functional designs.	Make products that are ugly or clunky.
Use a consistent design language.	Use a jumble of different styles.
Prioritize user-friendly design.	Focus on looks over usability.
Get feedback on the design.	Assume your design is perfect.
Design for accessibility.	Forget to design for disabilities.
Follow a clear brand identity.	Create a design that doesn't fit the brand.

SITUATION	SUGGESTED ACTION
Designers feel energized when visual and functional decisions align seamlessly across product experiences	Establish principles, refine patterns, and maintain coherence through consistent collaboration across teams
Teams take pride when interfaces feel intuitive and visually balanced across devices and contexts	Test layouts, adjust spacing, and refine interactions to support clarity and comfort for customers
Product managers appreciate designs that communicate purpose clearly without unnecessary complexity or distraction	Review intent, validate comprehension, and ensure visual choices reinforce core value across workflows
Engineers enjoy implementing designs that balance elegance with technical feasibility across systems	Collaborate early, clarify constraints, and refine solutions supporting both beauty and performance
Teams celebrate progress when prototypes demonstrate harmony between aesthetics and usability across journeys	Share demos, gather feedback, and iterate to strengthen cohesion and customer satisfaction
Leaders value design excellence when it elevates customer trust and product credibility across markets	Highlight impact, support investment, and reinforce standards promoting quality and consistency

The joy of building features that make a "real and measurable difference" for customers is the bliss of "impactful contribution." It's the feeling when a customer says, "this changed my life." This bliss is the "validation of effort." Daily, it manifests as positive feedback and rising "retention metrics." It provides a sense of "professional meaning." This bliss is the reward for staying "customer-obsessed." When you know that the "code you wrote" or the "model you trained" is actually saving someone time, reducing their stress, or helping them succeed, you achieve a deep and lasting sense of fulfillment that transcends the daily grind of product management.

DO'S FOR THE LEADER	DON'TS FOR THE LEADER
Prioritize features customers need.	Build features you think are cool.
Validate new features with users.	Make assumptions about feature value.
Use data to rank feature ideas.	Let a single person decide on features.
Focus on solving a key problem.	Build features that solve nothing.
Build features that add new value.	Just copy your competitors' features.
Listen to your sales team's needs.	Ignore your sales team's feedback.

SITUATION	SUGGESTED ACTION
Teams feel energized when features solve meaningful customer problems with measurable impact across workflows	Validate needs, design intentionally, and deliver improvements enhancing essential tasks and customer outcomes
Designers feel fulfilled when interactions help customers accomplish tasks more easily and confidently across contexts	Observe behavior, refine flows, and optimize usability through iterative testing and structured feedback
Product managers feel proud when features drive adoption and satisfaction across diverse customer segments	Track metrics, gather feedback, and adjust priorities based on real outcomes and evolving needs
Engineers feel motivated when building capabilities that improve performance or reliability significantly across systems	Identify bottlenecks, optimize architecture, and validate improvements through monitoring and testing
Teams celebrate when customers express appreciation for features improving daily experiences and workflows	Collect testimonials, analyze patterns, and reinforce decisions supporting customer value and satisfaction
Leaders feel confident when feature investments demonstrate clear alignment with strategic goals and long-term priorities	Review impact, assess reach, and prioritize initiatives delivering sustained value across markets

The satisfaction of successfully launching a product into a "new market" and seeing it flourish is the bliss of "strategic expansion." It's the feeling of "scaling your impact." This bliss is the "reward for courage." Daily, it looks like seeing your product being used by a "new demographic" or in a "new region." It provides a sense of "limitless growth." This bliss is the realization that your "vision" has a universal appeal. When your AI transformation allows you to solve problems for people who were previously "unreachable," you have achieved a major milestone. This bliss is the pinnacle of the growth journey, where your "local success" becomes a "global phenomenon," proving the power of your product.

DO'S FOR THE LEADER	DON'TS FOR THE LEADER
Create a smooth onboarding process.	Make it hard for new users to start.
Provide clear instructions and help.	Let users figure it out on their own.
Launch with a strong marketing plan.	Just release a product and hope for the best.
Get feedback from early adopters.	Ignore what your first users say.
Make it easy to see product value.	Make users search for the value.
Offer a product demo or tutorial.	Expect people to adopt it instantly.

SITUATION	SUGGESTED ACTION
Teams feel energized when early customers respond positively to initial product offerings and messaging	Gather insights, refine positioning, and adjust launch strategy based on real reactions from representative customer segments
Designers feel proud when localized experiences resonate with cultural expectations and norms across regions	Adapt patterns, validate assumptions, and refine interfaces to support regional needs and preferences effectively
Product managers feel accomplished when launch metrics confirm strong demand across targeted customer segments	Analyze performance, identify opportunities, and optimize strategy using early data and validated customer behavior
Engineers feel satisfied when systems scale reliably under new market conditions and increased demand	Monitor load, address issues, and strengthen infrastructure to support growth and ensure consistent performance
Teams celebrate when partnerships accelerate adoption and strengthen credibility across new markets	Build relationships, coordinate efforts, and reinforce alignment with expectations from local partners and stakeholders
Leaders feel confident when expansion demonstrates strategic fit and long-term potential across regions	Review outcomes, assess risks, and plan next steps based on validated traction and market signals

The satisfaction of designing a product that is "respectful of and useful for diverse cultures" is the bliss of "inclusive excellence." It's the feeling of "connecting the world." This bliss happens when your localized AI feels like it was "made specifically" for a culture different from your own. Daily, it manifests as "high adoption" in international markets and a "positive global reputation." It provides a sense of "universal empathy." This bliss is the reward for the hard work of "cultural adaptation." When you know that your product is helping people around the world in a way that respects their unique values and traditions, you have built a brand that is truly "world-class" and "human-centered."

DO'S FOR THE LEADER	DON'TS FOR THE LEADER
Adapt your product for local needs.	Use a single product for all cultures.
Research cultural norms and laws.	Use an insensitive or offensive design.
Translate your content accurately.	Use poor or literal translations.
Offer local payment options.	Force everyone to use the same payment.
Test your product with local users.	Assume it will work everywhere.
Respect local holidays and customs.	Send promotions on local holidays.

SITUATION	SUGGESTED ACTION
Teams feel inspired when cultural insights shape product decisions across regions and diverse communities	Conduct research, engage experts, and integrate findings into design choices supporting authentic customer experiences
Designers feel fulfilled when interfaces reflect local norms without compromising usability or clarity	Adapt visuals, refine language, and validate patterns with representative users across cultural contexts
Product managers feel confident when features accommodate cultural expectations across multiple markets	Review requirements, assess differences, and prioritize inclusivity across experiences and workflows
Engineers feel motivated when building systems supporting localization and regional customization effectively	Implement frameworks, test variations, and ensure flexibility across environments and customer groups
Teams celebrate when customers recognize cultural respect embedded in product experiences and interactions	Gather feedback, analyze sentiment, and refine approaches supporting authenticity and trust
Leaders feel proud when cultural sensitivity strengthens brand trust and global relevance across markets	Highlight impact, support investment, and reinforce commitment to inclusive design practices

The pride that comes from building a product that is "technically sound, reliable, and elegant" is the bliss of "technical mastery." It's the feeling of "clean code" and "stable models." This bliss is the "death of the 3 AM alert." Daily, it means your system "just works," allowing you to focus on the "next big thing" rather than fixing the "last big mess." It provides a sense of "internal strength." This bliss is the realization that "integrity" is the best long-term strategy. When your AI infrastructure is so well-built that it handles "scale and complexity" with ease, you have achieved a level of engineering excellence that few can match, providing a rock-solid foundation for the future of your company.

DO'S FOR THE LEADER	DON'TS FOR THE LEADER
Build for quality and long-term use.	Build products that break quickly.
Offer a good warranty or guarantee.	Offer no support for broken products.
Use high-quality materials.	Use cheap parts to save money.
Design for easy repair.	Make products impossible to repair.
Test for long-term wear and tear.	Only test for short-term functionality.
Release timely software updates.	Abandon products after the sale.

SITUATION	SUGGESTED ACTION
Engineers feel proud when systems operate reliably under heavy load without performance degradation	Monitor metrics, optimize architecture, and strengthen components to ensure consistent behavior across environments
Teams feel accomplished when technical decisions reduce long-term maintenance burdens and operational risk	Standardize patterns, document rationale, and reinforce practices that improve durability and simplify future enhancements
Designers appreciate engineering excellence when interactions feel smooth, responsive, and stable across devices	Collaborate early, clarify constraints, and refine designs to support performance expectations
Product managers feel confident when technical foundations support rapid iteration without compromising stability	Align priorities, protect core systems, and invest in improvements enabling sustainable delivery
Teams celebrate when customers notice improved speed, reliability, and overall product quality across workflows	Track feedback, analyze patterns, and reinforce decisions that strengthen customer trust and satisfaction
Leaders feel assured when engineering investments create long-term resilience and strategic advantage	Evaluate outcomes, assess risks, and prioritize initiatives that enhance reliability and technical strength

BONUS CHAPTER

Ai6x6™ Transformation Blueprint

Product Diversity

"There is a huge business case for diversity. You will be making products for people you don't understand, you don't interact with. If you don't have an inclusive, diverse workforce, it makes you myopic."

Inga Beale, Former CEO of Lloyd's of London. A lack of diversity leads to myopia (short-sightedness), which is fatal in the fast-paced, customer-centric world of digital transformation.

Product Transformation relies heavily on diversity to fuel breakthrough innovation and guarantee widespread market adoption. When the teams responsible for the product lifecycle—including ideation, design, and development—reflect a broad mix of demographics, skills, and lived experiences, they are positioned to identify and solve problems that are often entirely invisible to a homogenous team. This cognitive diversity is the most effective tool for overcoming **feature tunnel vision**, the failure to see use cases outside one's immediate frame of reference. Diverse teams naturally force a wider consideration of accessibility issues, cultural relevance, and unexpected product applications, resulting in digital offerings that are more robust and universally viable. By valuing and integrating these varied viewpoints from the initial design phase, the company avoids launching products that only serve a niche group and instead develops solutions that are genuinely revolutionary and scalable. Diversity in product teams is the quality assurance step that ensures the new digital portfolio is not just technically feasible, but truly innovative, accessible, and poised for **exponential market success**.

In today's interconnected world, diversity is a fundamental pillar of strength and innovation. Often simplified to visible differences, true diversity is a rich tapestry of attributes shaping how individuals perceive and contribute. Understanding this multifaceted nature is crucial for fostering environments where every voice thrives. This essay explores diversity through six critical dimensions: **demographic, cognitive, experiential, geographic, functional, and perspective diversity**. By examining these distinct facets, we can better appreciate how a holistic approach to diversity enriches teams and communities, driving robust solutions.

First, **Demographic Diversity** encompasses varied inherent and acquired characteristics. This includes age, bringing diverse generational insights, and gender, race, ethnicity, sexual orientation, ensuring a wide range of lived experiences. Socioeconomic status, religion, disability, and nationality further contribute to a richer collective understanding. Valuing demographic diversity ensures equitable representation and leverages unique insights from varied personal journeys.

Secondly, **Cognitive Diversity** highlights how people think, process information, and approach problems. This includes varied thinking styles like analytical vs. intuitive approaches, and different problem-solving or learning methods. When a group embraces cognitive diversity, it tackles problems from multiple angles, leading to more innovative solutions and a thorough understanding of complex issues, free from a single dominant thought process.

Next, **Experiential Diversity** focuses on the breadth of life and work experiences individuals bring. This covers varied work histories across industries and roles, providing a rich pool of professional knowledge. Diverse educational backgrounds, from different disciplines or institutions, also offer distinct frameworks. Beyond formal credentials, personal life experiences like travel and cultural immersion contribute significantly, equipping teams with a wide array of skills and adaptable approaches.

Geographic Diversity emphasizes the impact of place on an individual's perspective. This includes differences from regional or local backgrounds, shaping values and priorities. National backgrounds introduce distinct cultural norms, while urban versus rural upbringings offer varying insights into community dynamics and societal needs. These varied geographical lenses provide a broader understanding of global and local markets and behaviors.

Furthermore, **Functional Diversity** pertains to the variety of professional roles, specialized areas, and departmental expertise within a team. This dimension recognizes contributions from different functions like engineering, marketing, or finance, each bringing specific domain knowledge. Variations in hierarchical roles and industry experience also inform cross-pollination of best practices. Leveraging functional diversity ensures all necessary technical skills are present, fostering comprehensive problem-solving.

Finally, **Perspective Diversity** encompasses the fundamental differences in viewpoints and mental models individuals hold, often influenced by other diversity forms. This includes varied ideological stances, cultural interpretations, and distinct value systems. Understanding diverse stakeholder perspectives—from customers to employees—is crucial for holistic decision-making. Ultimately, perspective diversity allows for problems to be framed in multiple ways, challenging assumptions and encouraging empathetic, effective outcomes.

By intentionally cultivating and integrating these six dimensions—demographic, cognitive, experiential, geographic, functional, and perspective—organizations and societies can harness the profound power of true diversity. This comprehensive approach not only fosters equity but also enhances creativity, resilience, and innovation, demonstrating that a multitude of viewpoints is the clearest path to collective success.

1. DEMOGRAPHIC DIVERSITY

Embracing **demographic diversity** within product teams, from ideation to launch, means intentionally cultivating a talent pool that brings varied life experiences to product design, development, and market understanding. This goes beyond mere representation; it's about leveraging the inherent value that comes from product professionals with varied ages, genders, racial and ethnic backgrounds, physical abilities, socio-economic upbringings, and sexual orientations. These visible and non-visible differences are crucial for identifying diverse user needs, preventing unintended biases in product features, ensuring inclusive design, and developing products that resonate with a global and varied customer base. A demographically diverse product team brings distinct insights into market segments, cultural sensitivities, and potential blind spots in product strategy, which are crucial for effective product transformation. When product teams lack this breadth, they risk developing offerings that are narrow in appeal, fail to address the needs of significant user populations, or perpetuate existing inequalities. True demographic diversity fosters more innovative, inclusive, and ultimately, more successful products, better equipped to address complex market challenges and connect authentically with diverse users and communities.

Topic	Tangible steps	Challenges you may face
AGE DIVERSITY IN PRODUCT INNOVATION		
Varied age groups bring different perspectives on user needs, technology adoption, and market trends. Enhances innovation and broadens product appeal.	Form cross-generational product development teams. Conduct user research with diverse age groups. Implement reverse mentorship for product leadership on emerging tech.	Resistance to new product ideas from established teams. Overcoming biases favoring specific age groups in target markets. Managing different approaches to product lifecycle.
GENDER DIVERSITY IN PRODUCT DESIGN		
Balanced gender representation influences product features, user experience, and market adoption. Diverse gender perspectives lead to more inclusive and appealing designs.	Set targets for gender balance in product design & UX teams. Implement inclusive design guidelines. Conduct gender-focused user testing and feedback loops.	Unconscious biases in user persona creation. Lack of female product leaders. Resistance to changing established product design norms.
RACIAL & ETHNIC DIVERSITY IN MARKET FIT		
Diverse racial & ethnic backgrounds foster critical insights into cultural relevance, market acceptance, and ethical product positioning. Improves global product fit and brand trust.	Develop cultural sensitivity training for product marketing. Establish product ERGs for diverse insights. Prioritize diverse representation in product market research.	Systemic bias in target market assumptions. Difficulty fostering psychological safety to discuss cultural nuances. Insufficient understanding of cultural implications for product features.

Cognitive diversity within product teams is paramount for navigating the inherent complexities of product transformation. It acknowledges that true innovation and robust problem-solving in product development stem not just from varied experiences or demographics, but fundamentally from differences in how people think, process information, and approach design and market challenges. A product team composed solely of analytical thinkers might excel at optimizing existing features but could struggle with identifying novel user needs or designing truly disruptive products. Conversely, a team too reliant on intuitive thinking might miss critical data points or overlook practical constraints in product implementation. By fostering a mix of analytical, creative, systemic, and neurodiverse cognitive styles across all product roles, organizations are better equipped to challenge assumptions in product strategy, identify unseen opportunities, predict unforeseen market shifts, and design more comprehensive and resilient product offerings. It's about building a collective product intelligence where different cognitive strengths complement each other, leading to more thorough product evaluations, more innovative product solutions, and a greater capacity for adaptive learning in the face of evolving user demands. This richness in thought ensures that product decisions are thoroughly vetted from multiple angles, preventing blind spots and fostering intellectual agility crucial for sustained product transformation success.

Topic	Tangible steps	Challenges you may face
ANALYTICAL VS. INTUITIVE THINKING IN PRODUCT STRATEGY		
Balancing data-driven & gut-feeling approaches in product strategy shapes roadmap & risk. Leads to nuanced decision-making, combining market rigor with insightful foresight.	Design product strategy frameworks requiring data/qualitative input. Encourage product managers to articulate reasoning. Form diverse product teams combining cognitive styles.	Friction between styles. Difficulty quantifying intuition in product. Culture valuing one style over another hinders product collaboration.
SYSTEMIC THINKERS IN PRODUCT ECOSYSTEMS		
Product professionals seeing interconnectedness of product lines & long-term implications drive comprehensive product transformation. Ensures holistic, sustainable product changes.	Provide systems thinking training for product architects/managers. Assign systemic thinkers to cross-product initiatives. Use "product journey mapping" to identify potential downstream impacts.	Difficulty identifying/developing systemic thinking in product roles. Resistance to long-term product views. Overwhelm from too much interconnected product information.
CREATIVE PROBLEM-SOLVERS IN PRODUCT IDEATION		
Product professionals with imaginative thinking inspire novel solutions to complex user problems. Crucial for breakthrough product innovations & disruptive strategies.	Establish product innovation labs or hackathons. Implement brainstorming for unconventional product ideas. Create time for blue-sky product thinking sessions.	Risk aversion stifling new product ideas. Difficulty translating creative product ideas into actionable development plans. Perceived lack of "seriousness" for certain product functions.

Experiential diversity among product professionals is a powerful catalyst for successful product transformation, providing a rich tapestry of practical insights drawn from varied career paths, industries, educational backgrounds, and personal life stages. Product professionals who have navigated different organizational cultures, worked across diverse sectors (e.g., startup to corporate, public to private), or gained international exposure bring a broader repertoire of product solutions and a more nuanced understanding of complex product challenges. This breadth of experience prevents insular thinking in product development, encourages the adoption of "best practices" from seemingly unrelated product fields, and fosters resilience by drawing on lessons learned from past successes and failures in varied contexts. For instance, a product professional with crisis management experience can bring invaluable composure and clear-headedness during periods of intense product recalls or market disruptions. By intentionally cultivating a product team with diverse professional and personal journeys, an organization gains a formidable collective wisdom, enhancing its ability to foresee product opportunities, mitigate product risks, and execute product transformation initiatives with greater adaptability and effectiveness. It promotes a learning culture where varied practical product wisdom is valued and applied to new strategic product challenges.

Topic	Tangible steps	Challenges you may face
CROSS-INDUSTRY INSIGHTS IN PRODUCT DEVELOPMENT		
Product professionals from different industries bring fresh perspectives & transferable solutions to product hurdles. Broadens product solution set.	Recruit product talent with diverse industry backgrounds. Organize peer-learning forums for cross-industry product insights. Encourage participation in external product management communities.	Difficulty translating product insights to new contexts. Resistance to "outsider" product approaches. Lack of understanding of new industry product nuances for integration.
STARTUP VS. CORPORATE PRODUCT DEVELOPMENT		
Impact of product professionals from agile startup vs. structured corporate environments on product development. Blending these fosters innovation & stability.	Integrate startup product talent into corporate innovation units. Create joint ventures pairing corporate/startup product teams. Foster internal "intrapreneurship" programs for products.	Culture clash between agile & traditional product structures. Difficulty integrating fast product decisions. High expectations for immediate product results.
GLOBAL VS. LOCAL PRODUCT LAUNCH EXPERIENCE		
Product professionals with diverse international & local market exposure balance global product standardization & regional adaptation. Optimizes product stratcgy.	Implement global product rotation for managers. Establish regional advisory boards for local product knowledge. Encourage dual reporting lines for international product launches.	Complex balance of global vs. local product needs. Resistance from local product teams to global mandates. Misinterpretation of cultural nuances in product marketing.

Geographic diversity within product teams is indispensable for organizations operating in a globalized world, especially during periods of significant product transformation. Product professionals deeply rooted in varied regions, countries, or even urban versus rural settings bring critical local market understanding, cultural nuances, and insights into unique regulatory environments and logistical challenges of product distribution. Without this localized product perspective, global product strategies risk being homogenous, culturally insensitive, or simply impractical in execution. For example, a product professional experienced in emerging markets might bring crucial insights into scalable, low-cost product solutions, while someone from a highly regulated economy understands the complexities of product compliance. This diversity ensures that product transformation initiatives are not just conceptualized centrally but are designed with built-in adaptability for regional product variations. It empowers the organization to navigate geopolitical risks affecting products, understand localized customer behaviors for products, and foster stronger relationships with diverse regional product stakeholders. Product professionals with varied geographic exposure contribute to product strategies that are globally coherent yet locally relevant, maximizing impact and minimizing missteps in a fragmented product world.

Topic	Tangible steps	Challenges you may face
REGIONAL PRODUCT MARKET UNDERSTANDING		
Product professionals from different regions offer unique insights into local product market dynamics, consumer behaviors, & competition. Vital for global product strategies.	Recruit product talent with local market knowledge. Mandate global product strategies be reviewed/adapted by local product teams. Support employee travel/immersion in diverse regions for product insights.	Overlooking product nuances for standardization. Difficulty adapting to rapid changes in local markets. Resistance from central product teams to delegate.
NAVIGATING GLOBAL PRODUCT REGULATORY LANDSCAPES		
Product professionals familiar with diverse international legal/compliance frameworks are crucial for multi-national product launches. Ensures adherence & mitigates product risks.	Create global product compliance guidelines for product teams. Invest in continuous education on international product regulations. Consult local legal experts for every product launch.	Keeping up with rapidly changing product laws. Reconciling conflicting requirements. Risk of non-compliance due to inadequate local product understanding.
LOCAL INFRASTRUCTURE AND PRODUCT DISTRIBUTION		
Product professionals understanding local infrastructure (logistics, digital access) & resources impact product distribution. Optimizes local product implementation & delivery.	Empower local product teams for infrastructure investments impacting products. Develop centralized knowledge of regional product capabilities. Foster local service provider partnerships for products.	Underestimating varied local infrastructure impact on products. Inaccurate resource assumptions for product distribution. Difficulty scaling global product solutions.

5. FUNCTIONAL DIVERSITY

Functional diversity within product teams is critical for executing comprehensive and interconnected product transformations. It ensures that product strategic and operational decisions are informed by the deep expertise and unique perspectives of various professional disciplines, rather than being confined to a single departmental silo. A product team comprising experts from finance, technology, human resources, operations, marketing, and legal functions can collectively identify product opportunities, anticipate product challenges, and design product solutions that are holistic, practical, and sustainable. For instance, a finance analyst ensures fiscal viability of product investments, while a technology specialist champions digital product adoption, and an HR generalist focuses on people-centric change management for product teams. Without this blend, product transformation efforts might optimize one area at the expense of another, leading to fragmented implementation or unintended negative consequences in product usage. By fostering cross-functional collaboration and valuing diverse functional expertise, an organization can break down traditional barriers, enhance teamwork, and create truly integrated product strategies that leverage the strengths of every part of the business, ensuring that the product transformation is not just conceptually sound but also executable across all organizational facets.

Topic	Tangible steps	Challenges you may face
FINANCE-LED PRODUCT ROI MEASUREMENT FOR TEAMS		
Product professionals with strong financial backgrounds prioritize cost optimization & ROI as primary product transformation drivers. Ensures fiscal responsibility in product projects.	Appoint finance professionals to product project teams to define ROI. Establish clear financial metrics & ROI targets for product initiatives. Implement rigorous cost-benefit analyses for all investments.	Over-emphasis on short-term product cuts. Resistance if financial goals overshadow product innovation. Difficulty projecting ROI for intangible product benefits.
TECHNOLOGY-DRIVEN PRODUCT INNOVATION		
Product professionals with deep tech expertise champion digital product adoption & disruptive innovation. Essential for driving tech-enabled product change.	Elevate technologists to strategic roles in product projects. Foster "tech literacy" for product teams. Create cross-functional innovation teams combining product tech/business experts.	Resistance to new product technologies. Rapid evolution challenges product skills. Difficulty integrating new product tech with legacy systems and traditional product workflows.
HR'S ROLE IN PRODUCT TEAM DEVELOPMENT		
HR professionals focus on cultural shifts, talent management, & employee well-being during product transformation. Ensures human capital is central to product success.	Position HR teams as strategic partners in all product change initiatives. Develop robust change management/communication plans for product adoption. Implement well-being/reskilling programs for product teams.	Business units resist prioritizing people over product technology. Difficulty measuring HR ROI in product initiatives. Challenges addressing employee anxiety/resistance to new product processes.

Perspective diversity within product teams is arguably one of the most subtle yet profound drivers of successful product transformation. It goes beyond demographics, experience, or function, delving into the varied worldviews, core values, ethical frameworks, and risk tolerances that shape how product professionals interpret information and prioritize product objectives. A product team with diverse perspectives challenges inherent biases in design, avoids groupthink in strategy, and considers a wider spectrum of product possibilities and potential consequences. For example, balancing product professionals who prioritize short-term product gains (e.g., quick feature releases) with those focused on long-term product sustainability (e.g., platform stability), or those who are risk-seeking with product experimentation with those who are risk-averse with product safety, leads to more robust and balanced product strategic decisions. This diversity ensures that product transformation is not just about efficiency or growth in features, but also about ethical responsibility, product accessibility, user well-being, and genuine stakeholder value from products. It encourages product professionals to approach product problems from fundamentally different angles, leading to more resilient product strategies, better user engagement, and a more adaptive and morally grounded organization capable of navigating complex and ambiguous future product landscapes. It cultivates a culture of constructive challenge and holistic product-driven decision-making.

Topic	Tangible steps	Challenges you may face
CUSTOMER-CENTRIC VS. PRODUCT-CENTRIC VIEWS		
Product professionals balance internal product focus with external customer needs. Ensures products solve customer problems & resonate with market.	Implement "Voice of Customer" programs for all product professionals. Conduct customer immersion days for product teams. Establish KPIs balancing product features with customer satisfaction.	Internal resistance to shifting from product-centric. Difficulty capturing diverse customer needs. Product professionals struggle balancing product desires vs. feasibility.
SHORT-TERM VS. LONG-TERM PRODUCT VISION		
Product professionals' differing time horizons influence project pace, scope, & success of product initiatives. Balancing these ensures both immediate product gains & future product viability.	Develop multi-year product roadmaps with short-term product milestones. Assign product teams with complementary focuses. Use scenario planning to visualize impacts of product strategies.	Pressure for immediate product results. Product professionals may prioritize short-term product gains. Difficulty articulating long-term product value to all stakeholders.
RISK-SEEKING VS. RISK-AVERSE APPROACHES IN PRODUCT DEVELOPMENT		
Varied risk tolerance impacts pursuit of disruptive vs. incremental product improvements. A healthy balance drives calculated product growth and controlled product experimentation.	Foster culture where calculated product risks are encouraged. Establish "fail fast, learn faster" practices for product projects. Create product project portfolios balancing high-risk/reward with safer bets.	Decision paralysis when product risk tolerances conflict. Dominance of extreme product risk views. Difficulty setting clear, unified product risk parameters across teams.

BLUEPRINT SERIES

SECTION G

Moral Compass

Ai6x6™ Transformation Blueprint

Story 1 — The Nuremberg Trials and Ethical Accountability (1945–1946): After World War II, the world confronted atrocities committed under the guise of legal authority. The Nuremberg Trials established that individuals and institutions are accountable for actions that violate fundamental human rights. Legal frameworks were strengthened by moral reasoning, not replaced by it. Judges balanced evidence with ethical principles to define new global norms. The trials demonstrated that morality must guide power. They became a foundation for modern ethical governance.

Story 2 — The Tuskegee Syphilis Study Exposure (1932–1972): For decades, researchers withheld treatment from Black men to observe disease progression, violating consent and dignity. Public exposure triggered national outrage and forced reforms in medical ethics. Oversight boards, informed consent, and patient rights became mandatory. Machines and data had played roles in tracking outcomes, but human ethics had been absent. The scandal showed how systems fail when morality is excluded. It reshaped global standards for ethical research.

Story 3 — The Environmental Movement and Earth Day (1970): Industrial growth created pollution that threatened ecosystems and public health. Scientists provided data, but moral arguments mobilized action. Citizens demanded accountability, and governments introduced regulations to protect air, water, and wildlife. The movement reframed environmental protection as a shared ethical duty. It demonstrated how moral clarity can transform policy, industry, and public behavior. The cultural shift established sustainability as a long-term societal obligation.

Story 4 — The Anti-Apartheid Global Solidarity Movement (1950s–1994: Apartheid enforced racial segregation and oppression in South Africa. Activists, governments, and institutions worldwide united to challenge the regime through sanctions, protests, and moral pressure. Human stories amplified the injustice, while global networks coordinated action. The movement showed how collective moral conviction can influence political systems. It proved that ethical solidarity can reshape international norms. The end of apartheid became a landmark in moral mobilization.

Story 5 — The Rise of Corporate Social Responsibility (1990s–2000s): Globalization exposed labor abuses, environmental harm, and opaque supply chains. Consumers demanded transparency, and companies adopted codes of conduct, sustainability reporting, and ethical sourcing. Machines tracked compliance, but humans defined the standards. CSR shifted from public relations to strategic responsibility. Organizations learned that long-term value depends on ethical behavior. It marked a turning point where morality became integral to corporate identity.

Story 6 — The AI Ethics Movement (2015–2025): As AI systems influenced decisions in hiring, policing, healthcare, and finance, concerns grew about bias, transparency, and accountability. Researchers, policymakers, and civil society pushed for ethical frameworks to guide development. Machines provided capability, but humans defined boundaries. The movement emphasized fairness, explainability, and human oversight. It demonstrated that emerging technologies require moral governance as much as technical innovation. AI ethics became a global priority shaping the future of intelligent systems.

LESSONS FROM A PRODUCT TRANSFORMATION PERSPECTIVE

Product maturity in moral contexts emerges when organizations design offerings that respect human rights, minimize harm, and promote fairness. Over time, teams learned that products reflect the values of their creators—ethical blind spots become embedded in features, data flows, and user experiences. Mature product cultures integrate ethical review into design, testing, and deployment. They consider long-term societal impact, not just short-term utility. Machines enable scale, but humans ensure responsibility. Maturity is achieved when products embody ethical intent, balancing innovation with safeguards that protect users and communities. Ethical design becomes a hallmark of trustworthy products.

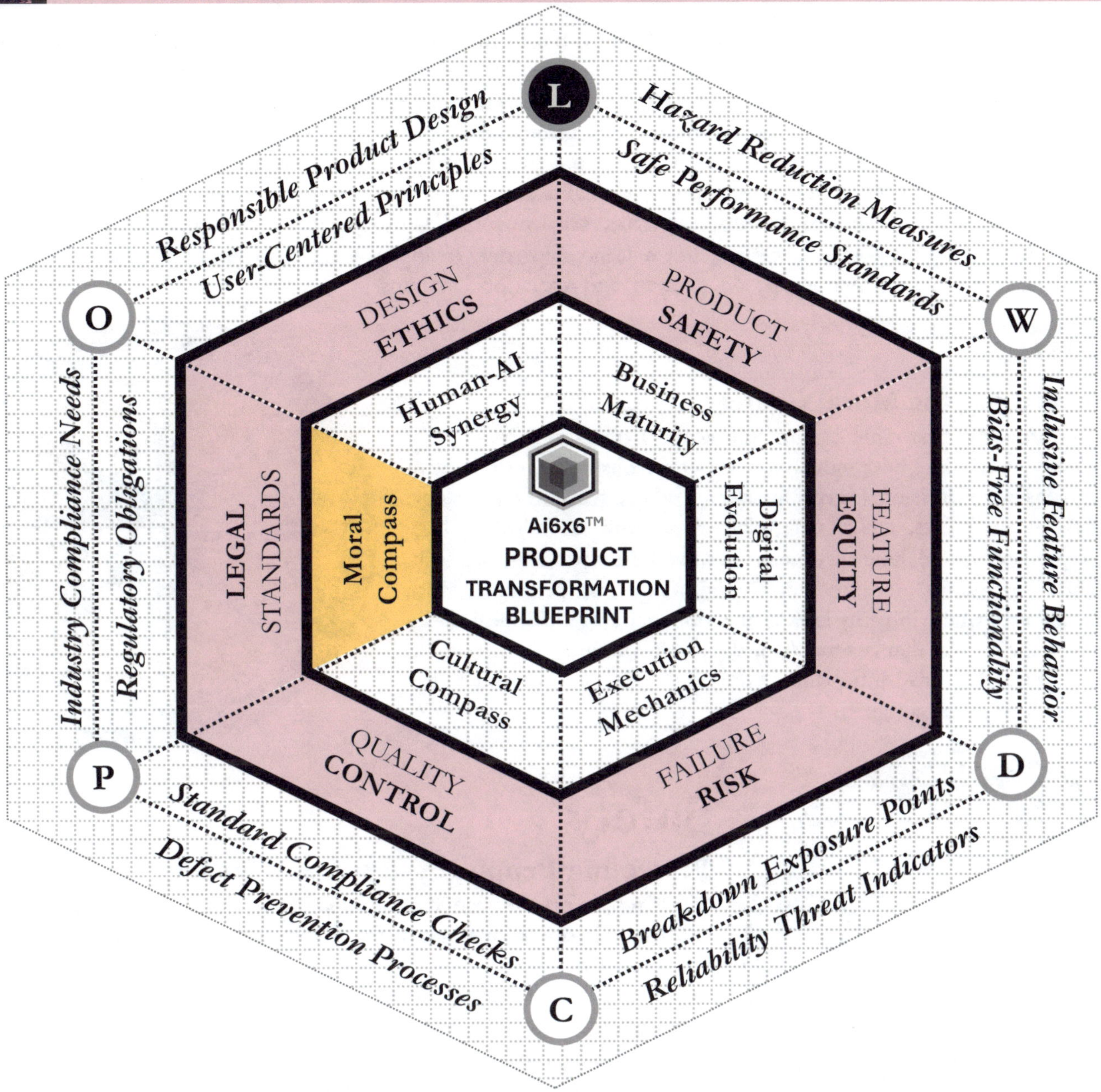

The **Ai6x6 Research Lab** identified the Moral Compass as the most vital lens because it serves as the final guardian of the product's honor and the user's safety. This is the ultimate filter for every decision, ensuring the drive for innovation never outpaces the bedrock of **Design Ethics** and **Respect**. We chose this because a product's legacy is built on its **Reliability** and **Safety**, not just its novelty. It is vital because it provides the moral authority required to lead through the ethical "fog" of autonomous agents and automated liability.

This lens is the heart's final check—a rigorous audit of **Failure Risk**, **Equity**, and **Quality Control**. It ensures the transformed product is not only a titan of performance but a beacon of public trust. The Lab believes that in the 2026 landscape, the greatest risk to a product is not a technical bug, but a moral failure in its design or accessibility. By focusing on **Product Standards** and **Validation**, we provide the tools to navigate the heavy responsibility of creation. This lens is about the courage to be honorable when the temptation for speed is highest, ensuring the legacy of the product is one of integrity and human value.

Morality has always shaped how organizations behave, but in the age of AI, it becomes a structural requirement rather than an aspirational ideal. The **Moral Compass** lens forces organizations to confront a truth that is often ignored: intelligence—whether human or machine—does not create ethical behavior. It amplifies whatever moral foundation already exists. When an organization is grounded in integrity, clarity, and principled action, AI strengthens it. When the foundation is inconsistent, opaque, or misaligned, AI exposes those weaknesses with unforgiving speed. Morality is no longer a philosophical preference; it is the architecture that determines whether accelerated decisions, automated systems, and scaled consequences lead to progress or harm. This lens reframes morality as an enterprise-wide discipline, not a leadership trait. It applies equally to workforce practices, data governance, customer interactions, product design, and operational systems.

Ethics

Establishing the Moral Center

Ethics forms the first dimension of the Moral Compass. Ethics is not simply about avoiding misconduct; it is about establishing a moral center that guides every decision, every interaction, and every use of technology. Across all six pillars, ethics determines how data is collected, how employees are treated, how customers are served, how products are built, and how operations are executed. Organizations that prioritize ethics create cultures where integrity is expected, not encouraged. Ethical clarity becomes the anchor that ensures AI is deployed with intention, restraint, and respect. In this dimension, principled behavior becomes the foundation of moral authority.

Safety

Protecting People, Systems, and Society

Safety represents the second dimension. As AI accelerates decision-making and expands operational reach, the potential for harm increases. Safety is not limited to physical environments; it includes psychological safety for the workforce, data safety for customers, model safety for AI systems, and operational safety for mission-critical processes. Safety ensures that no individual, team, or automated system can cause harm through unchecked influence or unintended consequences. It is not about limiting innovation; it is about ensuring that innovation does not outpace responsibility. In this dimension, responsible protection becomes essential to moral operation.

Equity

Ensuring Fairness Across Stakeholders

Equity forms the third dimension. Influence in modern organizations is no longer tied solely to hierarchy; it flows through data pipelines, algorithmic recommendations, and automated workflows. Equity ensures that these systems do not privilege certain groups, perspectives, or outcomes at the expense of others. It applies to workforce advancement, customer treatment, product accessibility, and data representation. Equity is not about equalizing results; it is about ensuring that the system itself is not biased. In this dimension, fairness becomes a structural requirement rather than a social aspiration.

Risk

Navigating Uncertainty With Discipline

Risk represents the fourth dimension. AI-enabled decisions carry greater speed, greater reach, and greater uncertainty. Risk is not about avoiding uncertainty; it is about navigating it with discipline. Organizations must anticipate consequences, stress-test assumptions, and understand the risks embedded in both human and machine judgment. This applies to operational risks, product risks, customer risks, workforce risks, and data risks. Organizations that embrace this dimension act decisively without acting blindly. In this dimension, risk-aware decision-making becomes a moral responsibility.

Control

Defining Governance and Accountability

Control forms the fifth dimension. As organizations adopt AI, authority becomes more complex: who decides, who approves, who monitors, and who intervenes when systems drift? Control ensures that governance structures are clear, decision rights are defined, and accountability is transparent. It applies to data governance, product governance, operational governance, and workforce governance. Control is not about bureaucracy; it is about ensuring that power—human or machine—is exercised with clarity and accountability. In this dimension, structured oversight becomes the backbone of moral operation.

Legal

Upholding Lawful and Responsible Conduct

Legal completes the Moral Compass. In a world where data, automation, and intelligence shape every action, legal obligations become inseparable from organizational obligations. Compliance is not a constraint; it is a safeguard. Legal integrity ensures that decisions align with regulatory expectations, societal norms, and ethical standards across all six pillars. It protects the organization from missteps and reinforces public trust. In this dimension, lawful stewardship becomes the final test of moral credibility.

The Moral Compass lens reminds organizations that moral authority is not a byproduct of good intentions; it is the result of deliberate, disciplined choices. Morality must be engineered into systems, not assumed in people. When organizations prioritize ethics, enforce safety, ensure equity, manage risk, establish control, and uphold legal accountability, they create environments capable of navigating complexity with confidence and integrity. They build systems where people feel protected, customers feel respected, and society feels considered. They create the moral conditions for transformation to take root and endure. In this maturity, morality becomes an enterprise capability—one that strengthens trust, accelerates adoption, and stabilizes transformation. It becomes the differentiator that determines whether AI amplifies value or amplifies harm. Organizations that master the Moral Compass do not merely deploy intelligent systems; they deploy them responsibly. They create futures where intelligence and integrity advance together, ensuring that progress is not only possible but principled.

MORAL COMPASS ANALOGY – NURTURING THE FAMILY HOME

Family Values — ETHICS

The foundational moral code—religious, cultural, and social—that dictates the ground rules for "Right vs. Wrong." Establishing the core integrity of the home so that all who belong understand the non-negotiable standards of conduct.

Home Security — SAFETY

Integrated protection—cameras, sensors, and smoke detectors—that provide total awareness of the premises. Deploying proactive warning systems and delegation guardrails to protect the inhabitants from "hidden fires."

Balanced Contribution — EQUITY

Ensuring everyone helps with the cooking and upkeep, with expenses shared equitably by each earning member. Maintaining participation and representation so that no member feels excluded from the collective success.

The Insurance Shield — RISK

Comprehensive fire, flood, and auto coverage to provide long-term peace of mind for the entire household. Using stress testing and "what-if" scenarios to protect the family assets from unforeseen market storms.

Access & Keys — CONTROL

Ensuring all adults have keys and codes, with clear emergency rules and a defined line of command for disasters. Establishing rigid RACI controls to ensure that a human always has the final word on the most critical locks.

The Estate Will — LEGAL

Recording Wills, Trusts, and Medical Mandates with a legal custodian to safeguard the valuables for the future. Securing the organization's legacy through regulatory accountability, ensuring the estate is legally unassailable for generations.

Ai6x6™ Transformation Blueprint

Design Ethics

"If we teach machines to see the world, we must also teach them to value it."

Sylvia Earle, marine biologist and ocean conservation pioneer.

Design Ethics requires principled creation, and Responsible Product Design defines how products must be conceived with safety, clarity, and user benefit at the forefront. It establishes expectations that prevent harmful, manipulative, or negligent design choices. This matters because design decisions shape user experience long before a product is used. Responsible Product Design prevents ethical failures by ensuring features align with user needs rather than shortcuts or exploitation. It also reinforces fairness by ensuring all users receive thoughtful, respectful design treatment. When ignored, design becomes inconsistent and potentially harmful. Teams experience confusion as design priorities drift. Over time, Design Ethics deteriorates as user trust erodes. User Centered Principles ensure design decisions prioritize real human needs and responsible outcomes.

Product is the physical manifestation of leadership intent. Leaders sit with Harm-Aware Design Logs and Ethical Design Reviews to ensure the spark of innovation doesn't create an operational fire. This topic moves the product from "Task Automation" to "Intent Orchestration," ensuring that "Default Settings" and "Transparency Disclosures" respect the user. By analyzing "Ethical Trade-off Records" in a print format, you can see the long-term consequences of design choices. Ethical design is the practice of building a product that reflects the organization's soul, turning every feature into a statement of radical brand integrity. In the print blueprint, you can engross yourself in "User Respect Screens," seeing how your design choices influence your brand's legacy. This clinical focus on "Design Ethics" ensures that your "Product Transformation" is as innovative as it is ethical, preventing your innovation from creating "Kill Zones" in your own market and ensuring that every product release is a masterclass in brand-led execution.

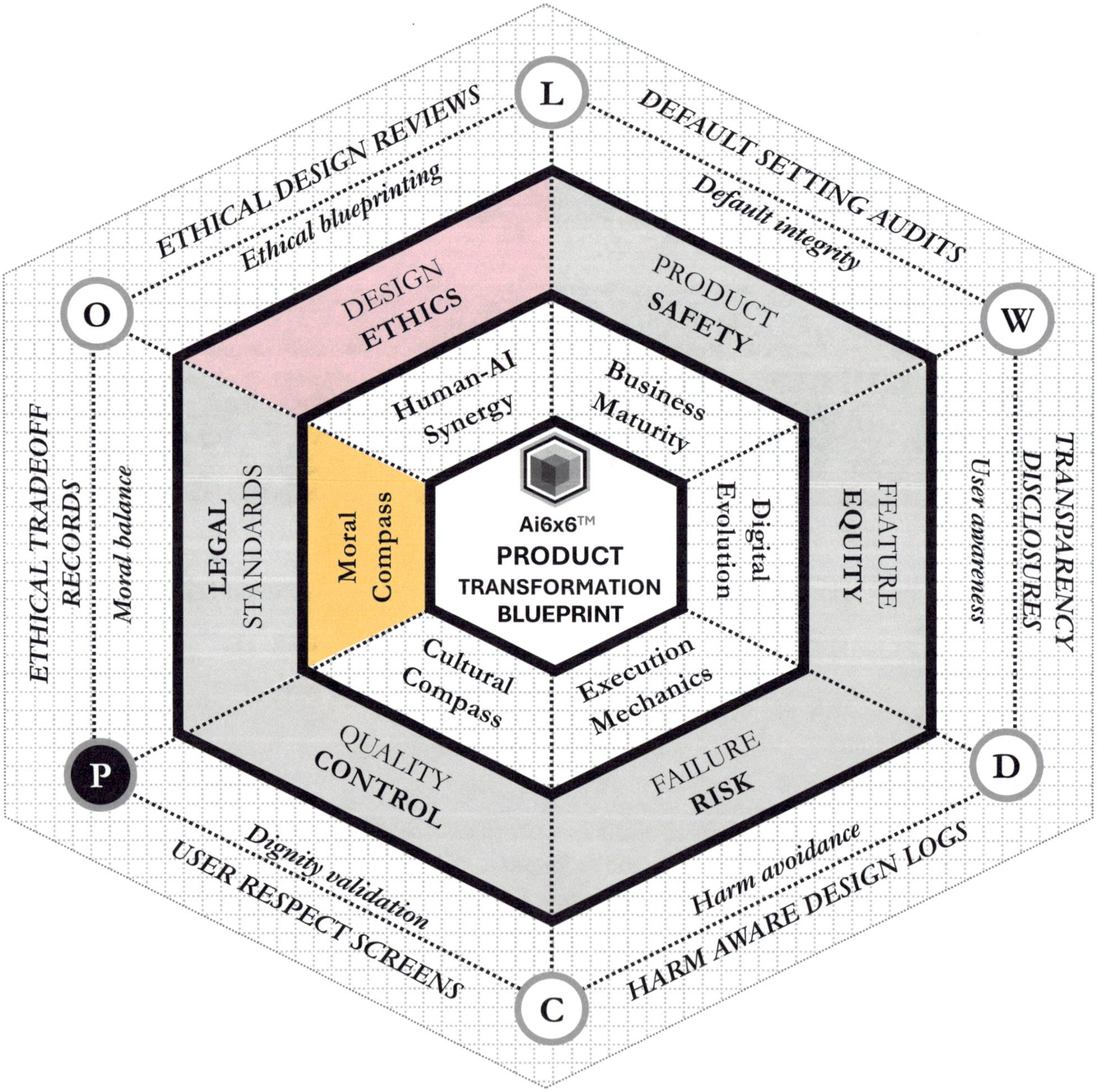

Ai6x6 DESIGN ETHICS FOR MORAL COMPASS | PRODUCT TRANSFORMATION

1. ETHICAL DESIGN REVIEWS

Design Ethics requires principled creation, and Ethical Design Reviews evaluate whether product decisions respect user dignity, autonomy, and well being. They examine assumptions, tradeoffs, and design choices that may introduce harm or manipulation. This matters because unethical design often hides inside convenience, efficiency, or growth incentives. Ethical Design Reviews prevent misuse by forcing teams to confront the moral implications of their decisions. They also reinforce fairness by ensuring all users—not just ideal or profitable ones—are considered. When ignored, products drift toward exploitative patterns that erode trust. Customers feel manipulated or disadvantaged. Over time, Design Ethics collapses as ethical boundaries dissolve.

When executed well, Ethical Design Reviews strengthen Design Ethics by embedding moral reasoning into the development process. Customers trust that the organization prioritizes their interests over short term gains. This improves loyalty, reduces risk, and enhances brand integrity. Reviews reinforce the principle that design is never neutral; it always shapes behavior. They remind leaders that ethical design requires discipline and foresight. Ultimately, they build an environment where products reflect responsibility, not opportunism.

MORAL COMPASS SCENARIO	LEADERSHIP ACTION
Mara discovers that a new onboarding flow nudges users into sharing more data than necessary. She realizes the design crosses into manipulative territory.	The leader removes coercive elements and aligns the flow with minimal-data principles. They also require ethical-design sign-off before launch.
Devin notices that a feature hides critical limitations behind expandable sections. He senses deceptive UX.	The leader surfaces limitations prominently and updates design guidelines. They also add transparency checks to all reviews.
Selah finds that a "recommended" option is preselected even though it benefits the company more than the user. She recognizes a conflict of interest.	The leader removes the preselection and tests neutral defaults. They also require conflict-of-interest screening in design reviews.
Mateo sees that error messages subtly blame users for system failures. He worries about psychological harm.	The leader rewrites messages with supportive, neutral language. They also add tone-of-voice checks to design audits.
Farah notices that a feature encourages excessive usage patterns. She senses potential for user harm.	The leader adds usage-limit controls and harm-mitigation prompts. They also require harm-aware evaluation for all engagement features.
Idris discovers that the design team skipped the ethical review due to time pressure. He realizes governance is being bypassed.	The leader enforces mandatory ethical reviews and blocks releases without approval. They also track compliance in the design governance dashboard.

2. DEFAULT SETTING AUDITS

Design Ethics requires protective baselines, and Default Setting Audits evaluate whether initial configurations support user safety, privacy, and fairness. They examine whether defaults nudge users toward beneficial or harmful outcomes. This matters because defaults influence behavior more than any other design choice. Default Setting Audits prevent exploitation by ensuring users are not pushed into risky or intrusive settings. They also reinforce fairness by protecting all users equally, including those who never adjust configurations. When ignored, defaults become silent mechanisms of manipulation. Customers unknowingly operate in unsafe or disadvantageous states. Over time, Design Ethics deteriorates as defaults prioritize organizational benefit over user welfare.

When executed well, Default Setting Audits strengthen Design Ethics by aligning initial experiences with user protection. Customers trust that the organization sets defaults with integrity. This improves satisfaction, reduces harm, and enhances ethical credibility. Audits reinforce the principle that safety and privacy must be the starting point, not an optional adjustment. They remind leaders that defaults carry moral weight. Ultimately, they build an environment where users begin every interaction protected by design.

MORAL COMPASS SCENARIO	LEADERSHIP ACTION
Nia sees that data-sharing defaults are set to "on" for all new users. She realizes this violates privacy-first principles.	The leader switches defaults to opt-in and updates onboarding flows. They also require justification for any non-protective default.
Corin notices that autoplay is enabled by default on child accounts. He fears harmful exposure.	The leader disables autoplay for minors and adds parental-control prompts. They also audit all child-facing defaults.
Elara finds that location tracking activates automatically after installation. She recognizes a consent breach.	The leader changes the default to off and adds clear activation prompts. They also require location-use explanations before enabling.
Mateo sees that default notification settings overwhelm users with constant alerts. He senses cognitive overload.	The leader reduces default notifications and categorizes them by importance. They also test user comprehension before rollout.
Soren notices that default visibility settings expose user profiles publicly. He worries about safety risks.	The leader sets profiles to private by default. They also add safety education during onboarding.
Hana discovers that default password requirements are weak. She recognizes a security vulnerability.	The leader strengthens password defaults and enforces MFA. They also add password-strength indicators.

3. TRANSPARENCY DISCLOSURES

Design Ethics requires clarity, and Transparency Disclosures explain how features operate, what data they use, and what tradeoffs they impose. They ensure that customers understand the mechanics behind the product experience. This matters because hidden design choices undermine informed decision making. Transparency Disclosures prevent confusion by revealing the logic behind key interactions. They also reinforce fairness by ensuring all users—not just experts—can understand how the product works. When ignored, opacity becomes a tool for manipulation. Customers feel misled or uninformed. Over time, Design Ethics collapses as transparency disappears.

When executed well, Transparency Disclosures strengthen Design Ethics by enabling informed participation. Customers trust that the organization communicates openly about design decisions. This improves confidence, reduces disputes, and enhances ethical alignment. Disclosures reinforce the principle that clarity is a moral obligation. They remind leaders that transparency builds trust. Ultimately, they build an environment where users understand the systems they rely on.

MORAL COMPASS SCENARIO	LEADERSHIP ACTION
Priya discovers that a feature collects behavioral data without clear disclosure. She realizes users are unaware of tracking.	The leader updates disclosures with explicit data-use explanations. They also require transparency reviews for all data-collecting features.
Rowan notices that algorithmic recommendations are presented as neutral suggestions. He senses users don't understand the underlying logic.	The leader adds "Why am I seeing this?" explanations. They also require algorithmic transparency for all recommendation systems.
Serena finds that product limitations are buried in footnotes. She recognizes a transparency failure.	The leader surfaces limitations prominently and tests comprehension. They also add limitation-visibility checks.
Omar sees that third-party integrations are not disclosed anywhere. He worries users are unaware of external data flows.	The leader adds integration disclosures and updates privacy notices. They also require third-party transparency in all features.
Lila discovers that subscription renewal terms are hidden behind multiple clicks. She senses deceptive design.	The leader surfaces renewal terms upfront. They also enforce renewal-transparency standards.
Andre notices that disclosures differ across platforms. He recognizes cross-platform inconsistency.	The leader standardizes disclosures and tests for parity. They also require platform-aligned transparency reviews.

4. HARM-AWARE DESIGN LOGS

Design Ethics requires anticipation of negative outcomes, and Harm Aware Design Logs document potential risks, vulnerabilities, and unintended consequences embedded in product decisions. They capture where design choices may create harm under real world conditions. This matters because harm often emerges from overlooked edge cases. Harm Aware Design Logs prevent damage by making risk visible early. They also reinforce fairness by ensuring vulnerable users are considered. When ignored, harm becomes predictable and unaddressed. Customers experience failures that could have been prevented. Over time, Design Ethics deteriorates as risk awareness fades.

When executed well, Harm Aware Design Logs strengthen Design Ethics by embedding foresight into the design process. Customers trust that the organization anticipates and mitigates harm proactively. This improves reliability, reduces liability, and enhances ethical responsibility. Logs reinforce the principle that harm must be identified before it occurs. They remind leaders that responsible design requires vigilance. Ultimately, they build an environment where risk is managed with discipline and care.

MORAL COMPASS SCENARIO	LEADERSHIP ACTION
Nadia sees that logs show repeated user frustration with a high-stakes workflow. She realizes emotional harm is accumulating.	The leader redesigns the workflow with clearer guidance. They also track harm-reduction improvements.
Trevor notices that logs highlight compulsive usage patterns. He senses addictive design.	The leader adds usage-limit controls and friction for excessive use. They also require harm-awareness checks for engagement features.
Aisha finds that users frequently misinterpret a warning message. She recognizes cognitive harm.	The leader rewrites the warning with clearer language and tests comprehension. They also add clarity scoring to harm logs.
Marco sees that logs show increased stress indicators during financial flows. He worries about emotional burden.	The leader simplifies financial steps and adds reassurance prompts. They also track stress-related harm metrics.
Evelyn notices that harm logs are not reviewed by product teams. She senses disconnect between insights and design.	The leader integrates harm logs into product planning. They also require harm-impact reviews before release.
Jonah discovers that harm logs only track negative outcomes, not near-misses. He recognizes blind spots.	The leader adds near-miss tracking and updates logging standards. They also require near-miss analysis in harm reviews.

5. USER RESPECT SCREENS

Design Ethics requires dignity, and User Respect Screens evaluate whether product interactions treat users with fairness, empathy, and consideration. They examine tone, flow, and the implicit messages embedded in design. This matters because disrespect—subtle or overt—erodes trust faster than functional failure. User Respect Screens prevent harm by identifying where design choices belittle, pressure, or manipulate users. They also reinforce fairness by ensuring all users receive respectful treatment. When ignored, disrespect becomes normalized and users disengage. Complaints rise as interactions feel demeaning or coercive. Over time, Design Ethics collapses as respect is replaced by exploitation.

When executed well, User Respect Screens strengthen Design Ethics by embedding dignity into every interaction. Customers trust that the organization values their humanity. This improves satisfaction, reduces friction, and enhances brand loyalty. Screens reinforce the principle that respect is a moral and strategic imperative. They remind leaders that design communicates values. Ultimately, they build an environment where respect is the default posture.

MORAL COMPASS SCENARIO	LEADERSHIP ACTION
Rhea discovers that error screens use condescending language. She realizes the tone undermines user dignity.	The leader rewrites screens with respectful, neutral phrasing. They also add tone-of-respect checks to UI reviews.
Omar notices that confirmation dialogs pressure users into choosing the company-preferred option. He senses coercion.	The leader redesigns dialogs with neutral choices. They also require respect-based decision framing.
Lila finds that loading screens display dismissive jokes. She recognizes cultural insensitivity.	The leader removes the jokes and updates content guidelines. They also require cultural-sensitivity reviews.
Andre sees that help screens assume high technical literacy. He worries about excluding novice users.	The leader rewrites help content with simpler language. They also test comprehension across skill levels.
Serena notices that system messages blame users for errors. She senses psychological harm.	The leader updates messages to focus on solutions, not blame. They also add empathy-based messaging standards.
Hugo discovers that respect screens are inconsistent across platforms. He recognizes fragmented UX.	The leader standardizes respect guidelines and enforces cross-platform parity. They also require respect audits.

6. ETHICAL TRADEOFF RECORDS

Design Ethics requires transparency in decision making, and Ethical Tradeoff Records document the compromises made between usability, safety, privacy, performance, and business goals. They reveal the rationale behind difficult design choices. This matters because tradeoffs shape user experience in profound ways. Ethical Tradeoff Records prevent hidden sacrifices by making decisions explicit. They also reinforce fairness by ensuring tradeoffs do not disproportionately harm specific user groups. When ignored, tradeoffs become invisible and unaccountable. Customers experience consequences without understanding why. Over time, Design Ethics deteriorates as decisions prioritize convenience over responsibility.

When executed well, Ethical Tradeoff Records strengthen Design Ethics by ensuring decisions are principled and transparent. Customers trust that the organization weighs their interests carefully. This improves credibility, reduces risk, and enhances ethical alignment. Records reinforce the principle that tradeoffs must be justified, not concealed. They remind leaders that design decisions carry moral weight. Ultimately, they build an environment where tradeoffs are made with integrity and accountability.

MORAL COMPASS SCENARIO	LEADERSHIP ACTION
Keisha sees that a feature prioritizes speed over accuracy without documenting the tradeoff. She realizes the decision lacks ethical justification.	The leader documents the tradeoff and tests user impact. They also require tradeoff records for all performance decisions.
Damon notices that a design choice improves engagement but increases cognitive load. He senses a harmful tradeoff.	The leader rebalances the design and documents the reasoning. They also add cognitive-load scoring to tradeoff reviews.
Liora finds that a privacy-reducing shortcut was approved informally. She recognizes governance failure.	The leader reverses the shortcut and formalizes the tradeoff process. They also require privacy-impact justification.
Mateo sees that accessibility was deprioritized to meet deadlines. He worries about exclusion.	The leader restores accessibility requirements and documents the tradeoff. They also enforce accessibility as a non-negotiable criterion.
Serena notices that a cost-saving decision reduces product safety margins. She senses unacceptable risk.	The leader rejects the decision and reinforces safety-first principles. They also require safety-impact scoring.
Omar discovers that tradeoff records are scattered across documents. He recognizes traceability issues.	The leader centralizes tradeoff records in a single repository. They also enforce version control.

Ai6x6™ Transformation Blueprint

Product Safety

"Trust is the bridge between human intention and machine action."

Fei-Fei Li, AI scientist known for pioneering computer vision research.

Product Safety requires proactive protection, and Hazard Reduction Measures define the safeguards that prevent users from encountering physical, digital, or operational harm. They establish the standards that ensure products function safely under expected conditions. This matters because unsafe products create immediate risk and long-term liability. Hazard Reduction Measures prevent incidents by identifying and mitigating potential dangers. They also reinforce fairness by ensuring all users receive equal protection regardless of context. When ignored, safety failures become more frequent and severe. Teams experience uncertainty about acceptable risk thresholds. Over time, Product Safety deteriorates as vulnerabilities accumulate. Safe Performance Standards ensure products operate reliably and predictably.

Auditing the product's armor is a clinical necessity for long-term survival. This topic uses Failure Containment Plans and Safety Test Reports to ensure the product remains a permanent, safe pillar of the brand's legacy. By mapping "Hazard Identification Logs," the leader takes responsibility for the product's impact in the real world. Safety isn't a "feature"; it's the foundation of "Product Transformation." A safe product is an antifragile product. By mastering "Safety Certification Files," the leader ensures that the organization isn't broken by the market's reaction to a preventable product failure or hazard. The print series allows you to physically audit your product safety against "Safety Monitor Dashboards," ensuring that your innovation engine doesn't outpace your ability to protect your users. This clinical focus on "Product Safety" transforms your product from a point of potential liability into a bedrock of customer trust and market defensibility, ensuring your brand remains relevant and respected in the competitive AI landscape.

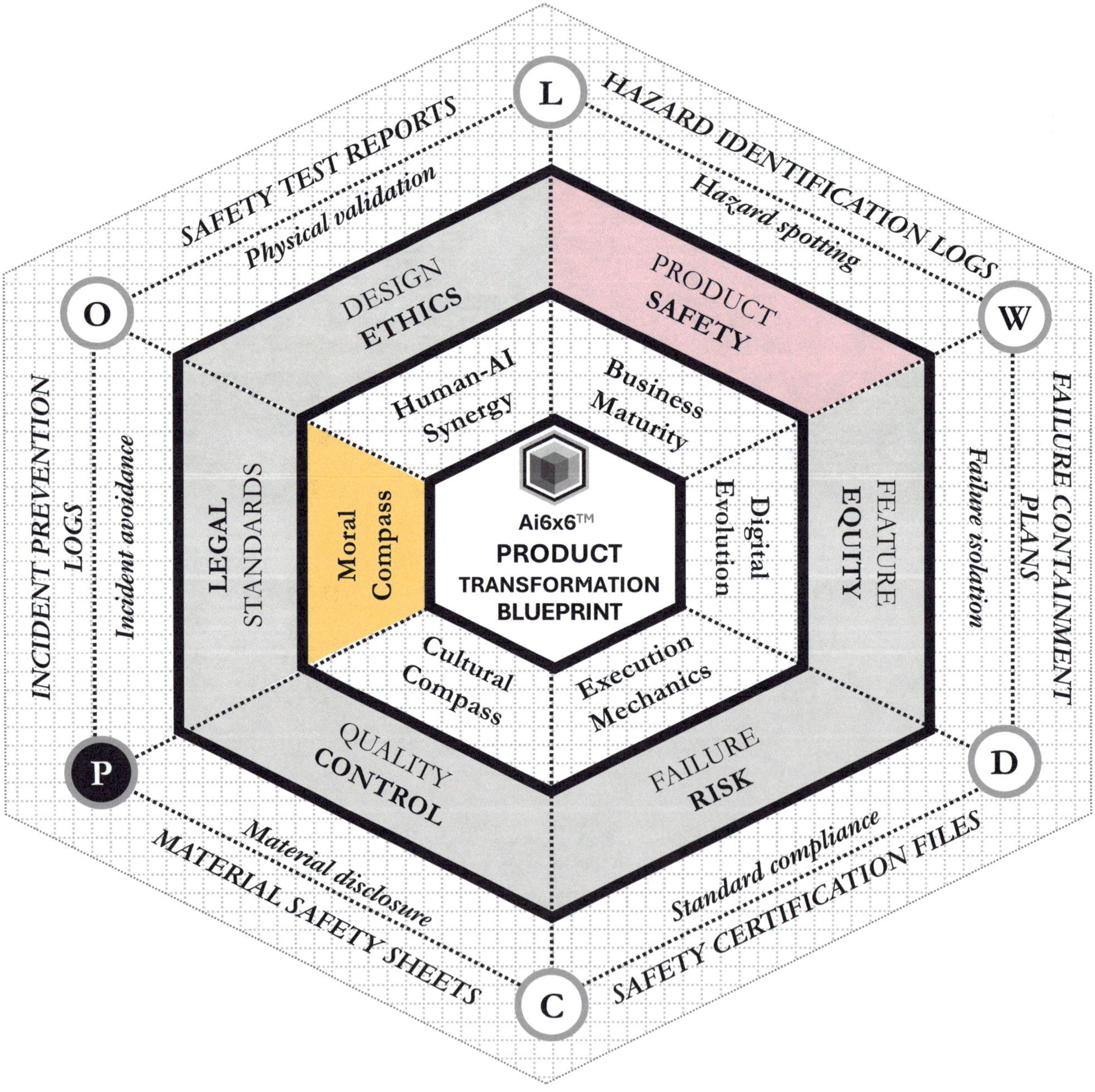

Ai6x6 PRODUCT SAFETY FOR MORAL COMPASS | PRODUCT TRANSFORMATION

1. SAFETY TEST REPORTS

Product Safety requires evidence based validation, and Safety Test Reports document whether a product performs safely under expected and extreme conditions. They capture structured results from stress tests, durability checks, environmental simulations, and failure scenarios. This matters because untested assumptions are one of the most common sources of product related harm. Safety Test Reports prevent risk by revealing vulnerabilities before customers encounter them. They also reinforce fairness by ensuring all users—not just ideal users—receive products that meet the same safety standards. When ignored, hazards remain hidden and customers face preventable injuries. Trust erodes as products fail unpredictably. Over time, Product Safety collapses as testing rigor disappears.

When executed well, Safety Test Reports strengthen Product Safety by embedding scientific discipline into product development. Customers trust that the organization validates safety claims with real data. This improves reliability, reduces liability, and enhances ethical responsibility. Reports reinforce the principle that safety must be proven, not assumed. They remind leaders that rigorous testing is a moral obligation. Ultimately, they build an environment where safety is verified through transparent, repeatable evidence.

MORAL COMPASS SCENARIO	LEADERSHIP ACTION
Mara discovers that a new hardware accessory fails thermal-stress tests at high ambient temperatures. She realizes the product could overheat in real-world conditions.	The leader halts production and orders a redesign of the thermal system. They also require expanded environmental testing before approval.
Devin notices that safety test reports show inconsistent results across manufacturing batches. He senses quality-control drift.	The leader initiates batch-level retesting and audits the production line. They also enforce tighter manufacturing tolerances.
Selah finds that a software safety test reveals a crash during emergency-stop actions. She recognizes a critical safety flaw.	The leader blocks release and assigns a rapid-response engineering team. They also add emergency-action tests to the standard suite.
Mateo sees that safety tests were skipped to meet a launch deadline. He realizes governance was bypassed.	The leader restarts the full safety-test cycle and delays launch. They also enforce mandatory sign-off from safety engineering.
Farah notices that test reports lack documentation of edge-case scenarios. She worries hidden risks remain.	The leader expands the test plan to include extreme and rare conditions. They also require scenario-coverage audits.
Idris discovers that third-party labs used outdated testing protocols. He recognizes compliance risk.	The leader demands updated certifications and switches to accredited labs. They also add vendor-protocol verification.

2. HAZARD IDENTIFICATION LOGS

Product Safety requires proactive detection of danger, and Hazard Identification Logs capture potential risks discovered during design, testing, manufacturing, or customer use. They document early warning signs before they escalate into failures. This matters because hazards often emerge gradually and are easy to overlook without structured tracking. Hazard Identification Logs prevent harm by ensuring risks are recorded, analyzed, and addressed. They also reinforce fairness by protecting all users from known dangers. When ignored, hazards accumulate silently and create systemic vulnerabilities. Customers encounter risks that could have been mitigated. Over time, Product Safety deteriorates as hazard awareness fades.

When executed well, Hazard Identification Logs strengthen Product Safety by creating a disciplined record of emerging threats. Customers trust that the organization monitors and responds to risks continuously. This improves safety performance, reduces incidents, and enhances ethical alignment. Logs reinforce the principle that danger must be documented to be eliminated. They remind leaders that vigilance is essential to responsible product stewardship. Ultimately, they build an environment where hazards are identified early and resolved decisively.

MORAL COMPASS SCENARIO	LEADERSHIP ACTION
Nia sees repeated reports of minor electrical shocks from a device. She realizes the hazard is escalating.	The leader initiates a root-cause investigation and issues a temporary usage advisory. They also add electrical-hazard checks to QA.
Corin notices that hazard logs show sharp-edge injuries from packaging. He senses a design oversight.	The leader redesigns packaging with safer materials. They also add packaging-hazard reviews.
Elara finds that hazard logs include multiple overheating complaints. She recognizes a thermal-management flaw.	The leader triggers a thermal-safety audit and updates cooling components. They also track overheating incidents post-fix.
Mateo sees that hazard logs are missing severity ratings. He worries critical issues are buried.	The leader adds severity scoring and triage workflows. They also require weekly hazard-review meetings.
Soren notices that hazard logs are not shared with product designers. He senses a disconnect between insights and design.	The leader integrates hazard logs into design reviews. They also require hazard-informed redesigns.
Hana discovers that hazard logs are stored in spreadsheets without access controls. She recognizes tampering risk.	The leader migrates logs to a secure system with audit trails. They also restrict editing rights.

3. FAILURE CONTAINMENT PLANS

Product Safety requires preparation for breakdowns, and Failure Containment Plans outline how the organization limits harm when a product malfunctions. They define steps for isolating failures, protecting users, and preventing escalation. This matters because even well designed products can fail under certain conditions. Failure Containment Plans prevent widespread damage by ensuring failures remain controlled and predictable. They also reinforce fairness by ensuring all customers receive the same level of protection during incidents. When ignored, failures spread unchecked and create cascading harm. Customers experience unpredictable danger and lose trust. Over time, Product Safety collapses as containment discipline disappears.

When executed well, Failure Containment Plans strengthen Product Safety by ensuring the organization is ready for the unexpected. Customers trust that failures will be managed responsibly. This improves resilience, reduces harm, and enhances brand credibility. Plans reinforce the principle that safety requires both prevention and response. They remind leaders that containment is a critical component of risk management. Ultimately, they build an environment where failures are controlled with precision and care.

MORAL COMPASS SCENARIO	LEADERSHIP ACTION
Priya discovers that a device failure cascades into system-wide shutdowns. She realizes containment boundaries are missing.	The leader redesigns the architecture with isolation layers. They also require containment testing for all critical components.
Rowan notices that the containment plan relies on manual intervention. He fears delays during emergencies.	The leader automates containment triggers and tests fail-safe responses. They also add redundancy for critical paths.
Serena finds that containment plans don't include cloud-service outages. She recognizes a major blind spot.	The leader adds cloud-failure scenarios and tests cross-region failover. They also require cloud-resilience reviews.
Omar sees that containment plans are outdated after a recent system overhaul. He realizes they no longer reflect reality.	The leader updates all plans and enforces version control. They also require updates after every architectural change.
Lila notices that containment plans don't include communication protocols. She worries teams won't know how to respond.	The leader adds communication steps and assigns escalation owners. They also run containment drills.
Andre discovers that containment plans assume perfect sensor accuracy. He senses unrealistic dependencies.	The leader adds degraded-sensor scenarios and tests fallback logic. They also require sensor-failure modeling.

4. SAFETY CERTIFICATION FILES

Product Safety requires external validation, and Safety Certification Files document compliance with recognized safety standards and regulatory requirements. They include test results, audit findings, and third party certifications. This matters because independent verification strengthens credibility and reduces bias. Safety Certification Files prevent shortcuts by requiring products to meet established benchmarks. They also reinforce fairness by ensuring all customers receive products that meet the same certified standards. When ignored, products enter the market without adequate oversight. Customers face risks that should have been prevented through certification. Over time, Product Safety deteriorates as compliance weakens.

When executed well, Safety Certification Files strengthen Product Safety by providing transparent proof of safety performance. Customers trust that the organization meets or exceeds industry standards. This improves confidence, reduces regulatory exposure, and enhances ethical alignment. Files reinforce the principle that safety must be validated by impartial experts. They remind leaders that certification is a safeguard, not a formality. Ultimately, they build an environment where safety is externally verified and consistently upheld.

MORAL COMPASS SCENARIO	LEADERSHIP ACTION
Nadia discovers that certification files for a new product are incomplete. She realizes the product cannot legally ship.	The leader halts distribution and completes all required testing. They also enforce certification-file audits.
Trevor notices that certification files reference outdated standards. He senses compliance risk.	The leader updates all certifications to current regulations. They also require annual standards reviews.
Aisha finds that third-party certification documents are missing authenticity seals. She recognizes potential fraud.	The leader verifies documents with the issuing body and replaces invalid files. They also add authenticity checks to procurement.
Marco sees that certification files are scattered across multiple systems. He worries about traceability.	The leader centralizes all certification files in a secure repository. They also enforce version control.
Evelyn discovers that certification files don't include regional variants. She recognizes international compliance gaps.	The leader adds region-specific certifications and updates distribution rules. They also require regional compliance reviews.
Jonah notices that certification renewals are overdue. He senses operational negligence.	The leader renews all certifications and adds automated renewal reminders. They also assign certification owners.

5. MATERIAL SAFETY SHEETS

Product Safety requires clarity about physical components, and Material Safety Sheets document the properties, hazards, and handling requirements of materials used in products. They provide essential information about toxicity, flammability, reactivity, and safe disposal. This matters because customers and workers must understand the risks associated with product materials. Material Safety Sheets prevent harm by ensuring dangerous substances are identified and managed responsibly. They also reinforce fairness by giving all stakeholders equal access to safety information. When ignored, hazardous materials become invisible and unregulated. Customers face exposure risks they cannot anticipate. Over time, Product Safety collapses as material transparency disappears.

When executed well, Material Safety Sheets strengthen Product Safety by making material risks visible and actionable. Customers trust that the organization discloses hazards honestly. This improves safety awareness, reduces incidents, and enhances ethical responsibility. Sheets reinforce the principle that transparency is essential to material safety. They remind leaders that responsible product design includes responsible material disclosure. Ultimately, they build an environment where material risks are clearly communicated and properly managed.

MORAL COMPASS SCENARIO	LEADERSHIP ACTION
Rhea discovers that a new adhesive used in manufacturing lacks a Material Safety Sheet. She realizes workers are exposed without guidance.	The leader obtains the sheet and updates safety protocols. They also require MSDS verification before material approval.
Omar notices that existing sheets are outdated and reference discontinued chemicals. He senses compliance gaps.	The leader updates all sheets and audits material inventories. They also enforce annual MSDS reviews.
Lila finds that sheets are stored only in English. She recognizes accessibility issues for multilingual teams.	The leader provides translated versions and updates distribution rules. They also require multilingual safety documentation.
Andre sees that sheets are not accessible on the production floor. He worries workers cannot reference them quickly.	The leader posts physical and digital copies at all workstations. They also require accessibility checks for safety materials.
Serena discovers that sheets don't include emergency-response steps. She senses incomplete safety guidance.	The leader updates sheets with emergency procedures. They also require emergency-response training.
Hugo notices that sheets are not linked to procurement workflows. He recognizes risk of unvetted materials entering production.	The leader integrates MSDS checks into procurement. They also require approval gates for new materials.

6. INCIDENT PREVENTION LOGS

Product Safety requires continuous improvement, and Incident Prevention Logs track actions taken to reduce the likelihood of future failures. They document interventions, design changes, and process adjustments that address known risks. This matters because prevention is the most effective form of safety. Incident Prevention Logs prevent recurrence by ensuring lessons learned translate into concrete improvements. They also reinforce fairness by ensuring all customers benefit from enhanced safety measures. When ignored, organizations repeat the same mistakes and expose customers to avoidable harm. Trust erodes as incidents recur. Over time, Product Safety deteriorates as prevention discipline weakens.

When executed well, Incident Prevention Logs strengthen Product Safety by creating a structured record of safety enhancements. Customers trust that the organization learns from experience and acts responsibly. This improves reliability, reduces harm, and enhances ethical alignment. Logs reinforce the principle that prevention is a continuous obligation. They remind leaders that safety evolves through iteration. Ultimately, they build an environment where safety improves with every cycle of learning.

MORAL COMPASS SCENARIO	LEADERSHIP ACTION
Keisha sees that logs show repeated near-misses involving the same component. She realizes a failure is imminent.	The leader replaces the component and investigates root causes. They also track near-miss trends.
Damon notices that prevention logs are filled with vague descriptions. He senses poor documentation discipline.	The leader standardizes logging templates and trains teams on specificity. They also audit log quality monthly.
Liora finds that logs show rising incidents during night shifts. She recognizes staffing or training gaps.	The leader increases supervision and retrains night-shift teams. They also monitor incident rates by shift.
Mateo sees that prevention logs are not reviewed by engineering. He worries insights aren't informing design.	The leader integrates logs into engineering sprints. They also require design updates based on prevention insights.
Serena discovers that logs don't include environmental conditions. She senses missing context.	The leader adds environmental-factor fields and updates logging standards. They also analyze correlations with incident patterns.
Omar notices that prevention logs are siloed across departments. He recognizes fragmented visibility.	The leader centralizes logs and enforces cross-department access. They also assign prevention-log owners.

Ai6x6™ Transformation Blueprint

Feature Equity

"Ethics is the anchor that keeps innovation from drifting into harm."
Mary Barra, CEO of General Motors, known for industrial transformation leadership.

Feature Equity requires balanced functionality, and Inclusive Feature Behavior ensures product features work consistently across diverse users and contexts. It establishes expectations that prevent exclusion, bias, or unequal performance. This matters because feature behavior directly shapes user access and satisfaction. Inclusive Feature Behavior prevents inequity by ensuring features respond fairly to all users. It also reinforces trust by demonstrating that functionality is designed with inclusivity in mind. When ignored, disparities emerge in how features behave. Teams experience misalignment as functionality becomes inconsistent. Over time, Feature Equity deteriorates as gaps widen. Bias Free Functionality ensures features operate without favoring or disadvantaging specific groups.

Every feature is a promise that must be kept for everyone. Accessibility Feature Maps and Inclusive Feature Reviews help a leader see where those promises might break. This topic prioritizes "Feature Fairness" in the 90-day roadmap, ensuring that innovation doesn't leave segments of the market behind. By engrossing yourself in "Performance Equity Reports," you identify where the product's "Why" is failing its "How." Equity in product design is a competitive moat. It ensures that your innovation engine scales "Bliss" rather than "Silos," making your product a tool for democratization rather than a source of exclusion. In the print series, you can map "Feature Fairness Logs" against your "Customer Representation Maps," seeing the whole map of product inclusivity. This clinical focus allows you to move innovation capital to the most critical "Survival" and "Growth" features, ensuring that your "Product Transformation" is as broad as it is deep, securing your future against the predators who hunt for silos in feature accessibility.

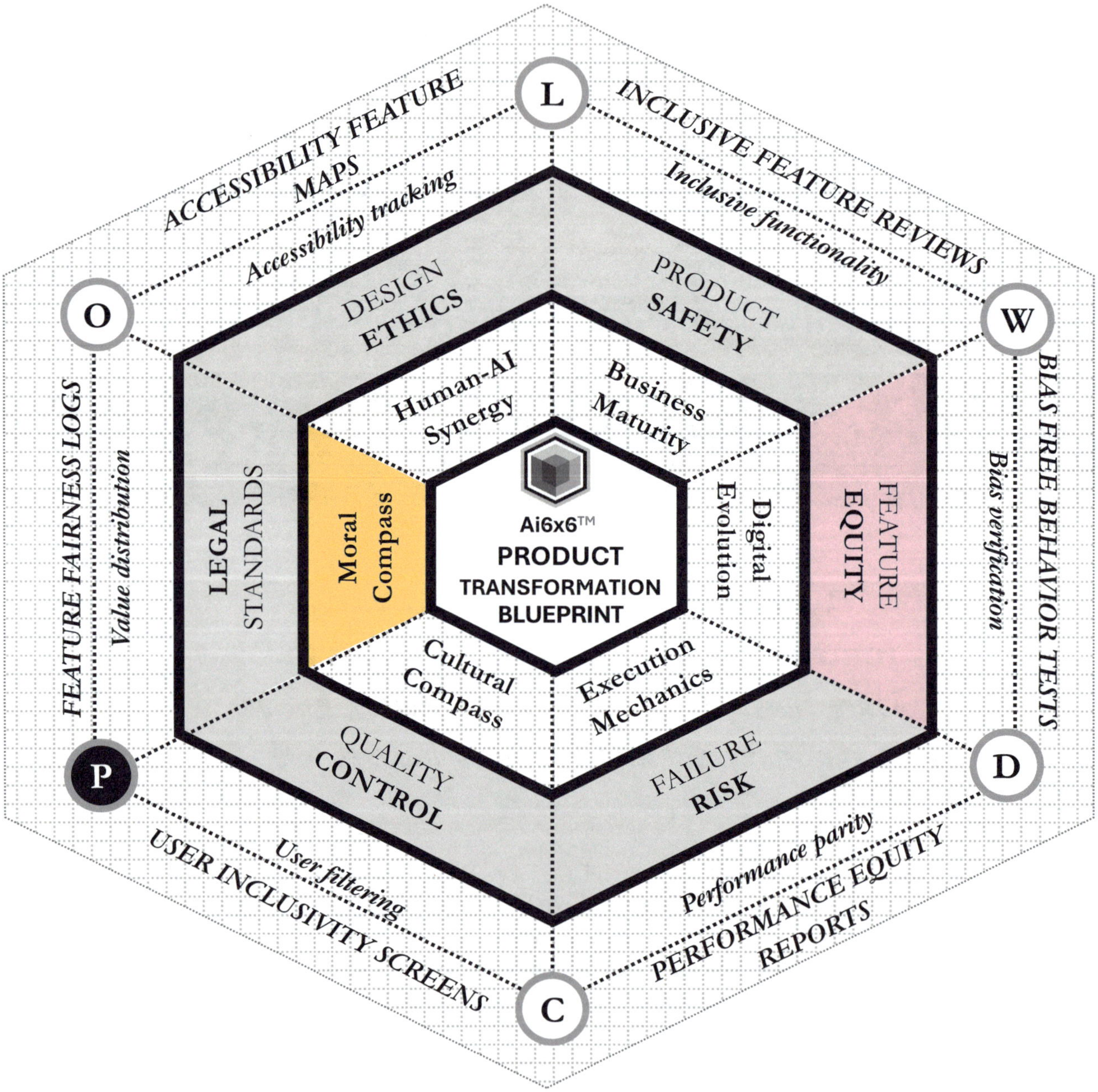

Ai6x6 FEATURE EQUITY FOR MORAL COMPASS | PRODUCT TRANSFORMATION

1. ACCESSIBILITY FEATURE MAPS

Feature Equity requires universal usability, and Accessibility Feature Maps document how product features support customers with diverse physical, cognitive, and sensory needs. They reveal where accessibility is strong, where gaps exist, and where design choices unintentionally exclude users. This matters because accessibility failures often stem from incomplete visibility rather than intentional neglect. Accessibility Feature Maps prevent inequity by making inclusion a measurable design requirement. They also reinforce fairness by ensuring all customers—not just those with typical abilities—can benefit from the product. When ignored, accessibility becomes inconsistent and reactive. Customers with disabilities face unnecessary barriers. Over time, Feature Equity deteriorates as exclusion becomes embedded in the product.

When executed well, Accessibility Feature Maps strengthen Feature Equity by ensuring accessibility is integrated into every feature, not added as an afterthought. Customers trust that the organization recognizes and supports diverse needs. This improves usability, expands reach, and enhances ethical alignment. Maps reinforce the principle that accessibility is a core design obligation. They remind leaders that inclusion must be intentional and visible. Ultimately, they build an environment where every feature is accessible by design.

MORAL COMPASS SCENARIO	LEADERSHIP ACTION
Mara discovers that accessibility features are scattered across menus with no unified structure. She realizes users with disabilities cannot easily find essential tools.	The leader consolidates all accessibility features into a single, intuitive hub. They also require accessibility-map updates for every release.
Devin notices that the feature map excludes voice-control pathways. He senses a gap for users with mobility impairments.	The leader adds voice-control mapping and tests it with real users. They also require multimodal accessibility coverage.
Selah finds that accessibility features differ across platforms. She recognizes cross-platform inconsistency.	The leader standardizes accessibility features and enforces parity. They also add platform-specific accessibility audits.
Mateo sees that the map doesn't include compatibility with assistive technologies. He worries about integration failures.	The leader adds AT-compatibility mapping and validates with screen readers and switch devices. They also require AT-integration testing.
Farah notices that the map lacks cognitive-accessibility features. She senses exclusion of neurodiverse users.	The leader adds cognitive-support features and updates design guidelines. They also require cognitive-accessibility reviews.
Idris discovers that the map hasn't been updated after major UI changes. He realizes users are navigating outdated information.	The leader refreshes the map and enforces version control. They also require updates after every UI overhaul.

2. INCLUSIVE FEATURE REVIEWS

Feature Equity requires intentional representation, and Inclusive Feature Reviews evaluate whether product features serve diverse customer groups equitably. They examine assumptions, workflows, and design patterns that may privilege one group while disadvantaging another. This matters because exclusion often originates in early feature decisions. Inclusive Feature Reviews prevent inequity by ensuring diverse perspectives shape feature development. They also reinforce fairness by ensuring no customer group is overlooked or underserved. When ignored, features become optimized for narrow use cases. Customers outside those cases experience frustration and disengagement. Over time, Feature Equity collapses as exclusion becomes systemic.

When executed well, Inclusive Feature Reviews strengthen Feature Equity by embedding diversity into the feature lifecycle. Customers trust that the organization considers their unique contexts and needs. This improves satisfaction, broadens adoption, and enhances ethical responsibility. Reviews reinforce the principle that inclusion must be engineered, not assumed. They remind leaders that equitable features require deliberate scrutiny. Ultimately, they build an environment where every feature reflects the full spectrum of customer needs.

MORAL COMPASS SCENARIO	LEADERSHIP ACTION
Nia sees that a new feature assumes high reading proficiency. She realizes it excludes users with lower literacy.	The leader redesigns the feature with simplified language and visual cues. They also require literacy-level testing.
Corin notices that a feature uses gestures not recognized across cultures. He senses global usability risk.	The leader replaces ambiguous gestures with universal interactions. They also add cultural-variation testing.
Elara finds that a feature requires rapid response times. She recognizes exclusion for users with motor impairments.	The leader adds adjustable timing controls. They also require motor-accessibility validation.
Mateo sees that color-only indicators are used for critical states. He worries about color-blind accessibility.	The leader adds shape and text indicators. They also enforce non-color-dependent design rules.
Soren notices that the feature review team lacks diverse representation. He senses blind spots.	The leader expands the review panel with diverse testers. They also require demographic coverage in reviews.
Hana discovers that inclusive reviews are skipped for minor features. She realizes small exclusions accumulate.	The leader mandates inclusive reviews for all features, regardless of size. They also track compliance in the design governance dashboard.

3. BIAS-FREE BEHAVIOR TESTS

Feature Equity requires neutrality, and Bias Free Behavior Tests evaluate whether product features behave consistently across demographic groups. They detect patterns where algorithms, rules, or interactions produce unequal outcomes. This matters because bias often emerges unintentionally through data, design, or logic. Bias Free Behavior Tests prevent inequity by revealing where features treat customers differently without justification. They also reinforce fairness by ensuring all users receive consistent behavior. When ignored, biased features create harm, distrust, and reputational damage. Customers feel misrepresented or disadvantaged. Over time, Feature Equity deteriorates as bias becomes embedded in product behavior.

When executed well, Bias Free Behavior Tests strengthen Feature Equity by ensuring features operate with impartiality. Customers trust that the product treats them fairly. This improves credibility, reduces risk, and enhances ethical alignment. Tests reinforce the principle that fairness must be validated, not assumed. They remind leaders that bias is a technical and moral failure. Ultimately, they build an environment where product behavior is equitable across all users.

MORAL COMPASS SCENARIO	LEADERSHIP ACTION
Priya discovers that a predictive feature behaves differently for users with non-standard dialects. She realizes linguistic bias is embedded.	The leader retrains the model with diverse language samples. They also add dialect-specific fairness tests.
Rowan notices that behavior tests show slower response times for older devices. He senses device-based inequity.	The leader optimizes performance for low-end hardware. They also add device-equity metrics.
Serena finds that a feature flags certain names as suspicious. She recognizes cultural bias.	The leader removes biased heuristics and rebuilds the logic. They also require cultural-neutrality testing.
Omar sees that behavior tests ignore intersectional groups. He worries about hidden disparities.	The leader adds intersectional test cases. They also require intersectional fairness reporting.
Lila discovers that behavior tests rely on synthetic data that doesn't reflect real diversity. She senses skewed results.	The leader validates synthetic data and supplements it with real samples. They also require provenance documentation.
Andre notices that behavior tests are run only once per release cycle. He recognizes fairness drift risk.	The leader increases test frequency and automates execution. They also track fairness stability over time.

4. PERFORMANCE EQUITY REPORTS

Feature Equity requires consistent performance, and Performance Equity Reports evaluate whether features operate reliably across devices, regions, networks, and customer segments. They reveal disparities in speed, accuracy, or stability that affect certain groups more than others. This matters because uneven performance creates unequal experiences. Performance Equity Reports prevent inequity by making performance gaps visible and actionable. They also reinforce fairness by ensuring all customers receive the same quality of service. When ignored, performance disparities widen and certain users face chronic disadvantages. Customers lose confidence in the product's reliability. Over time, Feature Equity collapses as inconsistency becomes normalized.

When executed well, Performance Equity Reports strengthen Feature Equity by driving uniform performance across all contexts. Customers trust that the product will work reliably for them. This improves satisfaction, reduces complaints, and enhances brand integrity. Reports reinforce the principle that performance is a fairness issue. They remind leaders that reliability must be equitable, not selective. Ultimately, they build an environment where every customer receives consistent, dependable performance.

MORAL COMPASS SCENARIO	LEADERSHIP ACTION
Nadia sees that performance varies significantly between high-bandwidth and low-bandwidth regions. She realizes users in rural areas are disadvantaged.	The leader optimizes performance for low-bandwidth environments. They also add region-level performance equity metrics.
Trevor notices that load times are slower for older devices. He senses hardware-based inequity.	The leader improves backward compatibility and tests across device tiers. They also require device-equity reporting.
Aisha finds that performance degrades for users with accessibility tools enabled. She recognizes an inclusion failure.	The leader fixes compatibility issues and adds AT-performance testing. They also track performance parity.
Marco sees that performance equity reports exclude non-English locales. He worries about linguistic inequity.	The leader adds locale-specific performance metrics. They also test performance across languages.
Evelyn notices that performance varies by time of day due to load spikes. She senses infrastructure inequity.	The leader improves load balancing and adds peak-hour performance monitoring. They also track time-based equity.
Jonah discovers that performance equity reports are not shared with engineering. He realizes insights aren't driving improvements.	The leader integrates reports into engineering sprints. They also require performance-equity remediation plans.

5. USER INCLUSIVITY SCREENS

Feature Equity requires respect for diverse identities, and User Inclusivity Screens evaluate whether features acknowledge, support, and represent the full range of customer backgrounds. They examine language, imagery, defaults, and assumptions that may exclude or misrepresent users. This matters because exclusion often appears subtly in how features frame identity. User Inclusivity Screens prevent harm by identifying where features fail to reflect real world diversity. They also reinforce fairness by ensuring all customers feel recognized and valued. When ignored, features communicate narrow or biased assumptions. Customers feel unseen or marginalized. Over time, Feature Equity deteriorates as inclusivity disappears from design.

When executed well, User Inclusivity Screens strengthen Feature Equity by embedding representation into every feature. Customers trust that the product respects their identity. This improves engagement, reduces friction, and enhances ethical alignment. Screens reinforce the principle that inclusivity is a design requirement, not a marketing gesture. They remind leaders that representation shapes user experience. Ultimately, they build an environment where every customer feels included.

MORAL COMPASS SCENARIO	LEADERSHIP ACTION
Rhea discovers that onboarding assumes users have perfect vision. She realizes inclusivity screens are missing.	The leader adds vision-friendly onboarding options. They also require inclusivity screens for all entry points.
Omar notices that inclusivity screens don't include language selection. He senses linguistic exclusion.	The leader adds multilingual options and tests comprehension. They also require language-inclusivity checks.
Lila finds that inclusivity screens ignore neurodiverse needs. She recognizes cognitive-accessibility gaps.	The leader adds cognitive-support modes. They also require neurodiversity-inclusive design reviews.
Andre sees that inclusivity screens are inconsistent across platforms. He worries about fragmented UX.	The leader standardizes inclusivity screens and enforces parity. They also add cross-platform inclusivity audits.
Serena notices that inclusivity screens don't include safety preferences. She senses risk for vulnerable users.	The leader adds safety-preference controls. They also require safety-aware inclusivity reviews.
Hugo discovers that inclusivity screens are hidden behind multiple steps. He realizes users won't find them.	The leader surfaces inclusivity screens during onboarding. They also require visibility scoring.

6. FEATURE FAIRNESS LOGS

Feature Equity requires continuous oversight, and Feature Fairness Logs document whether features behave consistently and equitably over time. They track fairness issues, interventions, and improvements across releases. This matters because fairness can degrade as products evolve. Feature Fairness Logs prevent regression by ensuring equity is monitored and maintained. They also reinforce accountability by making fairness performance visible. When ignored, inequities reappear unnoticed and harm customers. Trust erodes as features behave unpredictably. Over time, Feature Equity collapses as fairness becomes ungoverned.

When executed well, Feature Fairness Logs strengthen Feature Equity by creating a disciplined record of fairness performance. Customers trust that the organization monitors and protects equity continuously. This improves reliability, reduces risk, and enhances ethical responsibility. Logs reinforce the principle that fairness must be maintained, not assumed. They remind leaders that equity requires ongoing vigilance. Ultimately, they build an environment where fairness is sustained through structured oversight.

MORAL COMPASS SCENARIO	LEADERSHIP ACTION
Keisha sees that fairness logs show a feature performs worse for older adults. She realizes age-based inequity is emerging.	The leader redesigns the feature and validates improvements with older users. They also track age-equity metrics.
Damon notices that fairness logs reveal slower response times for non-English locales. He senses linguistic inequity.	The leader optimizes localization performance. They also add locale-specific fairness checks.
Liora finds that fairness logs show higher error rates for users with disabilities. She recognizes accessibility failures.	The leader fixes accessibility issues and adds AT-specific fairness testing. They also require accessibility parity.
Mateo sees that fairness logs are missing demographic metadata. He worries inequities are invisible.	The leader adds demographic tagging and updates logging standards. They also require demographic-impact reviews.
Serena notices that fairness logs show regressions after each major update. She senses release-cycle instability.	The leader adds fairness gating to release workflows. They also track fairness deltas post-release.
Omar discovers that fairness logs are siloed across teams. He recognizes fragmented governance.	The leader centralizes fairness logs and enforces cross-team visibility. They also assign fairness-log owners.

Ai6x6™ Transformation Blueprint

Failure Risk

"AI must help us repair the planet, not accelerate its decline."

Al Gore, former U.S. Vice President and climate advocate.

Failure Risk requires awareness of product vulnerabilities, and Breakdown Exposure Points identify where systems are most likely to fail under stress or normal use. They establish visibility into weaknesses that compromise reliability. This matters because product failures create operational, financial, and reputational damage. Breakdown Exposure Points prevent disruption by revealing where redesign or reinforcement is needed. They also reinforce fairness by ensuring all users receive dependable performance. When ignored, failures multiply and become harder to predict. Teams experience confusion as reliability declines. Over time, Failure Risk increases as weaknesses compound. Reliability Threat Indicators highlight early signs of potential breakdown.

Spotting the "Breakpoints" in the engine before they occur. By analyzing Edge Case Exposure Logs and Stress Test Results, leaders ensure that innovation velocity doesn't outpace the organization's ability to manage failure. This topic involves using "Failure Mode Logs" to run a clinical diagnostic on the product's resilience. It's about moving away from "Daydreaming with a Budget" to evidence-based innovation. Failure risk management allows the leader to "Build the road while running it," providing the "Failure Containment Plans" that ensure even when a product breaks, the organization remains antifragile and focused on growth. The print format is essential for this diagnostic; it lets you see the whole "Failure Map" at once, identifying where your product armor is thin. This clinical focus replaces the "million-dollar guess" of product reliability with a roadmap for resilience, ensuring that your innovation engine remains robust and your "Product Transformation" moves as fast as your thoughts, protected from the "Kill Zones" of unpredictable product failure.

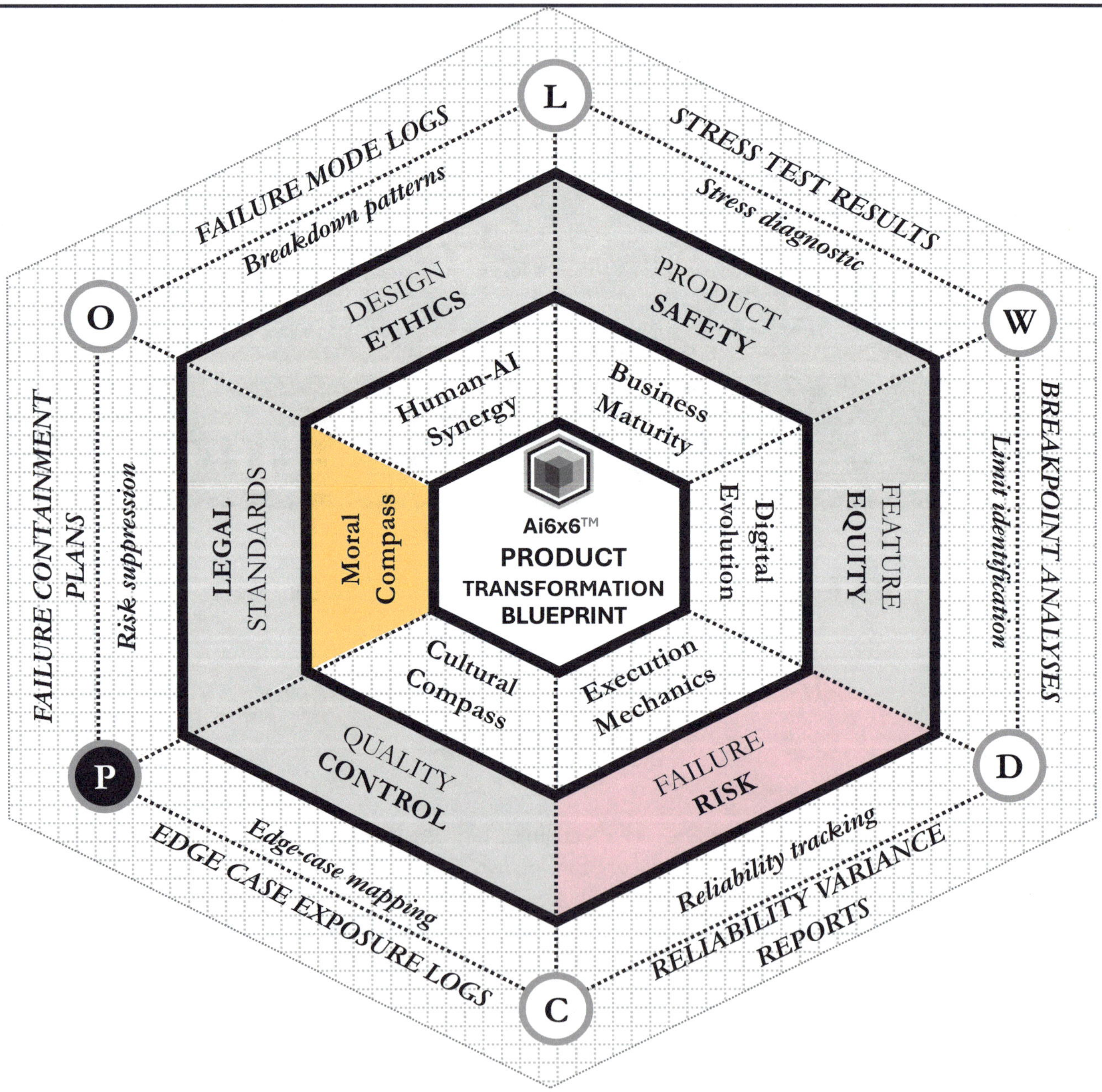

Ai6x6 FAILURE RISK FOR MORAL COMPASS | PRODUCT TRANSFORMATION

1. FAILURE MODE LOGS

Failure Risk requires structured visibility into breakdown patterns, and Failure Mode Logs document every known way a product can malfunction under real world conditions. They capture triggers, symptoms, severity, and the contexts in which failures occur. This matters because untracked failure modes create blind spots that lead to repeated harm. Failure Mode Logs prevent recurrence by making failure behavior explicit and analyzable. They also reinforce fairness by ensuring all customers—not just early adopters or technical users—benefit from improved reliability. When ignored, failures remain unpredictable and unaddressed. Customers experience inconsistent performance and lose confidence. Over time, Failure Risk escalates as the product becomes fragile.

When executed well, Failure Mode Logs strengthen Failure Risk management by creating a disciplined record of how the product behaves under stress. Customers trust that the organization understands and mitigates failure patterns. This improves reliability, reduces harm, and enhances ethical responsibility. Logs reinforce the principle that failure must be documented to be eliminated. They remind leaders that reliability is built through transparency. Ultimately, they build an environment where failure knowledge drives continuous improvement.

MORAL COMPASS SCENARIO	LEADERSHIP ACTION
Talia notices that a device intermittently freezes only when switching between two rarely used modes. She realizes the failure mode is buried in an obscure transition path.	The leader adds transition-state monitoring and assigns engineers to replicate the exact sequence. They also require mode-transition coverage in future testing.
Brant sees that logs show silent data corruption during long-running tasks. He recognizes a hidden integrity failure.	The leader halts affected workflows and adds checksum validation. They also require long-duration stress logging.
Mireya finds that failure logs spike only after firmware updates. She suspects a regression in the update pipeline.	The leader rolls back the firmware and audits the update process. They also add pre-deployment regression gates.
Jalen notices that failures occur only when users rapidly toggle settings. He realizes the UI thread is overloaded.	The leader throttles rapid-toggle behavior and optimizes thread handling. They also add rapid-interaction tests.
Soraya discovers that failures cluster around a specific hardware batch. She senses manufacturing variance.	The leader isolates the batch and runs component-level diagnostics. They also tighten supplier QA requirements.
Kade sees that logs show failures only when the device is used outdoors in cold weather. He recognizes temperature-sensitive behavior.	The leader expands environmental testing and updates hardware tolerances. They also add cold-weather performance metrics.

2. STRESS TEST RESULTS

Failure Risk requires proof of resilience, and Stress Test Results reveal how a product performs under extreme, abnormal, or boundary pushing conditions. They expose weaknesses that normal testing cannot detect. This matters because real world environments often exceed ideal operating assumptions. Stress Test Results prevent catastrophic failures by identifying where the product breaks under pressure. They also reinforce fairness by ensuring all customers—not just those in controlled environments—receive robust performance. When ignored, products fail unpredictably when pushed beyond nominal limits. Customers experience breakdowns that feel sudden and inexplicable. Over time, Failure Risk becomes systemic as resilience is never validated.

When executed well, Stress Test Results strengthen Failure Risk management by revealing the product's true durability. Customers trust that the organization tests beyond minimum requirements. This improves confidence, reduces liability, and enhances ethical alignment. Results reinforce the principle that resilience must be engineered, not assumed. They remind leaders that extreme conditions are part of real world use. Ultimately, they build an environment where products withstand stress with reliability and integrity.

MORAL COMPASS SCENARIO	LEADERSHIP ACTION
Liora sees that the system collapses when handling simultaneous uploads from multiple regions. She realizes the load balancer isn't distributing traffic correctly.	The leader reconfigures the balancer and reruns multi-region stress tests. They also add geo-load simulations to the test suite.
Camden notices that memory usage spikes uncontrollably during batch imports. He suspects a leak in the parsing engine.	The leader assigns a memory-profiling task force and patches the parser. They also add leak-detection automation.
Nalin finds that stress tests show severe latency when encryption is enabled. He recognizes a cryptographic bottleneck.	The leader optimizes encryption routines and tests hardware acceleration. They also add crypto-performance benchmarks.
Zoya sees that stress tests fail only during partial network outages. She realizes the system lacks graceful degradation.	The leader adds fallback logic and tests degraded-network scenarios. They also require resilience scoring.
Idran notices that stress tests reveal cascading failures between microservices. He senses dependency fragility.	The leader maps service dependencies and adds circuit breakers. They also require inter-service stress testing.
Freya finds that stress tests show UI lag only on high-resolution displays. She recognizes rendering inefficiency.	The leader optimizes rendering pipelines and tests across display tiers. They also add resolution-specific stress cases.

3. BREAKPOINT ANALYSES

Failure Risk requires understanding limits, and Breakpoint Analyses determine the exact thresholds at which a product stops functioning safely or effectively. They identify the boundaries between acceptable performance and failure. This matters because customers rely on predictable behavior, especially near operational limits. Breakpoint Analyses prevent harm by revealing where performance degrades before it becomes dangerous. They also reinforce fairness by ensuring all customers receive products with clearly defined limits. When ignored, products fail without warning and expose users to unnecessary risk. Customers feel blindsided by unpredictable behavior. Over time, Failure Risk escalates as limits remain unknown.

When executed well, Breakpoint Analyses strengthen Failure Risk management by defining the product's safe operating envelope. Customers trust that the organization understands and communicates performance boundaries. This improves safety, reduces misuse, and enhances ethical responsibility. Analyses reinforce the principle that limits must be measured, not guessed. They remind leaders that predictability is essential to user confidence. Ultimately, they build an environment where product limits are transparent and responsibly managed.

MORAL COMPASS SCENARIO	LEADERSHIP ACTION
Nadir discovers that performance collapses once user profiles exceed a certain size. He realizes the breakpoint is tied to unindexed fields.	The leader adds indexing and re-architects the data model. They also require profile-size breakpoint testing.
Selene sees that the system becomes unstable when more than three plugins are enabled. She suspects plugin-interaction conflicts.	The leader isolates plugin dependencies and adds compatibility constraints. They also require plugin-stack breakpoint analysis.
Corvin notices that the product fails when processing unusually large images. He recognizes a memory-allocation breakpoint.	The leader adds size-based throttling and optimizes memory handling. They also require large-asset breakpoint tests.
Amara finds that the app crashes when users exceed a certain number of saved items. She senses a scaling flaw.	The leader redesigns the storage layer and adds pagination. They also track scaling breakpoints across features.
Jiro sees that the system becomes erratic when network jitter exceeds a threshold. He realizes the breakpoint is environmental.	The leader adds jitter-tolerant buffering and tests unstable-network scenarios. They also require network-variability breakpoint mapping.
Tamsin notices that the product fails when too many background tasks run concurrently. She recognizes a scheduling breakpoint.	The leader optimizes task scheduling and adds concurrency limits. They also require background-task breakpoint analysis.

4. RELIABILITY VARIANCE REPORTS

Failure Risk requires consistency, and Reliability Variance Reports evaluate how product performance fluctuates across environments, devices, user behaviors, and time. They reveal where reliability is strong and where it degrades. This matters because inconsistent reliability creates unpredictable customer experiences. Reliability Variance Reports prevent inequity by identifying which customers face higher failure rates. They also reinforce fairness by ensuring all users receive dependable performance. When ignored, reliability gaps widen and certain customers experience chronic instability. Trust erodes as performance becomes inconsistent. Over time, Failure Risk becomes embedded in the product's reputation.

When executed well, Reliability Variance Reports strengthen Failure Risk management by driving uniform reliability across all contexts. Customers trust that the product will perform consistently for them. This improves satisfaction, reduces complaints, and enhances brand stability. Reports reinforce the principle that reliability must be equitable. They remind leaders that variance is a signal of deeper systemic issues. Ultimately, they build an environment where reliability is predictable and dependable.

MORAL COMPASS SCENARIO	LEADERSHIP ACTION
Rami sees that reliability varies drastically between two identical hardware models. He realizes firmware divergence is causing inconsistent behavior.	The leader unifies firmware versions and validates parity. They also add hardware-variant reliability tracking.
Elowen notices that reliability drops only during specific time windows. She suspects scheduled background jobs are interfering.	The leader reschedules jobs and tests time-based reliability. They also add temporal-variance monitoring.
Dax finds that reliability is lower for users with large datasets. He recognizes scaling-related variance.	The leader optimizes large-dataset handling and adds dataset-size variance metrics. They also test performance across dataset tiers.
Mirek sees that reliability varies by language locale. He senses localization-related defects.	The leader audits locale-specific code paths and fixes inconsistencies. They also add locale-variance reporting.
Saffron notices that reliability drops only when users enable advanced settings. She realizes edge-case configurations are untested.	The leader expands test coverage to include advanced-setting combinations. They also track configuration-based variance.
Halden finds that reliability differs between cloud regions. He recognizes infrastructure variance.	The leader aligns regional configurations and tests cross-region reliability. They also add region-specific variance dashboards.

5. EDGE CASE EXPOSURE LOGS

Failure Risk requires awareness of rare but impactful scenarios, and Edge Case Exposure Logs document unusual conditions that reveal hidden vulnerabilities. They capture scenarios that fall outside typical usage patterns but still occur in real life. This matters because edge cases often expose the most dangerous failures. Edge Case Exposure Logs prevent harm by ensuring rare conditions are analyzed and addressed. They also reinforce fairness by protecting customers whose use cases differ from the majority. When ignored, edge cases become sources of unpredictable and severe failures. Customers experience breakdowns that feel unfair and unexplained. Over time, Failure Risk escalates as rare scenarios accumulate unmitigated.

When executed well, Edge Case Exposure Logs strengthen Failure Risk management by expanding the organization's understanding of real world complexity. Customers trust that the product is tested beyond typical assumptions. This improves resilience, reduces harm, and enhances ethical alignment. Logs reinforce the principle that rare does not mean irrelevant. They remind leaders that edge cases reveal the true robustness of a product. Ultimately, they build an environment where even uncommon scenarios are handled responsibly.

MORAL COMPASS SCENARIO	LEADERSHIP ACTION
Kira discovers that edge-case logs show failures only when users input extremely long text strings. She realizes buffer limits are insufficient.	The leader adds input-length validation and tests extreme-length scenarios. They also require boundary-condition coverage.
Oren notices that failures occur when users rapidly switch between offline and online modes. He senses synchronization instability.	The leader redesigns sync logic and tests rapid-state transitions. They also add offline-mode exposure metrics.
Maelin finds that edge-case logs show errors when users upload unsupported file types. She recognizes missing validation.	The leader adds strict file-type checks and user-friendly error handling. They also require file-type exposure testing.
Jax sees that edge-case logs spike when users operate the product with outdated OS versions. He realizes backward compatibility is weak.	The leader improves legacy support and tests across OS tiers. They also add OS-version exposure tracking.
Suri notices that edge-case logs show failures only when users have extremely slow network speeds. She senses unhandled latency.	The leader adds latency-tolerant logic and tests low-bandwidth scenarios. They also require bandwidth-exposure mapping.
Thane discovers that edge-case logs reveal failures when users disable animations. He recognizes a UI-state dependency.	The leader decouples animations from core logic and tests animation-off states. They also add accessibility-mode exposure tests.

6. FAILURE CONTAINMENT PLANS

Failure Risk requires disciplined response, and Failure Containment Plans define how the organization limits harm when a product malfunctions. They outline steps for isolating failures, protecting users, and preventing escalation. This matters because even well designed products can fail under certain conditions. Failure Containment Plans prevent widespread damage by ensuring failures remain controlled and predictable. They also reinforce fairness by ensuring all customers receive the same level of protection during incidents. When ignored, failures spread unchecked and create cascading harm. Customers experience unpredictable danger and lose trust. Over time, Failure Risk collapses as containment discipline disappears.

When executed well, Failure Containment Plans strengthen Failure Risk management by ensuring the organization is ready for the unexpected. Customers trust that failures will be managed responsibly. This improves resilience, reduces harm, and enhances brand credibility. Plans reinforce the principle that safety requires both prevention and response. They remind leaders that containment is a critical component of risk management. Ultimately, they build an environment where failures are controlled with precision and care.

MORAL COMPASS SCENARIO	LEADERSHIP ACTION
Liora discovers that a subsystem failure silently corrupts downstream data before alerts trigger. She realizes containment boundaries are too late.	The leader adds upstream validation gates and early-warning triggers. They also require pre-propagation containment checks.
Bronn notices that a failure in one microservice causes retries that overload the entire cluster. He senses cascading amplification.	The leader adds circuit breakers and retry-rate limits. They also test cascading-failure containment.
Nessa finds that containment plans assume cloud availability. She recognizes that offline fallback paths are missing.	The leader adds offline-mode containment and tests degraded-connectivity scenarios. They also require offline-resilience reviews.
Jalen sees that containment plans don't include user-facing messaging. He worries customers will misinterpret failures as data loss.	The leader adds clear failure-state communication and tests user comprehension. They also require communication-layer containment.
Sorin notices that containment plans rely on manual overrides that require specialized knowledge. He senses operational fragility.	The leader simplifies override procedures and trains frontline teams. They also add automation for high-risk scenarios.
Helia discovers that containment plans don't include rollback paths for partial deployments. She recognizes deployment-stage exposure.	The leader adds granular rollback checkpoints and tests partial-deployment containment. They also require rollback-readiness checks.

Ai6x6™ Transformation Blueprint

Quality Control

"Morality is not a constraint on AI — it is its greatest potential."

Martha Nussbaum, philosopher known for work on ethics and human capability.

Quality Control requires disciplined oversight, and Standard Compliance Checks define the benchmarks products must meet before release. They establish the criteria that prevent defects, inconsistencies, or performance gaps. This matters because quality determines user satisfaction and long term trust. Standard Compliance Checks prevent errors by ensuring products meet defined expectations. They also reinforce fairness by ensuring all users receive the same level of product reliability. When ignored, defects become more frequent and damaging. Teams experience misalignment as quality becomes unpredictable. Over time, Quality Control deteriorates as standards weaken. Defect Prevention Processes ensure issues are identified and resolved before reaching users.

The distance between intent and execution is measured in quality. Defect Prevention Logs and Testing Protocol Records are the benchmarks of mastery, ensuring the engine scales sparks without the technical debt of the past. This topic uses "Release Validation Reports" to ensure that every ship is a "Masterclass" in execution. By studying "Performance Monitoring Dashboards" on paper, the leader can focus on the "clinical truth" of the product's health. Quality control is the "Blueprint" that replaces the "Million-Dollar Guess," ensuring that the product is a self-governing engine of market dominance and operational excellence. In the print series, you can physically flip between "Quality Standard Checklists" and "Defect Prevention Logs," engrossing yourself in the diagnostic that prevents brand-destroying quality gaps. Quality control is the clinical moat that ensures your "Product Transformation" is built on a foundation of mastery, ensuring that your innovation engine moves cleanly and consistently, transforming your strategic intent into a tangible, high-quality market reality.

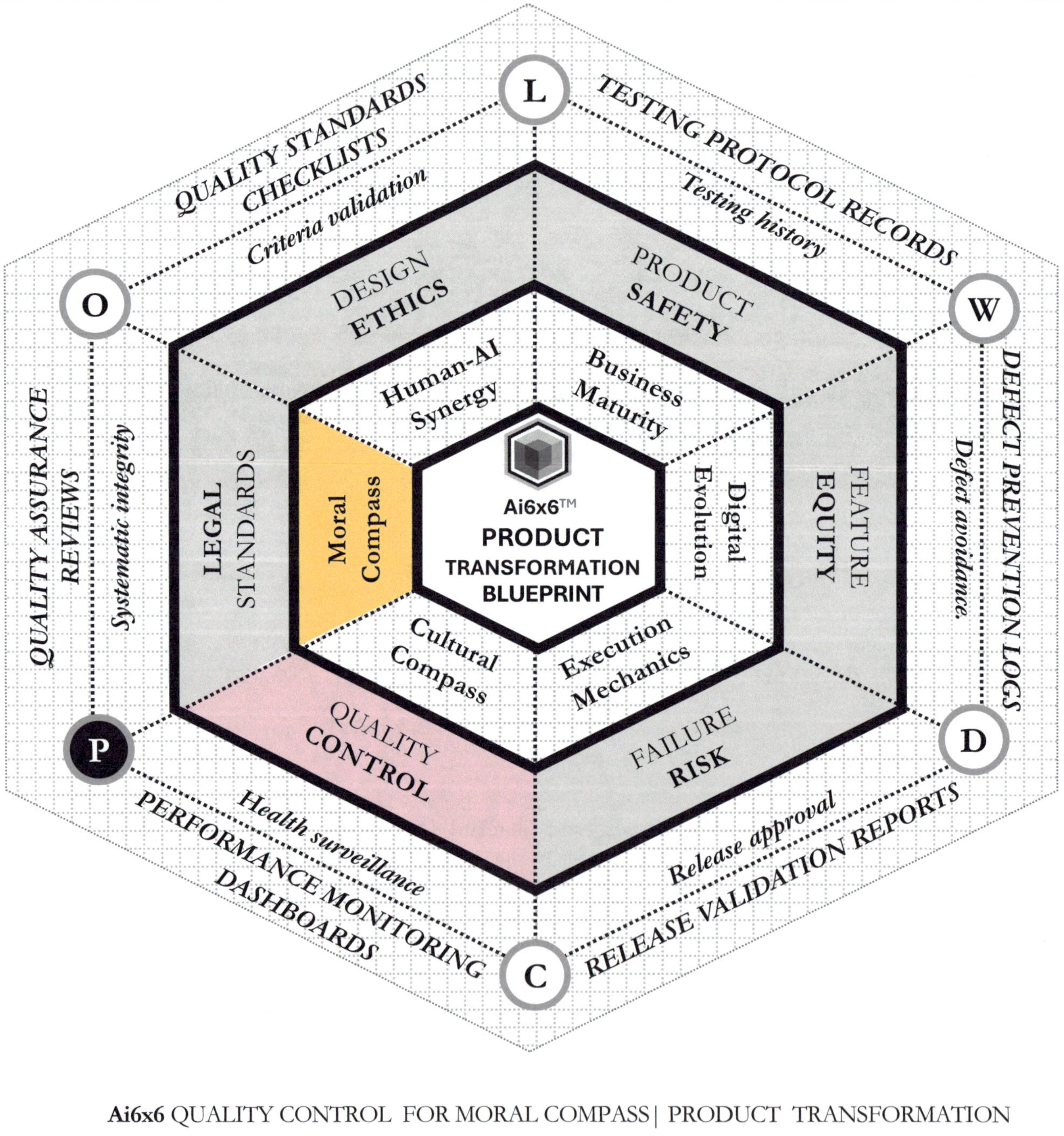

Ai6x6 QUALITY CONTROL FOR MORAL COMPASS | PRODUCT TRANSFORMATION

1. QUALITY STANDARDS CHECKLISTS

Quality Control requires clear expectations, and Quality Standards Checklists define the precise criteria a product must meet before release. They translate abstract quality principles into measurable requirements that guide engineering, testing, and validation. This matters because ambiguity in standards leads to inconsistent performance and unpredictable customer outcomes. Quality Standards Checklists prevent drift by ensuring every team evaluates quality using the same benchmarks. They also reinforce fairness by ensuring all customers receive products built to identical standards. When ignored, quality becomes subjective and uneven. Customers experience variability that erodes trust. Over time, Quality Control deteriorates as standards lose authority.

When executed well, Quality Standards Checklists strengthen Quality Control by creating a shared definition of excellence. Customers trust that the organization delivers consistent, reliable products. This improves satisfaction, reduces defects, and enhances brand credibility. Checklists reinforce the principle that quality must be intentional, not accidental. They remind leaders that clarity is the foundation of reliability. Ultimately, they build an environment where quality is predictable, measurable, and consistently upheld.

MORAL COMPASS SCENARIO	LEADERSHIP ACTION
Maren discovers that engineers are skipping half the checklist items during sprint crunches. She realizes quality gates are being treated as optional.	The leader enforces mandatory checklist completion with automated blockers. They also tie checklist adherence to release approval.
Dorian notices that the checklist hasn't been updated since the product added new modules. He senses outdated criteria are masking gaps.	The leader refreshes the checklist and aligns it with current architecture. They also require quarterly checklist reviews.
Selene finds that teams interpret checklist items differently across regions. She recognizes inconsistent quality standards.	The leader rewrites items with precise definitions and examples. They also train regional leads on unified interpretation.
Marcus sees that the checklist focuses only on functional quality and ignores resilience. He worries about brittle releases.	The leader expands the checklist to include resilience, recovery, and load criteria. They also require resilience scoring.
Fiona notices that the checklist is stored in a static PDF. She realizes teams are using outdated versions.	The leader migrates it to a version-controlled system. They also enforce update notifications.
Idris discovers that the checklist lacks security items. He senses a major oversight.	The leader adds security-critical checks and integrates them into CI pipelines. They also require security sign-off.

2. TESTING PROTOCOL RECORDS

Quality Control requires disciplined execution, and Testing Protocol Records document the exact procedures used to evaluate product performance. They ensure that tests are repeatable, consistent, and aligned with defined standards. This matters because inconsistent testing produces unreliable results. Testing Protocol Records prevent gaps by ensuring every test follows the same method. They also reinforce fairness by ensuring all product units undergo identical scrutiny. When ignored, testing becomes ad hoc and incomplete. Customers encounter defects that should have been detected earlier. Over time, Quality Control collapses as testing rigor disappears.

When executed well, Testing Protocol Records strengthen Quality Control by embedding structure into the evaluation process. Customers trust that the organization validates performance with precision. This improves reliability, reduces risk, and enhances ethical responsibility. Records reinforce the principle that testing must be systematic. They remind leaders that quality emerges from disciplined verification. Ultimately, they build an environment where testing is consistent, transparent, and dependable.

MORAL COMPASS SCENARIO	LEADERSHIP ACTION
Nia sees that protocol records show inconsistent test coverage between teams. She realizes testing discipline varies widely.	The leader standardizes test protocols and enforces minimum coverage thresholds. They also audit team-level compliance.
Corin notices that records don't include environment details. He senses reproducibility issues.	The leader adds environment metadata requirements. They also require environment-specific test validation.
Elara finds that exploratory testing sessions are undocumented. She recognizes lost insights.	The leader mandates structured exploratory logs. They also train testers on capturing qualitative findings.
Mateo sees that protocol records omit negative test cases. He worries about untested failure paths.	The leader expands protocols to include negative and adversarial scenarios. They also require negative-case coverage reporting.
Soren notices that protocol records are scattered across multiple tools. He senses fragmentation.	The leader consolidates records into a unified repository. They also enforce consistent naming conventions.
Hana discovers that protocol records don't include test-data provenance. She recognizes traceability gaps.	The leader adds provenance fields and validates historical data sources. They also require data-origin checks.

3. DEFECT PREVENTION LOGS

Quality Control requires proactive mitigation, and Defect Prevention Logs track actions taken to eliminate root causes of quality issues. They document patterns, interventions, and improvements across development cycles. This matters because prevention is more effective than correction. Defect Prevention Logs prevent recurrence by ensuring lessons learned translate into structural improvements. They also reinforce fairness by ensuring all customers benefit from enhanced quality. When ignored, defects reappear and undermine reliability. Customers experience repeated failures that erode trust. Over time, Quality Control deteriorates as prevention discipline weakens.

When executed well, Defect Prevention Logs strengthen Quality Control by creating a continuous improvement loop. Customers trust that the organization learns from mistakes and acts responsibly. This improves product stability, reduces complaints, and enhances brand integrity. Logs reinforce the principle that quality evolves through iteration. They remind leaders that prevention is a strategic investment. Ultimately, they build an environment where quality improves with every release.

MORAL COMPASS SCENARIO	LEADERSHIP ACTION
Priya discovers that prevention logs show repeated UI regressions after minor updates. She realizes the component library is unstable.	The leader stabilizes the library and adds regression tests for all UI components. They also track regression frequency.
Rowan notices that logs show recurring defects tied to one integration point. He senses architectural fragility.	The leader redesigns the integration boundary and adds isolation tests. They also require integration-specific prevention reviews.
Serena finds that prevention logs highlight frequent misconfigurations by new engineers. She recognizes onboarding gaps.	The leader updates onboarding materials and adds configuration-validation tools. They also track misconfiguration trends.
Omar sees that logs show defects caused by ambiguous requirements. He worries upstream clarity is missing.	The leader enforces requirement-clarity checks and adds definition-of-ready criteria. They also train PMs on precision writing.
Lila notices that prevention logs show failures only in multilingual builds. She senses localization-specific defects.	The leader adds localization-focused prevention tests. They also require locale-specific QA cycles.
Andre discovers that prevention logs are not reviewed during sprint planning. He realizes teams aren't learning from past issues.	The leader integrates prevention logs into planning rituals. They also require defect-prevention commitments each sprint.

4. RELEASE VALIDATION REPORTS

Quality Control requires final verification, and Release Validation Reports confirm that a product meets all quality, safety, and performance requirements before launch. They consolidate test results, compliance checks, and risk assessments into a single decision artifact. This matters because release decisions carry significant customer impact. Release Validation Reports prevent premature launches by ensuring all criteria are met. They also reinforce fairness by ensuring all customers receive products that have passed the same rigorous evaluation. When ignored, products reach the market with unresolved issues. Customers experience failures that damage trust. Over time, Quality Control collapses as release discipline erodes.

When executed well, Release Validation Reports strengthen Quality Control by creating a transparent, accountable release process. Customers trust that the organization does not compromise quality for speed. This improves reliability, reduces risk, and enhances ethical alignment. Reports reinforce the principle that readiness must be proven, not assumed. They remind leaders that launch decisions require evidence. Ultimately, they build an environment where releases are safe, stable, and trustworthy.

MORAL COMPASS SCENARIO	LEADERSHIP ACTION
Nadia sees that validation reports show untested fallback paths. She realizes releases are going out without resilience checks.	The leader blocks the release and adds fallback-path validation. They also require resilience sign-off.
Trevor notices that validation reports differ between staging and production environments. He senses environment drift.	The leader aligns configurations and adds environment-parity checks. They also require pre-release environment audits.
Aisha finds that validation reports show inconsistent performance metrics across platforms. She recognizes platform-specific gaps.	The leader adds platform-parity validation. They also test performance across device tiers.
Marco sees that validation reports are missing security-scan results. He worries about unverified vulnerabilities.	The leader integrates security scans into the validation pipeline. They also require security approval before release.
Evelyn notices that validation reports are overly optimistic because teams self-report results. She senses bias.	The leader assigns independent QA reviewers. They also require evidence-based validation.
Jonah discovers that validation reports don't include rollback readiness. He recognizes deployment-stage risk.	The leader adds rollback-validation steps. They also require rollback drills.

5. PERFORMANCE MONITORING DASHBOARDS

Quality Control requires real time visibility, and Performance Monitoring Dashboards track product behavior across live environments. They reveal trends, anomalies, and degradation that may not appear during testing. This matters because quality issues often emerge only under real world conditions. Performance Monitoring Dashboards prevent escalation by enabling early detection of performance drift. They also reinforce fairness by ensuring all customers—not just those who complain—benefit from continuous monitoring. When ignored, issues remain hidden until they cause widespread harm. Customers experience unpredictable failures. Over time, Quality Control deteriorates as visibility disappears.

When executed well, Performance Monitoring Dashboards strengthen Quality Control by enabling rapid intervention. Customers trust that the organization monitors performance with discipline. This improves stability, reduces downtime, and enhances brand reliability. Dashboards reinforce the principle that quality must be maintained continuously. They remind leaders that real world performance is the ultimate test. Ultimately, they build an environment where quality is actively protected after release.

MORAL COMPASS SCENARIO	LEADERSHIP ACTION
Rhea sees that dashboards show latency spikes only during specific user journeys. She realizes the bottleneck is path-dependent.	The leader isolates the affected journey and optimizes the slowest steps. They also add journey-level performance metrics.
Omar notices that dashboards lack anomaly detection. He senses slow-burn performance degradation.	The leader adds anomaly-detection models. They also require alert thresholds for early warnings.
Lila finds that dashboards aggregate metrics too broadly. She recognizes that critical micro-patterns are hidden.	The leader adds granular, component-level metrics. They also require drill-down capabilities.
Andre sees that dashboards don't include accessibility-mode performance. He worries about inequitable performance.	The leader adds AT-specific performance metrics. They also test performance with assistive technologies.
Serena notices that dashboards show inconsistent data due to sampling errors. She senses monitoring pipeline issues.	The leader fixes sampling logic and validates data integrity. They also add monitoring-pipeline health checks.
Hugo discovers that dashboards are not shared with engineering teams. He realizes insights aren't driving improvements.	The leader integrates dashboards into engineering rituals. They also require performance-improvement commitments.

6. QUALITY ASSURANCE REVIEWS

Quality Control requires independent oversight, and Quality Assurance Reviews evaluate whether processes, standards, and testing practices are followed consistently. They examine the integrity of the entire quality system, not just the product. This matters because quality failures often originate in process gaps rather than technical defects. Quality Assurance Reviews prevent systemic issues by ensuring teams adhere to established procedures. They also reinforce fairness by ensuring all products benefit from the same disciplined oversight. When ignored, quality systems degrade and inconsistencies multiply. Customers experience unpredictable outcomes that erode trust. Over time, Quality Control collapses as governance weakens.

When executed well, Quality Assurance Reviews strengthen Quality Control by validating the reliability of the entire quality ecosystem. Customers trust that the organization maintains high standards across every stage of development. This improves consistency, reduces risk, and enhances ethical responsibility. Reviews reinforce the principle that quality requires governance, not just testing. They remind leaders that oversight is essential to sustained excellence. Ultimately, they build an environment where quality is upheld through disciplined, independent review.

MORAL COMPASS SCENARIO	LEADERSHIP ACTION
Keisha sees that QA reviews focus only on functional correctness and ignore usability. She realizes quality is being defined too narrowly.	The leader expands QA criteria to include usability, accessibility, and resilience. They also train QA teams on holistic evaluation.
Damon notices that QA reviews are rushed at the end of sprints. He senses time-compression risk.	The leader shifts QA earlier in the sprint and enforces minimum review windows. They also add mid-sprint QA checkpoints.
Liora finds that QA reviews rely heavily on manual testing. She recognizes automation gaps.	The leader expands automated test coverage and retrains QA engineers. They also require automation targets per release.
Mateo sees that QA reviews don't include cross-browser testing. He worries about inconsistent user experiences.	The leader adds browser-matrix testing. They also require parity validation across browsers.
Serena notices that QA reviews are not documented thoroughly. She senses traceability issues.	The leader enforces structured QA documentation. They also add audit trails for all reviews.
Omar discovers that QA reviews don't include risk-based prioritization. He realizes critical areas may be under-tested.	The leader adds risk scoring to QA planning. They also require risk-aligned test allocation.

Ai6x6™ Transformation Blueprint

Legal Standards

"A sustainable world requires machines that optimize for life, not profit."
Vandana Shiva, environmental activist and author.

Legal Standards require adherence to industry rules, and Industry Compliance Requirements define the legal and regulatory expectations products must meet to enter the market. They establish the minimum obligations for lawful operation. This matters because noncompliance exposes organizations to penalties, recalls, and reputational harm. Industry Compliance Requirements prevent violations by ensuring products align with established regulations. They also reinforce integrity by demonstrating that compliance is a core design and operational priority. When ignored, legal exposure increases rapidly. Teams experience uncertainty as decisions drift outside regulatory boundaries. Over time, Legal Standards deteriorate as compliance becomes inconsistent. Regulatory Product Obligations ensure products meet all required legal criteria.

The global benchmark of mastery and the foundation of product sovereignty. Mastering ISO, UL, and FDA Regulatory Standards ensures that the product is not just innovative, but globally compliant and durable. This topic treats "Legal Standards" as the "Insurance Policy" against market extinction. By reviewing "Liability Law" and "CE Conformity" in a unified view, the leader secures the product's place in the international market. Legal compliance is the clinical proof of a product's integrity. It ensures that your innovation engine is protected from the "Sledgehammer" of litigation, allowing for permanent, sustainable market dominance. Using the print series, you can sit with these global standards away from distractions, seeing how they form the "Moral Compass" of your product strategy. This clinical mastery ensures that your "Product Transformation" is not just efficient, but legally unassailable and morally sound, providing the ultimate insurance policy against the market extinction that follows the failure to meet the sovereign quality and safety standards of the global industrial landscape.

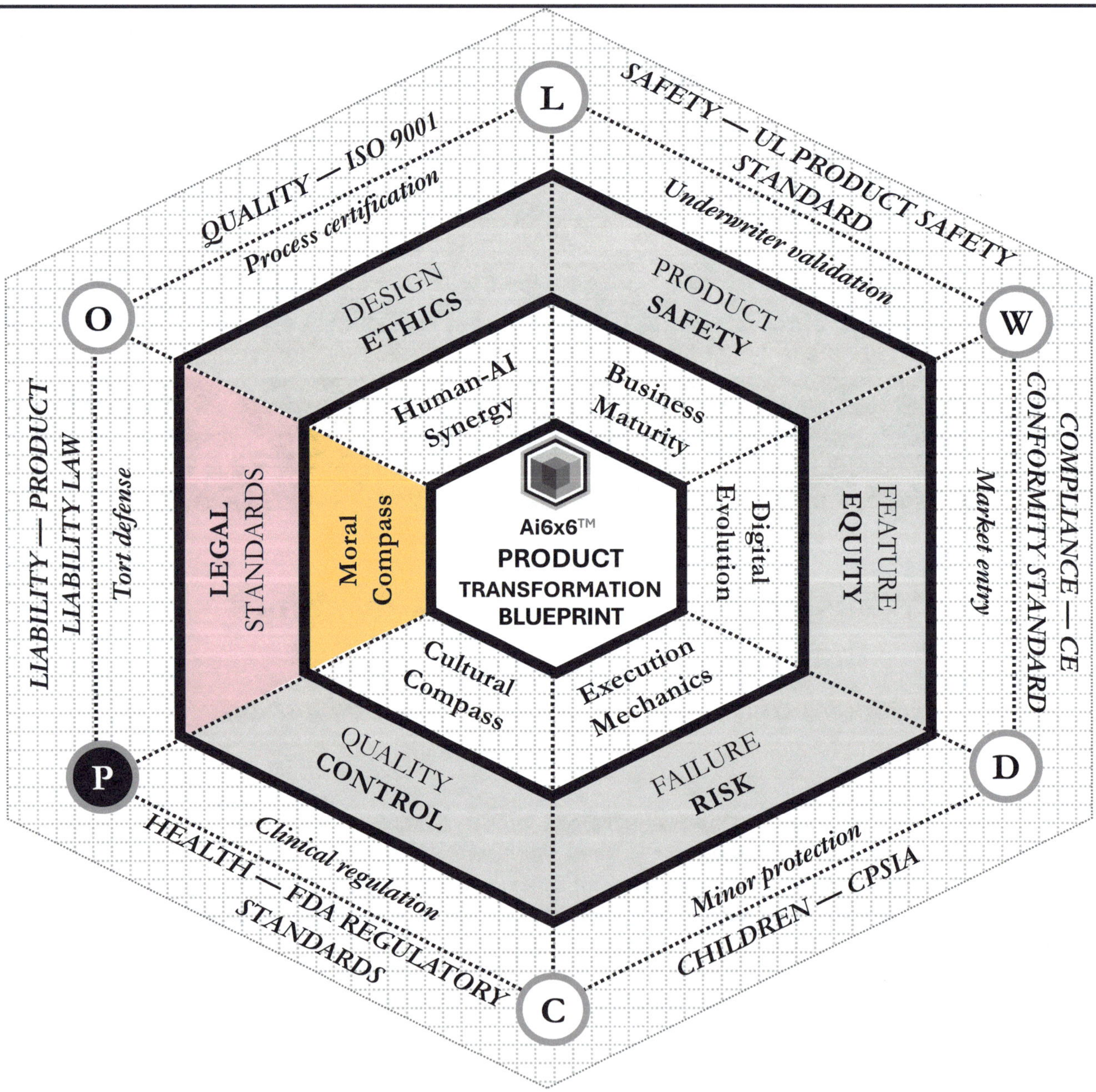

Ai6x6 LEGAL STANDARDS FOR MORAL COMPASS| PRODUCT TRANSFORMATION

1. QUALITY — ISO 9001

Legal Standards require disciplined quality management, and ISO 9001 establishes the global benchmark for consistent, repeatable, and auditable product quality. It defines the structures, processes, and documentation required to ensure products meet customer expectations reliably. This matters because quality failures often originate in weak systems rather than individual defects. ISO 9001 prevents inconsistency by enforcing standardized procedures across development and production. It also reinforces fairness by ensuring all customers receive products built under the same rigorous framework. When ignored, quality becomes unpredictable and dependent on individual teams or circumstances. Customers experience variability that erodes trust. Over time, Legal Standards deteriorate as quality governance collapses.

When executed well, ISO 9001 strengthens Legal Standards by embedding quality discipline into the organization's operating system. Customers trust that the product is produced under globally recognized controls. This improves reliability, reduces defects, and enhances brand credibility. Compliance reinforces the principle that quality must be systematic, not accidental. It reminds leaders that excellence requires structure and accountability. Ultimately, it builds an environment where quality is consistent, measurable, and continuously improved.

MORAL COMPASS SCENARIO	LEADERSHIP ACTION
Mara discovers that several teams are bypassing documented procedures and improvising their own workflows. She realizes the ISO process discipline has eroded.	The leader reinstates mandatory process-adherence audits and retrains teams on ISO requirements. They also tie compliance to release approval.
Devin notices that corrective-action entries are closed without evidence of verification. He senses superficial remediation.	The leader enforces verification-of-effectiveness steps and assigns independent reviewers. They also track closure quality.
Selah finds that internal audits are being scheduled irregularly. She recognizes a compliance-cycle breakdown.	The leader restores the audit calendar and assigns audit owners. They also require quarterly compliance reporting.
Mateo sees that document-control logs show outdated versions still in circulation. He worries teams are building from stale specs.	The leader centralizes document control and enforces version-locking. They also require version-awareness training.
Farah notices that ISO logs don't include supplier-quality issues. She senses a blind spot in the quality system.	The leader integrates supplier-quality tracking into ISO logs. They also require supplier-risk reviews.
Idris discovers that ISO logs are missing root-cause analysis for recurring issues. He realizes systemic problems are being ignored.	The leader mandates RCA documentation and assigns cross-functional teams to address patterns. They also track recurrence rates.

2. SAFETY — UL PRODUCT SAFETY STANDARD

Legal Standards require protection from physical harm, and UL Product Safety Standards define the requirements products must meet to ensure safe operation. They establish testing protocols, hazard thresholds, and certification criteria. This matters because customers expect products to function without exposing them to danger. UL standards prevent harm by validating that products meet recognized safety benchmarks. They also reinforce fairness by ensuring all customers receive products tested to the same rigorous criteria. When ignored, unsafe products reach the market and cause preventable injuries. Customers lose confidence in the organization's commitment to safety. Over time, Legal Standards collapse as safety oversight weakens.

When executed well, UL compliance strengthens Legal Standards by providing independent verification of product safety. Customers trust that the organization prioritizes their well being. This improves satisfaction, reduces liability, and enhances ethical alignment. Compliance reinforces the principle that safety must be validated, not assumed. It reminds leaders that rigorous testing is essential to responsible product stewardship. Ultimately, it builds an environment where safety is consistently protected through recognized standards.

MORAL COMPASS SCENARIO	LEADERSHIP ACTION
Nia discovers that UL certification files for a new device are incomplete. She realizes the product cannot legally ship.	The leader halts distribution and completes all required UL testing. They also enforce certification-file audits.
Corin notices that UL test results show borderline electrical-safety margins. He senses a risk of field failures.	The leader redesigns the circuitry to increase safety margins. They also require retesting before approval.
Elara finds that UL files reference outdated component specifications. She recognizes a documentation mismatch.	The leader updates all component records and revalidates compliance. They also enforce component-spec version control.
Mateo sees that UL files for international variants are missing. He worries about region-specific compliance gaps.	The leader adds region-specific UL documentation and updates distribution rules. They also require global compliance reviews.
Soren notices that UL files are stored across multiple drives. He senses traceability issues.	The leader centralizes all UL files in a secure repository. They also enforce structured naming conventions.
Hana discovers that UL test failures were never logged in the defect system. She realizes safety insights are siloed.	The leader integrates UL findings into defect workflows. They also require cross-team safety reviews.

Legal Standards require adherence to regional regulations, and CE Conformity ensures products meet the European Union's safety, health, and environmental requirements. It signals that a product complies with essential directives governing performance and risk. This matters because global markets demand consistent safety and compliance expectations. CE Conformity prevents regulatory violations by ensuring products meet EU standards before distribution. It also reinforces fairness by ensuring all customers in the region receive compliant products. When ignored, organizations face legal penalties, recalls, and reputational damage. Customers question the product's legitimacy and safety. Over time, Legal Standards deteriorate as compliance discipline erodes.

When executed well, CE Conformity strengthens Legal Standards by ensuring products meet international expectations. Customers trust that the product is safe, compliant, and responsibly manufactured. This improves market access, reduces regulatory risk, and enhances brand credibility. Compliance reinforces the principle that global distribution requires global responsibility. It reminds leaders that conformity is a prerequisite for trust. Ultimately, it builds an environment where compliance is embedded into every stage of product development.

MORAL COMPASS SCENARIO	LEADERSHIP ACTION
Priya discovers that the CE Declaration of Conformity references an outdated directive. She realizes the product is technically non-compliant.	The leader updates the declaration and revalidates against current directives. They also require annual directive reviews.
Rowan notices that CE technical files lack EMC test results. He senses a major compliance gap.	The leader completes EMC testing and updates the technical file. They also enforce EMC-test gating.
Serena finds that CE markings are applied before testing is complete. She recognizes premature labeling.	The leader halts labeling until all tests pass. They also enforce labeling-after-approval rules.
Omar sees that CE risk assessments are missing environmental-impact analysis. He worries about incomplete conformity.	The leader adds environmental-risk assessments and updates documentation. They also require environmental-impact reviews.
Lila discovers that CE files don't include supplier declarations. She senses supply-chain compliance gaps.	The leader obtains supplier declarations and integrates them into the file. They also require supplier-conformity attestations.
Andre notices that CE files are not updated after design changes. He recognizes a documentation lag.	The leader updates all CE records and enforces change-triggered updates. They also require design-change audits.

Legal Standards require special protection for vulnerable populations, and the Consumer Product Safety Improvement Act (CPSIA) establishes strict requirements for products intended for children. It mandates testing for hazardous substances, choking risks, and structural safety. This matters because children cannot assess or avoid danger on their own. CPSIA prevents harm by enforcing rigorous safety controls before products reach young users. It also reinforces fairness by ensuring all children receive the same level of protection. When ignored, unsafe products expose children to severe and preventable risks. Parents lose trust in the organization's integrity. Over time, Legal Standards collapse as child safety becomes compromised.

When executed well, CPSIA compliance strengthens Legal Standards by ensuring children's products meet the highest safety thresholds. Customers trust that the organization prioritizes child well being. This improves confidence, reduces liability, and enhances ethical responsibility. Compliance reinforces the principle that vulnerable users require heightened protection. It reminds leaders that children's safety is non negotiable. Ultimately, it builds an environment where products for children are designed and tested with exceptional care.

MORAL COMPASS SCENARIO	LEADERSHIP ACTION
Nadia discovers that children's products lack updated lead-content test results. She realizes CPSIA compliance is at risk.	The leader orders immediate testing and halts shipments. They also enforce annual lead-content testing.
Trevor notices that tracking labels on children's items are inconsistent. He senses traceability failures.	The leader standardizes tracking labels and audits all inventory. They also require label-integrity checks.
Aisha finds that small-parts warnings are missing from packaging. She recognizes a choking-hazard violation.	The leader updates packaging and adds mandatory warnings. They also require packaging-safety reviews.
Marco sees that CPSIA certificates are stored in email threads. He worries about missing documentation.	The leader centralizes certificates in a secure repository. They also enforce certificate-retention policies.
Evelyn notices that third-party labs used for testing are not CPSC-accepted. She senses invalid test results.	The leader switches to accredited labs and retests affected products. They also require lab-credential verification.
Jonah discovers that CPSIA attestation updates lag behind product revisions. He realizes compliance is drifting.	The leader synchronizes attestations with product changes. They also require revision-triggered compliance checks.

5. HEALTH — FDA REGULATORY STANDARDS

Legal Standards require protection of human health, and FDA Regulatory Standards govern the safety, efficacy, and labeling of health related products. They establish rigorous requirements for testing, validation, and post market surveillance. This matters because health products carry elevated risk when they fail. FDA standards prevent harm by ensuring products meet scientific and clinical thresholds before reaching customers. They also reinforce fairness by ensuring all users receive products that meet the same health protections. When ignored, unsafe or ineffective products enter the market and cause serious harm. Customers lose trust in the organization's credibility. Over time, Legal Standards deteriorate as health oversight weakens.

When executed well, FDA compliance strengthens Legal Standards by embedding scientific rigor into product development. Customers trust that the organization meets the highest health protection requirements. This improves safety, reduces risk, and enhances ethical alignment. Compliance reinforces the principle that health related products demand exceptional scrutiny. It reminds leaders that public health is a shared responsibility. Ultimately, it builds an environment where health products are safe, effective, and responsibly managed.

MORAL COMPASS SCENARIO	LEADERSHIP ACTION
Rhea discovers that a health-related feature was released without FDA classification review. She realizes regulatory oversight was skipped.	The leader pauses the feature and conducts a full regulatory assessment. They also require FDA-impact screening for all health features.
Omar notices that clinical-evaluation data is incomplete. He senses a risk of regulatory rejection.	The leader completes missing evaluations and updates the dossier. They also enforce clinical-data completeness checks.
Lila finds that labeling for a wellness device implies medical claims. She recognizes a regulatory boundary violation.	The leader rewrites labeling to remove medical implications. They also train teams on claim-classification rules.
Andre sees that FDA checklists are missing human-factors testing. He worries about usability-related safety risks.	The leader adds human-factors testing and updates the checklist. They also require usability validation for all regulated features.
Serena notices that adverse-event reporting procedures are unclear. She senses compliance risk.	The leader formalizes reporting workflows and trains teams. They also require periodic reporting drills.
Hugo discovers that FDA documentation is scattered across teams. He recognizes traceability gaps.	The leader centralizes all regulatory documents. They also enforce structured documentation governance.

Legal Standards require accountability for harm, and Product Liability Law establishes the legal framework governing responsibility for defective or dangerous products. It defines obligations related to design, manufacturing, warnings, and foreseeable misuse. This matters because customers must be protected when products fail. Product Liability Law prevents negligence by holding organizations accountable for harm caused by defects. It also reinforces fairness by ensuring customers have legal recourse. When ignored, organizations shift risk onto customers and evade responsibility. Trust erodes as failures go unaddressed. Over time, Legal Standards collapse as accountability disappears.

When executed well, adherence to Product Liability Law strengthens Legal Standards by ensuring organizations take responsibility for product safety. Customers trust that the organization stands behind its products. This improves confidence, reduces disputes, and enhances ethical responsibility. Compliance reinforces the principle that accountability is essential to trust. It reminds leaders that safety failures must be addressed, not denied. Ultimately, it builds an environment where customers are protected through clear legal safeguards.

MORAL COMPASS SCENARIO	LEADERSHIP ACTION
Keisha sees that risk reviews show a pattern of injuries linked to a specific feature. She realizes liability exposure is rising.	The leader redesigns the feature and conducts a hazard-elimination review. They also track injury-related risk metrics.
Damon notices that warnings are insufficient for high-risk use cases. He senses foreseeable-misuse exposure.	The leader adds explicit warnings and tests user comprehension. They also require misuse-scenario modeling.
Liora finds that risk reviews ignore international legal differences. She recognizes global liability gaps.	The leader adds region-specific liability assessments. They also require global legal reviews.
Mateo sees that risk reviews rely on outdated incident data. He worries about blind spots.	The leader updates incident datasets and re-runs risk scoring. They also enforce data-freshness requirements.
Serena notices that risk reviews don't include supplier-related failures. She senses supply-chain liability exposure.	The leader integrates supplier-risk analysis. They also require supplier-failure modeling.
Omar discovers that risk reviews are not shared with legal teams. He realizes cross-functional alignment is missing.	The leader integrates legal review into the risk-assessment process. They also require joint sign-off.

BONUS CHAPTER

Ai6x6™ Transformation Blueprint

The Materiality Paradox

"Design is a signal of human intention." -- **William McDonough**, the architect and co-author of *Cradle to Cradle* who pioneered the regenerative economy, defines the **Materiality Paradox**. His work forces designers to treat every product as a clear signal of their intention: to treat atoms as precious, infinite resources that must be designed for disassembly and return, rather than for a one-way trip to the landfill.

Traditional product design has been a linear journey from "Cradle to Grave," optimized for aesthetics and low manufacturing costs while ignoring the "embodied carbon" left in its wake. This one-dimensional focus on the point of sale has led to a global waste crisis and a brittle supply of rare materials. To achieve Net Zero, designers must master the **Materiality Paradox**. This demands a multi-dimensional view that treats every product as a temporary assembly of valuable atoms. It requires balancing the high-fidelity power of "Digital Twins" with their energy costs and prioritizing modularity and repairability over planned obsolescence. Traction comes when the "Return" is designed with as much precision as the "Sale," creating a truly regenerative and circular product ecosystem.

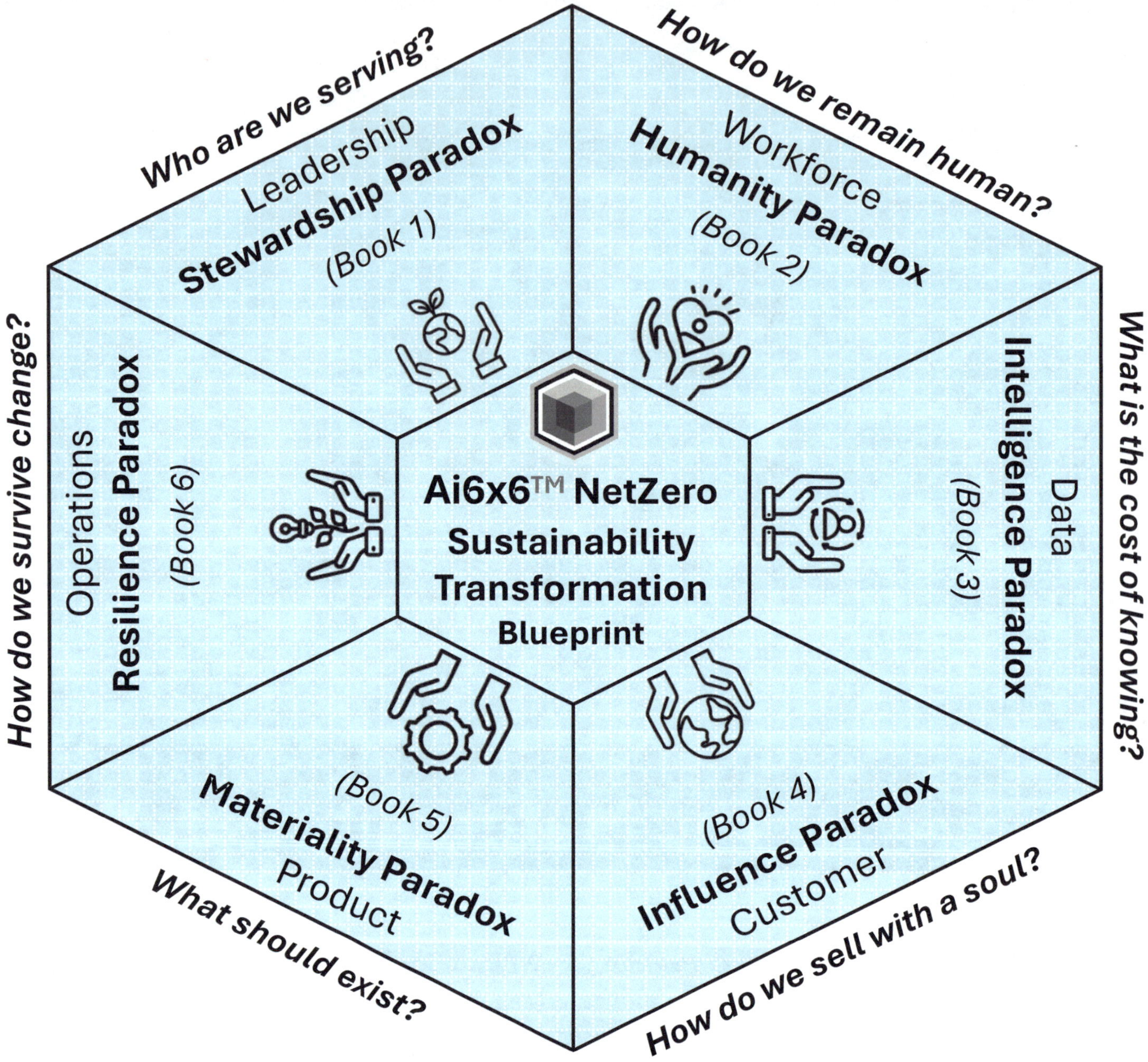

The **AI6x6 NetZero Sustainability Transformation Blueprint** redefines the path to Net Zero by replacing the traditional "check-the-box" compliance mindset with a rigorous framework of **paradoxical leadership**. In a world accelerated by AI, sustainability is rarely a choice between a clear right and wrong; instead, it is a constant navigation of thirty-six competing tensions across six organizational lenses. By framing the journey through these paradoxes—such as the conflict between **Digital Intelligence and Energy Consumption** or **Global Targets and Local Infrastructure**—the framework prepares leaders for the messy, non-linear realities of systemic change.

Rather than viewing these tensions as obstacles to be avoided, the **AI6x6 methodology** utilizes them as engines for innovation. It guides executives to move past the paralysis of "either/or" thinking, teaching them to embrace a "both/and" strategy that maintains fiduciary health while securing long-term planetary survival. This approach ensures that a company's moral compass remains calibrated even as technology and climate volatility shift the ground beneath them. Ultimately, the series provides a diagnostic roadmap that turns systemic friction into strategic advantage, empowering leaders to achieve radical Net Zero goals without sacrificing the operational resilience or human dignity required to thrive in the post-AI era.

THE MATERIALITY PARADOX is the fundamental tension of the physical world in a digital age: the struggle to create products that are both high-performance and zero-impact. As we move toward a Net Zero future, the very "stuff" our world is made of must be reinvented. We are caught between the need for durable, high-tech materials and the requirement that every product must eventually return to the earth or the factory without leaving a trace. This creates a conflict between the desire for "more" and the necessity of "nothing." To transform, we must stop designing for the point of sale and start designing for the point of return.

1. THE DIGITAL TWIN PARADOX (Virtual Testing vs. Real Power)

Using AI-driven "Digital Twins" allows engineers to test thousands of material combinations and designs in a virtual space, drastically reducing physical waste during the R&D phase. However, the paradox is that the massive compute power required to run these hyper-realistic simulations creates a significant carbon footprint before a single physical part is ever manufactured. Designers must balance the "virtual savings" against the "digital cost." The goal is to optimize the simulation process itself, ensuring that the environmental benefit of a perfectly designed product isn't erased by the energy-hungry servers that dreamed it into existence.

2. THE RECYCLABILITY-INFORMATION PARADOX (Easy to Recycle vs. Hard to Track)

To achieve a truly circular economy, every product needs a "Material Passport" that lists exactly what is inside it. The paradox is that as we make products more durable and complex to extend their life, the information required to safely track and eventually recycle them becomes overwhelmingly difficult to manage. A product with a hundred high-tech components is better for the user but a nightmare for the recycler. Success requires using blockchain and AI to make this data "invisible but accessible," ensuring that the complexity of a product's design doesn't become the reason it ends up in a landfill.

3. THE DEMATERIALIZATION PARADOX (Digital Services vs. Hidden Energy)

One of the most effective ways to save resources is to replace a physical product with a digital service, such as streaming music instead of buying CDs. The paradox, however, is that this "dematerialization" often hides a massive, energy-hungry infrastructure of server farms and the constant need for hardware upgrades to run heavier software. We swap physical waste for digital energy consumption. Product transformation requires a holistic view that counts the "hidden materials" of the cloud, ensuring that moving a service from the shelf to the screen actually results in a net win for the planet.

4. THE BIO-PERFORMANCE PARADOX (Natural Materials vs. Staying Strong)

Switching to plant-based or biodegradable materials is a moral victory for the environment, yet these materials often lack the extreme durability or fire-resistance of traditional synthetics. The paradox is that using "natural" materials might lead to a shorter product life, resulting in more frequent replacements and higher overall resource use. A brand's compass must navigate this trade-off by choosing the right material for the right lifespan. We must innovate to give biological materials industrial-grade strength, or accept that some products must be designed for a "short but circular" life rather than an "indestructible but toxic" one.

5. THE MODULAR-DESIGN PARADOX (Easy to Fix vs. Looking Good)

Designing for the "Right to Repair" means creating products with modular parts that anyone can swap out. The paradox is that this often clashes with the market's obsession with sleek, ultra-thin, "unibody" aesthetics that are inherently sealed and difficult to open. Customers want products that are easy to fix but also want them to look like seamless works of art. Solving this requires a new design language where "repairability" is marketed as a luxury feature. It is about proving that a product that can be taken apart and upgraded is more beautiful than one that is designed to be thrown away.

6. THE SOURCING-ETHICS PARADOX (Local Buying vs. Global Fairness)

A leader's moral compass often suggests buying materials locally to reduce the carbon footprint of shipping. However, the paradox is that many of the most sustainable or ethically sourced materials come from developing nations that rely on that trade for their economic survival. By "going local" to save carbon (the "E" in ESG), a company might be destroying social equity and fair labor opportunities abroad (the "S" in ESG). Transformation requires a sophisticated weighting system that balances transport emissions against the social impact of the global supply chain, ensuring that we don't save the air by hurting the people.

1. THE DIGITAL TWIN PARADOX (Virtual Precision vs. Physical Reality)

- **Aerospace:** An aviation firm uses high-fidelity virtual wind tunnels to simulate airflow, reducing the need for physical prototypes of a hydrogen-powered plane by 80%.
- **Consumer Goods:** A packaging company simulates "Drop Tests" virtually to calculate the minimum amount of virgin plastic required to maintain structural safety during transit.
- **Construction:** A developer creates a "Digital Twin" of a skyscraper to predict and optimize HVAC loads for the next 50 years, reducing operational carbon before breaking ground.
- **Fashion:** A luxury brand uses "3D Virtual Sampling" to fit garments on digital avatars, eliminating the carbon footprint of shipping physical samples back and forth to factories.
- **Electronics:** A hardware firm simulates chip heat-patterns to design passive cooling systems, removing the need for energy-hungry internal fans and heavy copper sinks.
- **Ocean Shipping:** A shipbuilder tests "Digital Hull" designs against 1,000 different wave patterns to maximize fuel efficiency before a single steel plate is cut.

2. THE RECYCLABILITY PARADOX (Complex Performance vs. Simple Sorting)

- **Beverages:** A global soda brand removes all plastic shrink-sleeves from its bottles, ensuring 100% optical sorting accuracy at municipal recycling plants.
- **Snack Foods:** A manufacturer moves to "Mono-material" bags that maintain freshness without the need for the unrecyclable foil-and-plastic layering found in traditional pouches.
- **Textiles:** A clothing brand creates "Cellulose-only" garments, including the thread and buttons, allowing the entire piece to be dissolved and spun into new fiber at end-of-life.
- **Energy Storage:** A tech firm designs "Salt-water" batteries for home storage that provide high performance without the toxic recycling challenges of lithium or cobalt.
- **Automotive:** A car manufacturer uses "Snap-together" dashboard components that avoid chemical glues, allowing a recycler to manually separate materials in under 60 seconds.
- **Furniture:** A desk manufacturer replaces mixed-material adhesives with mechanical joints and wood-dowels, ensuring every part can be easily sorted and recovered.

3. THE DEMATERIALIZATION PARADOX (Physical Objects vs. Digital Value)

- **Entertainment:** A gaming giant shifts its primary distribution from physical discs and plastic cases to a "Carbon-Aware" streaming service that optimizes data use.
- **Publishing:** A publisher replaces heavy, high-gloss coffee-table books with interactive "Augmented Reality" experiences that provide the same visual value with zero paper waste.
- **Automotive:** A brand moves all vehicle manuals and service logs into the dashboard's digital interface, eliminating 5 lbs of paper from every car produced.
- **Personal Finance:** A neobank eliminates physical credit cards entirely, issuing 100% "Digital Wallet" cards to reduce plastic production and shipping carbon.
- **Travel:** A tourism board replaces physical brochures and maps with high-resolution, GPS-enabled digital guides that update in real-time.
- **Art & Decor:** A gallery offers high-resolution digital "Art Subscriptions" for smart TVs, providing the aesthetic of rotating collections without the carbon of physical shipping.

4. THE BIO-PERFORMANCE PARADOX (Natural Materials vs. Industrial Strength)

- **Footwear:** A running brand develops a high-performance midsole made from "Mycelium" (mushroom root) that matches the bounce and durability of traditional synthetic foams.
- **Automotive:** A luxury EV maker uses "Pine-oil" based bio-plastics for interior trim, achieving a premium finish that is lighter and more durable than petroleum-based alternatives.
- **Outerwear:** A brand uses "Spider Silk" proteins to create a biodegradable rain-shell that provides extreme water resistance without the use of "forever chemicals" (PFAS).
- **Construction:** A startup uses "Hempcrete" blocks that are carbon-negative and provide superior natural insulation compared to traditional concrete or glass wool.
- **Personal Care:** A brand replaces plastic microbeads and synthetic exfoliants with crushed apricot seeds and biodegradable cellulose that performs identically in skincare.
- **Logistics:** A firm uses "Mushroom Packaging" instead of Styrofoam; it is grown from agricultural waste and provides the same shock-absorption for fragile electronics.

5. THE MODULAR-DESIGN PARADOX (Built to Last vs. Easy to Fix)

- **Smartphones:** A manufacturer designs a device with a "Modular Core," allowing users to click in a new camera or battery in under two minutes with a single standard screwdriver.
- **Appliances:** A washing machine brand creates a "Universal Motor" that can be easily removed and repaired, ensuring the rest of the steel machine lasts for 25 years.
- **Consumer Tech:** A laptop brand sells its hardware in "Open-Chassis" kits, allowing users to upgrade the motherboard and RAM while keeping the original screen and shell.
- **Furniture:** A modular shelving company ensures every part made in 2026 is backwards-compatible with units sold in 1990, preventing obsolescence.
- **Lighting:** An industrial lighting firm designs streetlights where the LED array can be swapped for newer, more efficient tech without replacing the entire aluminum pole.
- **Tools:** A power tool brand uses a "Single Battery Platform" across 50 different tools, reducing the total number of batteries and chargers a customer needs to own.

6. THE SOURCING-ETHICS PARADOX (Global Supply vs. Local Integrity)

- **Cosmetics:** A brand uses AI to trace the mica in its shimmer products back to individual, child-labor-free mines, publishing the verified audits on every package.
- **Electronics:** A hardware firm sources "Fair-mined" gold for its circuit boards, paying a premium to ensure local mining communities have safe water and fair wages.
- **Jewelry:** A brand shifts 100% to "Lab-Grown Diamonds" to eliminate the ecological destruction and human rights risks associated with traditional deep-crust mining.
- **Apparel:** A cotton retailer uses "Isotopic Mapping" to verify that their fiber was grown in high-standard regions, proving it wasn't sourced from forced-labor zones.
- **Chocolate:** A confectioner uses a "Direct-to-Farmer" digital payment system, ensuring that 100% of the sustainability premium reaches the grower without being lost to middlemen.
- **Furniture:** A brand uses "Satellite Monitoring" to guarantee that the timber for its tables did not contribute to illegal deforestation in protected rainforests.

Most comprehensive business transformation book series worldwide

Ai6x6™ Transformation Engine

The AI6x6 Transformation Engine™

Your Operating System for the AI Era

In an age of autonomous systems, fragmented strategy is a liability. The **AI6x6 Transformation Engine™** is a precision-engineered, 18-volume masterwork designed to move your enterprise from organizational inertia to scalable momentum. Built on a proprietary 6×6 geometry, this ecosystem unifies architecture, execution, and calibration into one high-performance unit.

The journey begins with the **Transformation Blueprint Series**, your **Intelligence Layer**. It codifies a "Navigation Deck" of Six Pillars and Six Lenses to anchor your strategic North Star. Next, the **Consulting Playbook Series** acts as the **Transmission Layer**, utilizing the IGNITE™ methodology to translate executive intent into high-torque, repeatable action. Finally, the **Maturity Diagnostic Series** serves as the **Sensor Layer**, providing a data-driven intake to quantify readiness and calibrate performance.

This isn't just a book collection; it is a rigid **Chassis** for institutional excellence. Secure the definitive architecture for sustained, high-stakes digital evolution. Drive the future with the Engine.

Transformation Series

To ensure this book series benefits your company holistically, we strongly recommend you buy the series and then distribute it to the appropriate transformation sponsors in each of these departments. Don't forget to browse through consulting offers at the end of the book.

WE NEED YOU!

WELCOMING OUR NEXT SET OF THOUGHT LEADERS

The remarkable success of the Ai6x6 Organization Transformation Series is built upon the **hard-won wisdom** of executive leaders who have successfully navigated the AI transformation within their industries, turning theory into a deeply practical, actionable guide. As we expand the series, we are extending a highly selective invitation to a new cohort of pioneers—individuals whose professional journey is at the **leading edge of AI strategy and execution**. We are looking for executives who possess a proven track record of rewriting business models, mastering unforeseen ethical and competitive challenges, and offering a **clear blueprint** for the next 3-5 years of the AI-Era.

Contributing to the Ai6x6 Series is a unique opportunity to **cement your professional legacy**, joining an exclusive, curated network of global authorities whose insights are distilled into executive-level knowledge cards, directly shaping the governance and innovation decisions of the world's most influential companies. Our selection process is based strictly on **pure expertise and demonstrated leadership** in the AI-Era, reinforcing our commitment to editorial integrity and authentic thought leadership—there is no monetary exchange involved. Select contributors will also receive an invitation to blog on our expert network microsite, further amplifying their platform. To confidentially inquire about contributing your expertise, please contact us via our dedicated portal at **www.Ai6x6.com/contact** or by scanning the adjacent QR code.

Contact

SUBSCRIBE TO OUR NEWSLETTERS: To stay ahead of the curve and receive continuous, executive-level intelligence on the rapid evolution of the AI-Era, we invite you to register for our **Ai6x6 Weekly Newsletter**. Each edition delivers curated insights directly from our thought leaders and highlights critical shifts in AI strategy, governance, and provides relevant excepts from our website and books. Keep your finger on the pulse of the future and ensure your company remains competitive by signing up on the same page

PLEASE REVIEW US ON AMAZON!

Product Blueprint

We know your time is valuable, but if the Ai6x6 Organization Transformation Series delivered the real-world, actionable strategies you needed, please consider leaving a brief review on Amazon. If we missed the mark, we would also like to know, and that is the only we we can improve. Your honest feedback is crucial; it helps fellow executives discover this unique framework and benefit from the insights of our thought leaders. Thank you for supporting our mission!

THE Ai6x6™ CERTIFIED TRANSFORMATION STRATEGIST

Certified for Thinking, Not Reciting

The **Ai6x6™ CTS (Certified Transformation Strategist)** is an elite designation for MBAs and strategic leaders with a minimum of **five years of professional experience**. This is not a test of your ability to regurgitate theory; it is a clinical evaluation of your ability to orchestrate high-stakes business transformation in real-time. It signals to the market that you are a strategist who can think on your feet, navigate ambiguity, and utilize a world-class clinical framework to deliver results.

The Non-Traditional Certification

We don't believe in "black box" testing. We believe in the "Open Book" reality of the modern boardroom. To earn your CTS, you must demonstrate mastery through a **60-minute Board-Level Defense** before a member of the Ai6x6™ Global Leadership Team.

$1000 per defense

The Preparation

Mastery of the Series: Candidates must familiarize themselves with all six volumes in the **Ai6x6™ Transformation Series**. You choose **two Retreat Exercises** from any two **Consulting Playbooks** within the series that best align with your expertise. Best practices for the presentation are provided on our website.

The 60-Minute Defense

1. **The Presentation (30 Minutes):** You have 15 minutes per exercise to present your strategic approach and expected outcomes. 60% of your score is for this exercise.
2. **The Situational Audit (30 Minutes):** Our leadership team will pose **six situational transformation scenarios**. You must demonstrate how to apply the **6 Lens Sequence™** to solve these complex frictions.
3. **The Open-Resource Advantage:** You are **encouraged** to refer to any of the six Ai6x6™ books during the exam. We test your ability to navigate and apply the framework under pressure, not your memory.

The Induction: Beyond the Credential

Passing the Board Defense is only the beginning. Successful candidates are immediately inducted into the **Ai6x6™ Executive Mentor Network**, unlocking:

- **Thought Leadership:** Opportunities to contribute to the Ai6x6™ books, knowledge articles and gain visibility among global industry leaders.
- **Engagement Pipeline:** Direct access to a network of future consulting engagements and strategic partnerships.
- **The Fast Track:** Exceptional participants may be invited to join our elite **Practitioner Program**, bypassing the standard entry requirements to lead their own white-label boutique practice.

SCHEDULE YOUR BOARD DEFENSE AT:

Ai6x6.com/certification

Consulting Certification

THE Ai6x6™ PRACTITIONER PROGRAM

Your Authority. Our Infrastructure. A Shared Legacy.

The **Ai6x6™ Practitioner Program** is designed for the high-level strategist—with proven industry leadership experience of over thirty years—who seeks a modern, standardized engine to deploy their operational wisdom. We provide the institutional infrastructure of a global consulting firm, delivered as a white-label solution that honors your professional identity and independence.

The "Un-Franchise" Model

From $2000 per month membership

We have re-engineered the consulting partnership for the modern era. Think of this as a **franchise without the friction or the high fees to get started**:

- **No Long-Term Commitments:** Maintain your agility with a model built for the independent architect.
- **No Loyalty Fees or Royalties:** You keep 100% of your engagement fees. We don't tax your success.
- **A Partnership of Mutual Trust:** Our relationship is powered by a transparent monthly membership, ensuring our incentives are perfectly aligned with your practice's growth.

The Foundations of Practice

All Practitioners gain immediate access to the **Ai6x6™ "Firm-in-a-Box"**:

- **White-Label Ecosystem:** Seamlessly integrate your own brand into our proprietary presentations, client pitches, and diagnostic audit templates.
- **The Clinical Engine:** Augment your strategic intuition with fact-based, quantitative scoring systems that replace subjective brainstorming with measurable transformation.

The White Glove Partnership

For those who wish to move from independent advisor to institutional authority, our **White Glove** tier offers a deep-tissue integration into the Ai6x6™ ecosystem:

- **Monthly 1-on-1 Strategic Check-ins:** Direct access to the **Ai6x6™ leadership team** to stress-test your pipeline, refine high-stakes offerings, or audit active client delivery.
- **Institutional Visibility:** Gain a global platform as a **Co-Author or Featured Expert** in upcoming volumes. We showcase your success stories on **Ai6x6.com**, cementing your legacy as a master of the 6 Lens Sequence.
- **The Bulk Purchase Program:** Exclusive access to deep discounts on the **Ai6x6™ Book Series**, enabling you to provide every client stakeholder with the "source code" for orchestration.

APPLY TO OUR PRACTITIONER PROGRAM AT:

Ai6x6.com/practitioner

Consulting Practitioner

THE Ai6x6™ STRATEGIC ROADMAP SERVICE

Your 3-Year Vision, Stress-Tested for the AI Era

The **Ai6x6™** book series has given you the "source code"—the clinical lenses, the vulnerability audits, and the diagnostic frameworks. But strategic orchestration at the highest level requires a cold, objective eye. If you are ready to move from self-diagnosis to radical transformation, we invite you to engage the **Ai6x6™ Roadmap Service.**

The 21-Day Transformation Engine

We don't believe in "billable hours" or six-month discovery phases. We believe in high-fidelity data and rapid execution. Our domain experts take your raw diagnostic output and synthesize it into a personalized, holistic blueprint for the next three years of your organization's life.

How It Works

1. **Submit Your Data:** Upload your completed Discovery Questionnaires, Vulnerability Audits, and Radar Charts directly to our secure portal or via email or via courier.
2. **Contextualize:** Provide your current 3-year roadmap and any business documentation that can provide more context on your line of business, your mission and success KPIs.
3. **The Clinical Review:** Our experts analyze your organizational DNA against the **6 Lens Sequence**, identifying hidden frictions and untapped leverage points. Assign a single point within your organization for asking questions.
4. **The Delivery:** Within **2 to 3 weeks**, you receive a personalized **Holistic Transformation Roadmap**, including our proprietary **Self-Disruption Strategy**.

Why Self-Disruption?

The greatest threat to your organization isn't your competitor; it is your own legacy success. Our roadmap doesn't just show you how to grow; it shows you how to cannibalize your outdated models before the market does it for you.

The Commitment

Packages from $3000

- **Speed:** 14–21-day turnaround.
- **Security:** Full NDA-backed engagement.
- **Precision:** Fact-based, quantitative analysis. No "fluff".

SECURE YOUR DIAGNOSTIC REVIEW:

Ai6x6.com/roadmap

Consulting Roadmap

THE Ai6x6™ C-SUITE GHOST PAIRING

For Senior Executives

Intelligence Off the Record and Under the Radar

The **Ai6x6™** series gave you the diagnostic frameworks. The Roadmap Service gave you the blueprint. But the reality of the C-Suite is a high-friction environment where internal politics often suppress the "Clinical Truth." If you are ready to operate with a dedicated, silent partner who provides the radical candor your inner circle cannot, we invite you to engage the **Ai6x6™ Strategic Ghost**.

The Shadow Cabinet

We do not believe in public accolades or "boardroom visibility." We believe in being the invisible intellectual infrastructure for the modern leader. This is a high-fidelity, secure partnership designed to be the objective "second brain" that your legacy organization is politically incapable of providing.

The Protocol: Subscribed Strategy

The Ghost Partnership is a precision-engineered monthly engagement. We don't appear on your org chart, your cap table, or your public calendar.

- **The Weekly Calibration:** 60 minutes of raw, unfiltered strategic synthesis—a "Cold Eye" review of your current friction points and tactical moves.
- **The Monthly Artifact:** One bespoke, high-stakes strategic asset—a Red Team analysis of your board deck, a Vulnerability Audit of a new initiative, or a competitor "Disruption Map."
- **The Direct Feed:** 24/7 access via a secure, encrypted WhatsApp channel for real-time decision support and "sanity checks."
- **The Annual Pivot:** A comprehensive year-end roadmap review to ensure your personal trajectory remains aligned with the 6 Lens Sequence.

The Shield: Total Secrecy

The Ghost operates under a dual-layer legal framework. A rigorous NDA and a specialized **Secrecy Agreement** ensure that our involvement remains entirely invisible. In a world of corporate noise, we are your only pure signal.

The Commitment

- **Access:** 4 hours of dedicated sync-up time per month.
- **Output:** 1 high-fidelity strategic artifact per month.
- **Security:** Full encryption and total anonymity.
- **Focus:** Your personal asymmetric advantage.

Subscriptions from $2500 per month

Consulting Ghost

SECURE YOUR SHADOW PARTNER :

Ai6x6.com/ghost

"Peace is its own reward."

Mahatma Gandhi

Ai6x6 Research Lab
Publishing Team

NEW YORK OFFICE
(Editorial Hub)

Editorial Lead: Marisa Thompson
Research Lead: Talia Rosen
Line Editor: Jordan Alvarez

LONDON OFFICE
(Design Hub)

Design Lead: Keiko Tanaka
Layout Specialist: Oliver Whitmore
Illustrator: Aisha Rahman
Design Team from Sumu Arts

KOLKATA OFFICE
(Support Hub)

Project Manager: Sohini Chatterjee
Production : Arindam Ghosh
Quality Lead: Priya Deshmukh
Fact Checker: Ravi Patel
Research Assistant: Ananya Sengupta
Research Assistant: Rohan Mukherjee
Project Assistant: Madhumita Roy
Production Assistant: Sagnik Dutta
Quality Assistant: Ishita Banerjee
Series Lead (Blueprint): Amit Verma
Series Lead (Playbook): Neha Reddy
Series Lead (Diagnostic): Karan Mehta

The End